EX LIBRIS

GIFT FROM DAVID MAYESKE
TO ROY, FOR HIS BIRTHDAY
— 2001 —

EDISON

THE UNIVERSITY OF CHICAGO PRESS
CHICAGO AND LONDON

EDISON

INVENTING THE CENTURY

Neil Baldwin

COPYRIGHTED BY BYRON N.Y. 1906

THOMAS EDISON

Excerpt on page 389 by George M. Cohan, "Thomas A. Edison, Miracle Man," © VosBurgh's Orchestration Service, New York, publishing agent for Light's Golden Jubilee committee.

Permission to quote from unpublished correspondence was granted by the following: Paul J. Christiansen and the Trustees of the Charles Edison Fund; The Henry Ford Museum and Greenfield Village; The Chautauqua Institution Library; and The New York Academy of Medicine Library.

Permission for quotations from other unpublished documents was made as follows: Special Collections, Vassar College Library, and Elizabeth Burroughs Kelley; The Henry and Albert A. Berg Collection, The New York Public Library, Astor, Lenox and Tilden Foundations; George Miller Beard Papers, Manuscripts and Archives, Yale University Library; Edison Electric Institute Library; The Trustees of Columbia University in the City of New York; The Collections of the Edison/Ford Estate, Fort Myers, Florida; and The Collections of the New Jersey Historical Society.

The University of Chicago Press, Chicago 60637
The University of Chicago Press, Ltd., London
Copyright © 1995, 2001 by Neil Baldwin
All rights reserved. Originally published 1995 by Hyperion
University of Chicago Press edition 2001
Printed in the United States of America
05 04 03 02 01   6 5 4 3 2 1

Library of Congress Cataloging-in-Publication Data

Baldwin, Neil, 1947–
    Edison : inventing the century / Neil Baldwin
      p.   cm.
    Originally published : New York : Hyperion, 1995
    Includes bibliographical references and index.
    ISBN 0-226-03571-9 (pbk. : alk. paper)
    1. Edison, Thomas A. (Thomas Alva), 1847–1931. 2. Inventors—United
States—Biography. 3. Electric engineering—United States—History. I. Title.

TA140.E35 B35   2001
621.3′092—dc21
[B]   00-053261

⊗ The paper used in this publication meets the minimum requirements of the American National Standard for Information Sciences—Permanence of Paper for Printed Library Materials, ANSI Z39.48-1992.

Once again, for Roberta

Each age writes the history of the past anew with reference to the conditions uppermost in its own time.

—**FREDERICK JACKSON TURNER**
An American Definition of History, 1891

A. B. Dick, the mimeograph manufacturer, saw Thomas Edison in Paris during the great Electrical Exhibition of 1889, and they renewed their old acquaintance over lunch. Mr. Dick told me, "Mr. Garrity, I can remember it just as if it occurred yesterday. Edison threw his arm over my shoulder as we were walking down the street from the restaurant, and said, 'Dick, I would give everything I own to be a young man like you again because there is so much I want to accomplish before I die. I won't get 1/10 of 1% of it done!' I was 33, and Mr. Edison was 42."

—**PATRICK A. GARRITY**
member of the "Edison Pioneers," 1918

The old nations of the earth keep on at a snail's pace. The Republic thunders past with the rush of the express train.

—**ANDREW CARNEGIE**
Triumphant Democracy, 1886

Inventors must be poets so that they may have imagination.

—**THOMAS A. EDISON, 1925**

PREFACE

I remember an epiphany one spring day, early in my research for this biography. It was about a decade ago, during the waning years of the previous century. I had been spending every Friday at the Edison National Historic Site in West Orange, New Jersey, seated at a plain wooden table from morning until dusk, delving into box after box of materials drawn from the three million documents in the archives. Directly across the hallway from the modest room I shared with curators George Tselos and Doug Tarr was the headquarters for the Edison Papers Project of Rutgers University, an ambitious, multivolume enterprise to document and microfilm the inventor's *oeuvre*.

Many of the gray Hollinger boxes I requested to be brought to me from the archives vault had already been surveyed and prepared for future filming by the Rutgers history of technology team. Selected items were flagged and annotated according to whether recto or verso should be photographed, or which direction the piece should be displayed. At a certain level in each box, I would come to a large sheet of paper that read, "Do Not Film," below which the Edison Papers Project had decided not to proceed.

After a while, I began to dig straight for that point first. The terrain below that sheet of paper was the undiscovered country, my intended destination. Of course I was interested in Edison's sketches for his many inventions, patent applications, notebooks filled with ideas, lists of parts needed and chemicals requisitioned and books ordered. But below the borderline of that sheet of paper, I found the precipitants, the residue of Edison's scientific work, odds and ends from an omnivorous mind—letters, diary entries, personal ephemera, doggerel verse, references to other people and places and subjects outside the strict "technological" purview.

. In an emblematic short story by Henry James, *The Aspern Papers*, a young scholar expends considerable energy trying to convince an elderly woman to allow him access to a locked cabinet wherein he believes reside the ultimate secrets of his subject. All I recall of this story now is the sustained compulsion to seek out and find what others have not, the conviction permanently implanted in the biographer's mind that there is always something new out there, some unseen scrap of paper (or maybe even *the other side* of a scrap of paper that no one ever bothered to turn over) that will provide an unrealized solution.

Toward what I mistakenly believed was the end of my research for this book, there came another epiphanic moment when Paul J. Christiansen, the (now deceased) president of the Charles Edison Fund casually mentioned to me that the complete correspondence of Mina Miller Edison, the inventor's second wife, had been preserved on more than twenty rolls of microfilm in an underground safe-deposit vault in a bank in East Orange, New Jersey. "But you probably wouldn't be interested in those letters," he offhandedly remarked, "because your book is about *Mr.* Edison."

I felt a surge in the center of my chest as I fought to keep my voice at a respectful and professional level, replying "Well . . . I know that's the case, but I probably should just take a look . . ." He shrugged, and told me to meet him the following day at the bank. A blizzard was raging. I fought my way through the snowdrifts and then downstairs. After Mr. Christiansen's key immediately broke in the lock, it took a locksmith three hours to pry open the filing cabinet–sized compartment sealed since Mina's death in 1947.

It turned out that as a dutiful daughter, she had written to one or another member of her family back home in Akron, Ohio without fail every single day of the forty-six years of her marriage.

I am often asked why I made the transition to Thomas Edison after having written biographies of William Carlos Williams, the poet, and Man Ray, the artist. To me, the progression is logical; the wellspring of the imagination does not distinguish between words or images or electrical circuits. Dr. Williams was insistent that "the poem is a machine made of words," and photographer Man Ray—*bricoleur extraordinaire*—was exceedingly adept with his hands.

Likewise, I see Thomas Edison as a truly modernist figure, because his most sophisticated invention of all was to craft himself into a cultural icon.

—N. B.

January 2001

PREFACE

Because you are about to read a big, American story with a household name at its centerpiece, I would like to caution you against looking *only* for the expected landmarks along the road. Yes, the diminutive, hard-of-hearing lad with a jaunty cap is teased by his smalltown schoolmates; the entrepreneurial teenager takes off to seek his fortune on the locomotives of the vast midwest and, after the Civil War, arrives nearly penniless in the teeming metropolis to make his fortune as a telegrapher. Yes, he finds success in grimy, gothic workshops, assembling a band of loyal workers to join him in a New Jersey hillside pastoral retreat, from which he unleashes the light bulb and the phonograph upon a startled world while sleeping less than four hours a night.

Yes, he marries a beautiful young Akron heiress, becomes a millionaire while still in his forties, moves into a mansion and builds the laboratory and factory complex of his dreams. Yes, he thumbs his nose at the bankers and "robber barons" and strikes an existential quest for inventions to transform middle-class American life. He makes movies. He spends a decade mining iron ore, and ensuing years manufacturing concrete. He serves his country nobly during wartime, then assumes the mantle of homespun philosopher and plunges into research on organic rubber, spending his waning years as a revered icon—one of the most popular figures of modern culture. He is showered with honors and medals, and dies a hero, a man for the ages—a wizard.

These are indeed true pieces in the mosaic of Thomas Alva Edison. They are the elements of a myth we continue to cherish, with good reason. But they are not the full portrait.

For here you will find a man who rose up from colonial ancestors gripped by a wanderlust, and inherited an intoxication for commerce. Here is *homo faber* in his most extreme form, a self-constructor defined by pathological immersion in labor. Here is a man struggling mightily to make the painful transition from shop-culture artisan to corporate manufacturer, ignoring industrial trends that ultimately led to defeat. Here is the contrarian, the publicity hound craving reflection and silence, the solitary thinker who unerringly hired the best, anonymous minds of a generation to feed his invention factories.

Here you will find a family man who raised six children by two wives, all of whom in various ways fought to benefit from his influence and tried to decipher his love, even as they ran from the painful, seductive spotlight of his fame.

And here you will come to understand a man who—contrary to the impression left by previous popular accounts—did not inhabit a hermetically sealed world of his own exclusive devising, but, rather, was the product of an era straddling two centuries, and became an enduring metaphor for its intellectual values. Here you will find Thomas Edison set into a context defined by Ralph Waldo Emerson, Walt Whitman, Henry Adams, Thorstein Veblen, John Dewey, Edward Bellamy, William Dean Howells, William James, and many other great American thinkers.

Invention, as an art, is unpredictable and intuitive. Thomas Edison was a pragmatic, authoritarian—and cranky—businessman; he was also a moody, ruminative writer equally at home in an abstract plane as seated at a workbench. As you will discover, not all his ideas automatically led to the production of utilitarian things. Not all of his inspired concepts found their way into the marketplace.

The drive to "make it new" sustained Thomas Edison into his ninth decade until he could not lift his head from the sickbed pillow—long beyond the time when he had reason to care about any vicissitudes of commercial success or failure.

In this respect, as a survivor, he remains larger than life; and ultimately, like the subject of any biography, larger than its pages will embrace.

EDISON

CHAPTER

1

IN THE BEGINNING, TWO DUTCH EDESONS—PRONOUNCED
with a long *A* at the start—arrived in America. The widow of a miller from
the Zuyder Zee with her three-year-old son, John, in tow, landed at Elizabeth Town, eight miles south of Newark (then called New Ark), New Jersey, in 1730.

More than a century and a half later, Thomas Alva Edison would be
honored to be claimed for membership by the Holland Society of New
York. He was insistently proud to be as much a Yankee, or "Janke" ("Little
John"), as his great-grandfather had been. Indeed, Dutch Creole was still
widely spoken in the northern colonies when the early Edesons arrived.
Even the stereotypes of sharp-dealing Yankee frugality and independence
have a role here, and Thomas Edison, master mythmaker, would have
deemed them appropriate for his ancestral roots.

The inventor's biography commences with a symbolic vignette, the
fatherless child setting foot in a New World. It continues with the conventional mystery that very little is known of John Edeson's formative years.
His mother never remarried, left a considerable estate to her only son, and
on October 10, 1765, in the First Presbyterian Church of Hanover, New
Jersey, John married Sarah Ogden. Of wealthy English descent, Sarah's family had emigrated from Hampshire to Southampton, Long Island, in 1640,

moving westward from there to Elizabeth Town, founded by her grandfather.

Reminiscing in self-consciously revisionist fashion—to Frank Lewis Dyer and Thomas Commerford Martin in preparation for their "complete, authentic, and authorized" two-volume epic, *Edison, His Life and Inventions,* personally sanctioned by the subject himself ("This book is published with my consent," the frontispiece pronounced over his trademark signature)— Edison in 1909 asserted that his great-grandfather's name was actually Thomas, *not* John, and that he had been a prosperous bank official in New York in the 1770s, a "stalwart patriot" who signed Continental currency. However, British control of Manhattan Island from 1776–84 would have made this a highly suspect, virtually treasonous activity. There was a "Thomas Edison" employed as a Chief Clerk in the Continental Congress during the period after the British evacuated Philadelphia in 1778, but he seems to have been of no family relation to the inventor. (A decade after the Dyer and Martin biography was published, Edison's younger daughter, Madeleine, had begun to assemble the rudiments of her own oral history of her father, taking notes on informal conversations with him. And once again, Edison regaled her with "my great-grandfather was a banker in Wall Street who signed Revolutionary money.")

John and Sarah Edeson settled as gentry in Caldwell, New Jersey, purchasing two hefty plots of land. The first parcel, now Essex Fells, was traversed by a stream, Canoe Brook, which John dammed to help power a mill. They further acquired fifty-three acres next to the Caldwell Presbyterian Parsonage, then known as the Horse-Neck tract of western Essex County, so named because of the pattern incised in the rural landscape by the meandering Passaic River. On this property, on the old New Ark Road, the Edesons built the house—still standing today—where the first seven of their ten children (five boys, five girls) were born.

During the Revolution, John Edeson, propertied and prosperous, a true Jersey Dutchman, remained loyal to the Crown, as did many in his wife's family. Male citizens in rebel-dominated territory were compelled to sign an "Oath of Abjuration and Allegiance," or be held to charges of treason. Tory John, as he was now nicknamed, stood fast and refused. Additional accusations, never substantiated, were brought against him: that he had been an informant and guide to the British army when it crossed New Jersey during the winter of 1776. On February 27, 1777, John Edeson was arrested and spent thirteen months and eighteen days in "close confinement" in Morristown. Only the intervening influence of his wife's family, the Ogdens, saved him from hanging. Although in extreme cases the property of many Loyalists was confiscated, this does not appear to have been John's penalty.

The political climate had become, to say the least, inhospitable for the Edeson family. In July 1783, they moved first to Staten Island, leaving behind, according to British records, "a farm of seventy-five acres, a house, a Negro slave, one black mare, fifteen sheep, three beehives and other goods to a value of 288 pounds."

"My great-grandfather was a rebel in the Revolution," Thomas Edison correctly told his daughter Madeleine on another, more lucid occasion, "and fled to Nova Scotia, where my father was born." Indeed, we next find the Edesons arriving in the Canadian Maritime province of Digby, in the Basin area of the harsh western coast of Nova Scotia, on May 29, 1784, escorted there from New York by a British naval vessel. Among more than forty thousand American Loyalists in political exile, casualties of the Revolution, the Edesons were awarded by the Crown a generous, unsurveyed land grant of five hundred acres, the "Wilderness Lots," hilly property ten miles out from the boundaries of Digby town, near Marshalltown.

Here they encountered the stark reality of pioneering. Like their compatriots, they had lived up until then in settled, familiar areas of North America. They had known comfort, and were forced to give up positions of importance and distinction to face a rough, makeshift existence. "I climbed to the top of a hill," the grandmother of a Loyalist immigrant wrote years later to one of her descendants, "and watched the sails disappearing in the distance, and such a feeling of loneliness came over me that, although I had not shed a tear all through the war, I sat down on the damp moss with my baby in my lap and cried."

The Digby Muster Roll of the day shows the spelling shifted to the more familiar "Edison," but four succeeding generations of the family, descendants living in Canada, tenaciously still speak of themselves with that long initial A sound.

The Edison children—Samuel, Adonijah, Mary, Margaret, Thomas, Sarah, Catherine (born in America); and Moses, Phoebe (the only child to die in childhood), and Marcellus (born in Canada)—grew fruitful and multiplied, a trait that proved itself astonishingly characteristic of the family, and the clan stuck together as farmers over the succeeding two generations. John and Sarah's oldest child, Samuel Ogden, married Nancy Stimson (or

Simpson), herself from a Loyalist refugee family, in 1792. They in turn had eight children: the sixth, Samuel Ogden Edison, Jr.—Thomas Alva's father—was born in Digby on August 16, 1804.

The frontier passion seizing America was mirrored in Canada by the growing awareness of the unspoiled appeal of the southern province of Ontario, especially to residents of the hardscrabble, farmed-out Maritimes. John Edison was approached by Colonel Thomas Talbot—former private secretary to John Simcoe, first governor of Upper Canada; and since 1801 the superintendent of the Canadian district of London, nestled between Lakes Erie and Ontario—who offered him the opportunity to purchase six hundred acres of virgin pine forest along the valley formed by the Big Otter Creek, spreading for two miles inland from Lake Erie in the new Talbot Tract (now Elgin County).

In the early spring of 1811, octogenarian John Edison led his extended family to yet another new home. They sailed out of the Bay of Fundy, down the east coast of the United States, landing in New York. They crossed over to Elizabeth Town, reversing steps taken decades earlier, and visited with the Ogden relatives. Purchasing wagons and oxen, they set forth on an overland trek of more than eight hundred miles—the popular turnpike route to the Canadian Northwest Territories, across northern New Jersey, eastern Pennsylvania, through western New York, back into Canada at Fort Erie, near Buffalo, then west across the northern edge of Lake Erie to Bayham, Ontario.

Besides the married sons and daughters of John and Sarah Edison, there were less than half a dozen families occupying the incipient community of Bayham. Colonel Talbot, a strict landlord, and something of a despot ("whose power," it was said of him at the time, was "infinitely more to be dreaded than that of the King of Great Britain") would not issue deed certificates to lots in the forest until certain "land duties" were fulfilled. Thus, we can presume that, within the first three years of their arrival, after which their names finally appear in Bayham survey maps, the Edisons would have been required to build log cabins for each family, of at least sixteen by twenty feet. It was prodigious work; the nearest sawmill was twenty miles north along the Creek at Otterville, and there were no roads of any kind in the territory, beyond blazed trails littered with stumps, impassable for weeks at a time during the summer and fall.

The Edisons lived in tents or brush teepees until their houses were ready. The size of the finished dwellings was limited of necessity by the pine and cedar growth available. Roofs were waterproofed with two layers of hollowed troughs made from basswood logs, or by sheets of elm bark lashed together in the Indian thatching method. Later on, shingles were split and shaved by hand from cedar, pine, or hemlock. Samuel Ogden Edison, Jr., revealed himself growing up to be particularly adept at this craft. Outbuildings might include a log barn for those tenants fortunate enough to own livestock; a hog and chicken pen; and smoke house, root cellar, and stone bread-baking oven.

Homesteaders were further required by Colonel Talbot to clear and plant ten acres of land per family. Given their late summer arrival, the Edisons must have brought much of their first winter's supply of food with them. Their New Jersey oxen were employed in springtime to work the land; seeds were sown among the tree stumps to be pulled later, huge pine roots piled in rows to make fences. Harvests were perennially threatened by hordes of wild turkeys and literally thousands of passenger pigeons, which had to be staved off with long poles, as gunpowder was too precious a commodity to waste. In the absence of nearby mills, grain was separated from chaff by allowing the wind to blow through the uncleaned mixture as it was poured slowly into a basket. An Indian method was used to make flour. After the center of a hardwood stump had been hollowed out by live coals, a heavy log was suspended a few inches above the grain, attached to a "spring pole" thirty feet long, balanced in a forked tree limb set in the ground. Two men would bounce the heavy log against the grain, using it as a leveraged mortar to crush the oats. Such "stump-mills" could supply several families with flour. The nearest stores were at the tiny town of Port Rowan, fifty miles away, in the Long Point settlement; obtaining such simple luxuries as tea and sugar required a day's trip by horseback.

Once Colonel Talbot's construction and cultivation stipulations were met, and, further, half the width of the normal road allowance of sixty-six feet was cut in front of each dwelling, each family received fifty acres of land free. Beyond that amount, the cost was three dollars per acre. The Colonel kept a penciled-in map recording these transactions, and was known to invalidate them autocratically if landowners opposed his Tory politics.

Tracking the extent of the Edisons' ultimate property domain in these early days is made complicated by the fact that the settlement's first land registry office was sacked and burned by American raiders during the War of 1812. However, by the time of the death of patriarch John Edison at the age of eighty-seven, Bayham Township was undeniably an Edison stronghold. Between them, John's children owned more than two thousand acres up and down the main east-west concession road, known appropriately as Nova Scotia Street. The 200-acre Lot No. 15, originally homesteaded by Captain Samuel Edison, Sr., and his wife, Nancy, later became the founding location of Shrewsbury town (named in honor of Colonel Talbot's Irish ancestor, the Earl of Shrewsbury), and, in 1824, was rechristened Vienna—where Thomas Edison as a boy came north often to visit his grandfather, who lived to the advanced age of ninety-eight.

Up until the early 1970s, Vienna local lore mistakenly held that young Thomas Edison actually lived in the village for a short time, and attended school there. During a family reunion in Port Huron, Michigan, Edison's boyhood days in Vienna were alluded to by visiting relatives who said that the inventor was probably unable to recall much about his occasional visits to the village. Edison astonished the gathering by producing pencil stub and paper from his rumpled suit-jacket pocket and sketching a detailed street plan of the town, marking his favorite playing spots. Peering back over six decades, he spoke of grandfather Samuel as a cranky, white-haired gent with a tall silk hat, who emphasized orders by rapping sharply on the floor with a gold-headed cane.

Captain Samuel maintained a healthy dose of the Edison feistiness about him. Even in middle age, hanging fast to his Loyalist roots, he signed up with the First Middlesex Regiment under Colonel Talbot to battle the Americans during the War of 1812, and participated in the British-Canadian victory at Detroit in August of that year. A decade later, a widower with eight grown children, Samuel remarried and had four sons and a daughter—Simeon Ogden, Mahlon Burwell, Fordice Warner, Charles Oscar, and Elizabeth Jane—by his second wife, Elizabeth Cook.

NANCY MATHEWS ELLIOTT EDISON

By this time, Vienna was a substantial place, prospering with lumbering and shipbuilding, tribute to its seafaring Nova Scotian heritage. Many Edisons went into this line of work, establishing at Estherville—where the Big Otter curved northwestward—the first shipyard in the region. The earliest boats built there were small single- or two-masted schooners designed for plying the Great Lakes, before the Erie and Welland Canals were opened to the sea. White oak trees dotting the hills along the Big Otter valley were perfectly curved for making ship's knees and stems. And the smooth white pine, soaring one hundred feet before their limbs jutted forth, made excellent spars.

By the time of his marriage on September 12, 1828, to Nancy Mathews Elliott, the ever eclectic, restless Samuel Ogden Edison, Jr., had temporarily abandoned the tailoring and carpentry trades in favor of tavern-keeping. The newlyweds were a striking couple. Samuel was "six feet three inches tall, sinewy, straight in nature, very homely. He could scarcely write his own name. Because he was born in Nova Scotia," a contemporary account acknowledges, "he was called a 'Bluenose.' "

SAMUEL OGDEN EDISON, JR.

Nancy, by contrast, was—fortuitously, it transpired—a well-educated, Bible-reading girl of seventeen. Born of English stock in the tiny rural settlement of German, on the westernmost edge of Chenango County in central New York state, she was the daughter of a Revolutionary war hero of the Continental Army from Claremont, New Hampshire, and granddaughter of a Baptist minister. "She was of medium height, stout, with large brown eyes, brown hair, of rather easy-going, mild disposition . . . The salt of the earth, [Nancy] was noted for her good example and the lessons conveyed by her in housekeeping and other domestic duties, how to appear and conduct oneself, and be courteous to everyone."

Over the next decade this historic generation of Canadian-American Edisons would have four children while still living in Vienna: Marion, born 1829; William Pitt, 1831; Harriet Ann, 1833; and Carlile, 1836.

Perhaps it was a combination of nationwide crop failure, economic uncertainty, and resentment against the quirks of Colonel Talbot's elitist, tax-hungry colonial oligarchy, powered by the grip of the "Family Compact" which guaranteed de facto authority to the old-line British landowners; or, perhaps it was the ferment of evening gossip in the Edison tavern, pitting advocates for democratic reform against entrenched monarchists in the Canadian government, coupled with his American wife's Jeffersonian passions, that led Samuel Ogden Edison, Jr., to take up the Liberal cause and join the Mackenzie Rebellion, thereby accomplishing an utter reversal in his family politics.

Building in the force of his rhetoric if not in the actual strength of numbers of followers throughout 1837, William Lyon Mackenzie, mayor of Toronto, launched his long-planned anarchist uprising aimed at overthrowing the Royal Canadian government, with the dream of replacing it with a gospel that had more in common with Andrew Jackson's populism. The British House of Commons had authorized the government of Lower Canada to spend money without legislative approval. It was taxation without representation all over again. Six decades after their ancestors had in effect refused to take part in the American Revolution, Mackenzie's radicals hoped to reenact a revolution on the Loyalists' new home ground, inspired by the model of the newly constituted, idealistically democratic United States. "Is then the country under the control of a lawless band of sworn villains?" Mackenzie asked, with soapbox drama. "If so, the citizens will have to form, not only political unions, but armed assocations. . . . My creed," he said, with remarkably Jeffersonian tones, "has been equality of each man before society."

Samuel Ogden Edison, Jr., was intoxicated by this spirit of reform and by the charismatic oratory of Mackenzie, the "fiery Scotsman." Local folklore claims that Edison secretly trained rebel volunteers in the woods around Vienna; we do know that he joined a march toward Toronto in early December with a file of Port Burwell and Vienna recruits, planning to meet up with the main contingent of farmers and "mechanics" under Mackenzie, assembling in a tavern on Yonge Street north of the metropolis. But the revolt was terribly short-lived, for word reached the provincial stragglers that the militia had easily crushed the uprising in Toronto and was now heading southwest to meet them. Abandoning homes and with a price on their heads, Mackenzie and many of his followers fled to the United States, where all along the American border, sympathizers with the cause of "Canadian freedom" harbored fugitives, blatantly violating American neutrality laws. Nine members of Samuel's Vienna brigade were not quick enough; they were captured at Bayham where the Talbot Street Bridge crossed the Otter Creek, and hanged in Middlesex jail in London, Ontario.

Young Samuel Edison was among those who fled during that demoralizing, frigid winter. Again, tales of heroism abound: the lanky, muscular rebel kisses his wife and little ones a furtive good-bye and strikes off on foot through the snowy woods, making for the safety of the border eighty miles away, a trip lyrically reminiscent, for some biographers, of the famous journey of Daniel Boone, when, escaping from the Shawnees, he traversed one hundred and sixty miles of forest in four days, eating but one meal the entire time.

The militia were on Samuel's heels. Arriving at the old Vienna homestead, they were met by his crafty stepmother, who pretended to be extremely nervous, roaming from the main house to the root cellar and back, conveying the impression that Samuel might be hidden in the area. Family lore has it that the soldiers, taken in by Elizabeth Cook Edison's ploys, remained in the village until nightfall, then raided the cottage, hoping to surprise the rebel. By this time, he was many miles away.

Did Samuel hop a lumber barge bound diagonally across Lake Erie to Cleveland? Did he risk taking the Sarnia bridge, only to be recognized by an old friend and retreat into hiding again? Or did he, as is most likely, scramble across the windswept, icebound, serpentine St. Clair River where it narrows to less than a mile in width before opening into vast Lake Huron?

CHAPTER

2

WITH SAMUEL EDISON'S BREATHLESS ARRIVAL ALONE IN THE sawmill town of Port Huron at the bitter end of 1837, less than one year after Michigan had achieved statehood, he brought his family odyssey full circle—back to America.

Known as the gateway to the Northwest frontier, the four villages—including a Chippewa reservation—that would collectively become Port Huron at the time of Samuel Edison's solitary arrival were enmeshed within "an almost impassable marsh of cedars," according to Lieutenant Samuel P. Heintzelman of the nearby Fort Gratiot military post. "You cannot see the earth for the moss, fallen timbers, and roots," he lamented, and in wintertime the temperatures could plummet to seventeen degrees below zero.

The fort was built at the crucial junction of the Black and St. Clair rivers during the War of 1812 for the purpose of securing the strategic "outlet of Lake Huron," assuring United States naval vessels safe passage into the Great Lakes. In the 1820s, Fort Gratiot was used as a mission school, where agriculture and reading and writing in English were taught to European immigrants and neighboring Indians. It was structurally upgraded in 1830, complete with a ten-foot-high log palisade with bastions, but six years later, military posts along the Great Lakes were abandoned as a cost-cutting measure, since it was felt by the secretary of war that regional defenses should be handled by local militia. In the wake of the Mackenzie Rebellion, and in fear of further disturbances, the fort was regarrisoned in the spring of 1838.

Despite this apparently inhospitable setting, and the unsettling vicissitudes of military protection and abandonment, almost one thousand people called the area home, and the environs could hardly be called uncivilized. Several large farms had been laid out, or "platted," in the usage of the day. With statehood, work had begun on a railroad which was to run all the way to Chicago. There was even a town pump, bakery, newspaper (The Lake Huron *Observer*), Baptist church, and hotel, nobly named the Columbia, under construction when Samuel Edison came on the scene. The Black River Steam Mill, first in the Northwest Territory, was also built at Port Huron, complete with a neighboring boardinghouse for the workers. Local historians claimed that Samuel Edison stayed there for some months, or possibly with relatives, before moving on to Detroit; to Peru, Ohio, in Vredenburgh township; and, finally, in 1838, northeast through Norwalk to Milan, Ohio.

This pivotal phase of Edison family history was not clarified by Thomas Alva when he sat down in 1908 to write the first chapter in longhand of his authorized memoirs; his account began in medias res, during his own childhood years in the 1850s. The retrospective descriptions he does provide portray Samuel-in-exile rather desultorily "wandering the southern shore of Lake Erie earning a living here and there by lumbering." Wanderlust sets yet another deep-seated mythic tone in the Edison saga, as we shall see, but in fact Samuel was too pragmatic for gratuitous meandering at a time when his main concern was to find a suitable situation from which to send for his wife and children.

Milan at this precise moment made great sense as a home for a young, entrepreneurial tradesman. At the edge of the Western Reserve in northeastern Ohio, the lands lying between Lake Erie and the forty-first parallel on a flat-topped, glacially sculpted hill overlooking the Huron River, Milan—first settled by Moravian missionaries—came into existence at the turn of the nineteenth century as part of the "Firelands" compensation claims filed by the state of Connecticut after the American Revolution. Those Connecticut residents who had lost property to the British during the War were given equivalent plots to settle in a 500,000-acre strip running south of Lake Erie through Indiana into Illinois.

The Huron had always been an artery of commerce, but in the mid 1820s, was only navigable as far as Abbott's Bridge, some three miles north of Milan. A schooner originating in Abbott's Bridge town could sail the length of Lake Erie to Buffalo, then navigate the Erie Canal and the Hudson River to discharge cargo at New York City. The Milan town fathers envi-

MILAN, OHIO

sioned the undeniable advantages of a viable, deep waterway for barges, connecting their village to Abbott's Bridge, and thereby tapping into a vastly broader economic network, changing inland Milan into a virtual seaport.

Planning, capitalization, and construction of the Milan Canal took more than fifteen years. The Huron River was dammed; a towpath was built through marshlands, requiring tons of fill; then, aside from digging the deep draw ditch required, two locks would be needed, because the land fell off seven feet in the three miles from Milan to Abbott's Bridge. Hired labor came not from the professionals who had set the standard with New York's Erie Canal, but from local farmers manning shovels and two-handled scoops behind oxen in shale and muck.

On July 4, 1839, the Canal opened, momentously. The 150-ton schooner *Kewannee*, greeted by a crowd of five hundred people, arrived with great fanfare, fireworks, and music, "under the salute of cannon, lying along side of the dock." On behalf of "the Ladies of Milan," the schooner's Captain Moran was presented with a twenty-eight-and-two-half-star American flag, and told by Miss Maria Butman of "the esteem in which [we] hold your Class of the American Nation. Nothing can be more gratifying to us, than to bid you as Commander of this first Schooner that ever visited our town, a hearty welcome."

After remarks in acceptance, the Captain ran the flag up to his masthead, and invited any of those willing to take the short journey up through the locks and back again, whereupon his passengers disembarked and retired to the Eagle Tavern for a round of toasts, "accompanied by more cannon and loud cheers," according to the enthusiastic Norwalk *Reflector* of July 9, 1839.

No less than thirteen "Regular Toasts" and ten "Volunteer Toasts" were proposed, including tributes to "the state of Ohio . . . internal Improvements: the first and best are Canals . . . the Spirit of '76 . . . the Memory of George Washington . . . and to The Citizens of the County generally, who have mingled with us in celebrating this day. May health and happiness attend them."

It is difficult to imagine that Samuel Ogden Edison, Jr., and his reunited family were *not* witness to the raucous festivities on this golden day, the cusp of an unprecedented spurt of prosperity, the transformation of a quiet backwater into a nexus for trade. With financial aid from his Canadian relatives and a loan from Captain Alva Bradley, an American friend and ship owner from Vermilion, Ohio, Samuel had managed to set up a modest shingle mill and feed grain establishment in Milan. Bradley's ships shuttled back and forth between Port Burwell in Ontario and Abbott's Bridge, carrying Canadian cedar in three-foot double lengths to be split in Samuel's mill. By importing the raw "bolts," as they were called, Samuel avoided paying duty that would have been charged had the shingles been manufactured in Canada. A month before the canal opened, Samuel sent for Nancy and the children. Ferried over by Captain Bradley, they lived in rented quarters in town. Bradley subsequently spent nine winters boarding with the Edisons when the lakes were impassable, and, for a time, was engaged to Nancy Edison's niece, Ann Dunham.

With the birth of Samuel Ogden Edison III on March 5, 1840, the Edison clan was in dire need of a proper house. They purchased a prime lot in Nancy's name from Ebenezer Merry, one of the town's elder founders, builder of Milan's first water-powered grist mill, "a friend of the poor, protector of orphans and benefactor of the needy." On his hillside tract—virtually a hundred yards from the new canal basin where the ships unloaded their cargo—Samuel constructed a three-story home in Greek Revival style, a popular architectural mode of the day. The ground floor kitchen opening upon a sloping flower garden, and the first- and second-floor bedrooms, boasted unobstructed views of the incessant canal traffic. Built to last—the home is still standing on the bluff today, unchanged—the lower-level walls were twenty-two inches thick, of handcut sandstone, mined from area quarries. Window sills and lintels were likewise handcut and sawed stone.

Upper walls were two layers of brick laid in Flemish bond. Floors through-out were wooden-pegged locust wood. The roof was clad in cedar shingles, of course, hand-tapered and fashioned by Samuel, with a maul, frow, and drawknife which can still be viewed at the home.

It was a cozy place. The sitting room, also called the everyday parlor, was reserved for receiving ordinary callers who would be seated in matching horsehair side chairs by a wood-burning stove. The adjoining bedroom was furnished with a straw-mattressed rope bed graced by a Jacquard loom cov-erlet. The formal parlor was used only on family occasions. In the downstairs kitchen, Thomas Edison recalled in later years, his mother kept a birch rod "disciplinary switch" readily available behind the heirloom Seth Thomas clock "for his benefit."

The Edisons were clearly in the right place at the right time. "A few years since," the Milan *Tribune* conceded, "this was a dull inland village, doing but a comparatively trifling amount of retail business, without pre-tensions, and almost without hopes; but, excepting Cleveland, it is now probably the first town in value of its exports between Buffalo and Detroit. . . . The coast is clear, the skies are bright, and we see no room for despon-dency or misgivings. . . . If Milan is true to herself, she must be *the* town for successful trade."

"Who would have believed, six years ago," the *Tribune's* rhapsodizing continued, "when the citizens of Milan were struggling along with their canal improvement, amid the jeers and taunts of their neighboring towns, a laughing stock and byword over the county, that in her future commercial history the record would be made of the shipment of 25,000 bushels of wheat in a single day?" The town was touted as "the greatest grain port in the world," rivalling Odessa, Russia.

With the canal in full operation, farmers coming north preferred rolling their wheat the shorter distance to Milan; and, for that matter, schooner companies could bring their vessels right alongside the main wheat supply. The annual harvest trek drew wheat growers from as far as one hundred miles south, averaging one hundred wagons a day, reaching down to the national highway crossing Franklin and Muskingum counties, and embracing in all a region of fifteen counties. At peak times, six-horse-team rigs were lined up for three miles waiting to unload. On some days, more than fifteen ships cleared the port. Wheat flowed into Milan faster than the vessels could ship out, giving rise to a thriving warehouse industry astride the river. As a boy, Thomas Edison was known to play, flirting with danger, along the rims of the grain storage bins down the hill from his house, where once, he said, he lost his balance and fell in. Besides wheat, corn, and oats, hogs, cattle, lumber, barrel staves, potash, whiskey, and wine were also shipped out of Milan. Awaiting transactions, the farmers would unload at the canal basin, then climb the hill to shop in town, staying for a couple of days to pick up supplies and socialize with other merchants in Milan's taverns or at the Exchange Place hotel or the Milan Inn.

In addition to blacksmith shops, harness repair shops, and general stores springing up to meet the merchants' needs during the flush times of the 1840s, Milan enjoyed a shipbuilding boom as well. The free white oak timber growing in abundance in the area was ideal for ship construction, and the industry brought many more skilled laborers to town, including wrought-iron workers, sailmakers, and carpenters. Between 1841 and 1867, Milan yards produced more than seventy-five lake schooners, and also became involved in the lucrative business of constructing revenue cutters for the United States government.

Even while economic security came to the Edisons of Milan, brutal winters brought tragedy: Young Carlile died at six in 1842. Little Samuel Ogden died the following year just before another daughter was born, Eliza, who lived only until she was three, dying as Nancy, diabetic, nervous, clad in black, was pregnant with her seventh and last child. She spent three hours a day during her *accouchement* knitting a "thousand-shell pattern" coverlet for him. Born in the depth of winter, the boy was delivered at home in the first-floor birthing room by Dr. Leman Galpin, who lived just across the street and a few houses down, and did not have far to come when Samuel Ogden knocked on his door on the snowy night of February 11, 1847.

"Thomas" for his grand-uncle; "Thomas," again, for his uncle; and "Alva" for family ally Captain Bradley, was a frail infant with a large head; Dr. Galpin feared brain fever (for lack of a better term), and the newborn was watched carefully. A devout Presbyterian, his mother prayed for him. But Thomas Alva was a survivalist, first and always.

It is difficult to extract fact from fancy in recounting Thomas Edison's earliest childhood, because nearly all the anecdotal information we have comes down through the imprimatur of Edison himself. Rather than be stymied by Samuel's terse disclaimer, insisting his son "had no boyhood days," it is instructive to view what apocrypha remain as the younger Edison's building a case for the formative years of a child who must have been destined to become an inventor.

When "Little Al," as he was nicknamed by all, was two years old, his sister, Marion, married Homer Page, a local farmer, who owned a big spread of land in Huron, Ohio, three miles out of town. On one of many family visits to the Pages, Al wandered off to explore the grounds, as he was known to do, but after a while, Marion became concerned. "Where's Al? I haven't seen him all afternoon!" she asked her husband as he came in from his chores. "Well, by cracky, neither have I," replied her homespun husband, according to the officially sanctioned memoir *So Rich A Life*, by Arthur Palmer.

THOMAS EDISON WITH
SISTER HARRIET

Homer went to the barn first, "knowing how Al loved to commune with the livestock. . . . He spotted Al over in a corner. 'What in tarnation are y' doin' a settin' there in that box with all the straw? Ain't that the goose's nest? Did y' fall in an' make a nice om'let and yer afeerd t' git up? Wait till y' sister sees the seat o' y' pants.'

" 'No, no, Homer, I didn't fall in,' replied Little Al, 'I saw baby chickens come out of eggs the old hen was sitting on so I thought I could make little gooses come out of the goose eggs if I sat on them. If the hens and geese can do it, why can't I?'

" 'WHAT!' exploded Homer, leaning against a roof support and laughing so hard the uproar scared all the poultry in the place. As they ran cackling and squawking about he regained some of his composure. 'Well, mebbe it's b'cause you're not a hen or a goose. Heh, mebbe, y'are sort of a goose at that,' he managed to eject between guffaws as the disgusted experimenter reluctantly acknowledged defeat in his first challenge to nature."

Back at the house, Al's big sister was more understanding. "Mother-like," Palmer tells us, and, significantly, in language without dialect, "Marion drew Al to her side to assuage his disappointment and soothe his hurt with kind, sympathetic words, 'It's all right, Al, you did a very smart thing even if it didn't work. If no one ever tried anything, even what some folks say is impossible, no one would ever learn anything. *So you just keep on trying,'* she presciently intoned, *'and maybe some day you'll try something that will work.'* "

According to another popular account, *Young Edison: The True Story of Edison's Boyhood*, by W. E. Wise, Al was "inquisitive as a young red squirrel . . . a nervous little question-box. . . . Before he could talk straight, he ran down to the Milan shipyards and watched the men building boats for Great Lakes shipping. There, he asked the shipbuilders hundreds of questions. Why could you see a hammer hit a board before you could hear it, if you were at a distance? Why did you have to fit joints carefully? What was pitch made of? Some of the questions the men could answer and some they couldn't. One of them jokingly said to Samuel Edison, 'It would save time to hire a man especial to answer your young one's questions.' " And if his father could not answer his questions either, Al would turn to him and ask "Why *don't* you know?"

Stories of Al's precocious powers of observation and retention also abound: how he would gaze for hours out his bedroom window and watch the procession of grain wagons rumbling to and from the docks; how he would memorize the rowdy songs of the lumber gangs, and build "little plank roads" of his own design with debris collected from the nearby sawmills; how he would laboriously copy in his foolscap tablet the signs over village shops.

The tales of Al's youthful curiosity consistently stress the consequences of his compulsion to discover phenomena or validate nascent theories through direct experience: investigating a bumblebees' nest in the corner of a pasture, he was attacked by an angry ram. Exploring new ways to shorten a skate strap, the tip of his middle finger was cut off by an errant axe. Deciding birds could fly because they ate worms, he mixed mashed worms with water and convinced a neighborhood girl to drink the concoction. She got sick, and he got "switched."

As a boy in Milan, Thomas Edison saw his first telegraph. And two modern inventors lived there contemporaneously with the Edisons: Zenas King, builder of iron bridges; and Samuel Winchester, designer of one of the first hydrogen-inflated balloons. But it is not known whether Al ever met or came under the early influence of these two gentlemen.

In a home where Al's closest siblings, William Pitt and Harriet, were teenagers, Father Samuel emerges as the tolerant guardian, allowing Al to tag along with him on Saturday nights, buying his son peppermint sticks to suck on while the sociable townsmen convened to chat on street corners, then ambling over to the square to hear the band concert; or driving with horse and buggy into the countryside to select choice timber for his mill.

On the subject of his mother Edison was unfailingly reverent throughout his life. He spoke of her in hushed, serious tones, without embellishment, tempered by the fact that although she did not die young, Nancy Elliott Edison ironically missed seeing her son reach the pinnacle of success at midlife—from which he looked back upon her as the informing spirit: "I did not have my mother very long," Edison told an interviewer, "but in that length of time she cast over me an influence which has lasted all my life. . . . I was always a careless boy, and with a mother of different mental caliber I should have probably turned out badly. But her firmness, her sweetness, her goodness, were potent powers to keep me in the right path.

"My mother was the making of me," the inventor declared. "She was so true, so sure of me; and I felt I had someone to live for, someone I must not disappoint."

Telling words, if they are meant deliberately to tap into the logical imperative that great men *must* have great mothers; or, equally plausible, if they are meant to justify Thomas Edison's lifelong, compulsive drive toward solutions to the most complex problems. And what greater invention can there be than one's own *personae*?

Little Al grew up with the railroads, and railroad fever struck Ohio as it did the rest of the nation. By 1836, the first few miles of functional rail lines laid by the Erie and Kalamazoo connected Toledo with Adrian, Michigan— a town Edison would eventually visit as a teenage telegrapher. These first cars were drawn by horses over oak rails. Steam locomotives were placed in service the following year.

Local businessmen throughout Ohio wanted to get in at the beginning of this profitable trend, which often required buying canal lands outright from the state in order to establish a new right of way. As a consequence, Ohio's 299 miles of track in 1850 increased tenfold over the decade.

This uncoordinated, speculative hysteria challenged the new canal system, still under construction. By aiding the epidemic spread of the railroad network, the state essentially undercut its own waterway investment—buying instead into a faster, cheaper, more flexible mode of transportation, one that was not subject to a five-month winter freeze, when canal boats lay idle.

Inevitably, the conflict came home to Milan. In 1848 the Cleveland and Ohio Railroad sought right of way through the town. The Milan Canal Company rejected the appeal outright. Local farmers refused to accept the newfangled railroad. Engine-stack sparks would burn grasses in their fields; the shrill whistles scared livestock, they said; smoke and soot sullied everything; poorly maintained tracks of uneven grade resulted in fatal wrecks— the protests went on and on. And so the railroad rerouted through the neighboring town of Norwalk, four miles to the south, bypassing the proud, independent citizens of Milan.

Two years later, the Junction Railroad was projected, to run from Cleveland westward to Toledo via Sandusky. This incursion required bridging the Huron River, and once again, the Milan Canal Company, fearing that such a bridge would interfere with free and uninterrupted use of their canal, "resolved that [they] would resist the building of any bridge or any other structure across the navigable waters of the Huron River, between the basin of the canal at Milan and Lake Erie . . . as a violation of the Ordinance of 1787, establishing the North West Territory." It was fierce, rugged (if retrograde) American individualism at its best—but the railroad won the day in court, and built the bridge anyway.

The even more grandiose invasion of the New York, Chicago, and St. Louis Railroad subsequently cut right across the town of Abbott's Bridge, Milan's northern neighbor, with a gigantic trestle requiring one million feet of lumber to construct.

Milan's shortsighted local pride was its swift undoing. As the inexorable railroads simply forged on—they could destroy as well as build—the town precipitously lost commercial viability as a hub for trade. Numbers tell the story: In 1847, Milan's banner year as a grain port, 917,800 bushels of wheat were shipped through its bursting warehouses. With the railroad completed, the amount shipped plummeted by three-quarters, to 258,778 bushels. Grain terminals went up in other towns on the railroad lines; wagon roads with their once glorious Conestoga parade leading to Milan were now diverted to markets nearer the source of agricultural production.

Samuel Edison could read the writing on the wall. It was time for the Edisons to pick up and move on once again.

CHAPTER

3

IT WAS TO THE THUMB OF MICHIGAN THAT SAM EDISON RE-
turned, to the familiar terrain of Port Huron. During the intervening years, the town's sawmills had flourished. The Black River was glutted with logs from shore to shore, and the riverjacks worked like slaves. Port Huron was becoming a city: streets formerly paved with river sand were now cobble-stoned, some even lighted with gas lamps. Sewers were dug, and the first dry goods store planned.

In the spring of 1854 the Edisons made the fifty-mile trip to Port Huron on the steamship *Ruby* northeast from Detroit, through Lake St. Clair. Thomas Alva was only seven years old, and paid half-fare—seventy-five cents. The family moved into a two-story frame home on ten acres, south of Fort Gratiot, adjacent to the Soldier's Cemetery. Surrounded by a stand of pines, it perched on a low rise by the St. Clair River, facing north toward Lake Huron.

Familiar to locals, The White House, also called The House in the Grove, was built in 1840 by Reuben Hyde Walworth, the last chancellor of New York, for his daughter, Mary Elizabeth. Her husband, Edgar Jenkins, was the Fort Gratiot sutler—storekeeper, postmaster, and fishery lessee. Mary did not relish living on the grounds of the fort proper; her father was able to obtain a special license from the secretary of war to construct her house offsite. Sam Edison bought the place from its second owner, Port Huron lawyer and lumber merchant Bethuel Farrand.

It was nearly square with columned balconies shading porches at the front and rear. Inside, a central hall ran north and south connecting four high-ceilinged rooms—master bedroom, kitchen, parlor, and sitting room, according to Edison's firmly detailed memory sketch—with huge windows offering broad river and lake views. There were six bedrooms upstairs, a cellar with a wooden floor and fireplace complete with oven (unusual for the period), and storage rooms below. Outbuildings included a carriage house, barn, icehouse, woodshed, and well. A vegetable garden and pear and apple orchards covered the grounds. An 1850s "Bird's Eye View of Port Huron" reveals the Edisons' house at the very edge of a pine grove, near the army's parade ground, with the one-hundred-foot-high wooden "Edison Tower" next to it. The tower was a brainstorm of the ever-resourceful Samuel, and although ridiculed by the townspeople as "Edison's Tower of Babel," he considered it "a folly that paid." Any visitor willing to provide the 25-cent fee could climb to the top, enjoy the sights and sounds of the bustling river traffic, and taste the incessant, bracing wind.

It was a big house for a diminishing family. Marion had stayed behind at her farm in Milan; big brother William Pitt became a partner in a livery stable in town soon after the move; and Harriet Ann ("Tannie") got married. Father Samuel was out and about, continually dabbling in his usual diverse mixture of revenue-producing pursuits. Aside from his foray into the tourist trade, he ran a grocery store, dealt real estate, sold timber, and converted most of his lush acreage into a thriving truck farm.

Rather than leave Thomas Alva to march to the beat of his own drummer, Nancy, ever the devout Presbyterian, enrolled her child, albeit briefly, in the Family School for Boys and Girls. Established by the Reverend George Engle, the institution was advertised in the Port Huron *Commercial* as "a family of young persons pursuing a course of thorough instruction, upon Christian principles. . . . The rules of a well-regulated Christian family are the only restrictions." As a matter of fact, the Reverend's niece was the head teacher, and his children, Willis and Mary, were among Thomas Edison's fifteen classmates. Along with the traditional subjects, for an additional fee of $20 per semester, one could receive lessons in "piano and melodeon." For a short time, the boy also attended a one-room public school in Port Huron, forty pupils ranging in age from five to twenty-one.

Al was not a well child. He suffered a bout of scarlet fever soon after the family's arrival in Port Huron, and was subject constantly to colds and upper respiratory afflictions. He began to notice a deterioration in his hearing and may have suffered variable hearing loss as a result of fluid retention in his middle ear. This chronic problem, in turn, may have led to attention-deficit disorders, for Al was most decidedly not a natural student, according to the prevailing definition; or, more tellingly, he did not respond well to rigidly systematic teaching methods.

Nineteenth-century educational philosophers, proudly grounded in republicanism, Protestantism, and capitalism, considered the best schools to be analogous to factories, places where efficiency, manipulation and mastery, promptness and industry, were valued. School was supposed to be a place of morality and discipline above all else; children learned to be deferential, restrained, and obedient. Character formation and the careful exercise of intelligence were paramount. Silence and punctuality were at the top of every standard list of classroom virtues. Indeed, public education was established to alleviate "crime and poverty, poor work habits, and idle youth" by cutting off the potential sources for these social problems at the earliest possible stage in cognitive life. "The idle boy is almost invariably poor and miserable," declared Wilson's *Third Reader* (1860), while "the industrious boy is happy and prosperous."

One of the first strategies developed by early school advocates to keep standards high was, therefore, the "discovery" of learning problems. As opposed to eighteenth-century cultural mores, according to which if the child did not learn, it was assumed to be the fault of the methodology of instruction, learning problems were established (or, perhaps, created) by mid-nineteenth-century American pedagogues as a defensive response, a way to weed out "difficult" students. It was also believed that a lack of inherited abilities was to blame when students did not succeed. In either case, the failure was assumed to have emanated from the children and their family backgrounds, not from the teacher.

This is the milieu in which we must consider the oft-quoted diagnosis of the legendary, abusive "Mr. Crawford," Thomas Edison's public school master, who—angered by the lad's inattentive, "dreamy," distracted behavior, frustrated by his tendency to drift off during recitations, to draw and doodle in his notebook instead of repeating rote lessons—cuffed and ridiculed Al in front of his motley classmates.

Teachers saddled with disaffected students like Edison were judged by how many of their pupils were promoted from one grade to the next, and they needed to rationalize the actions of children who were "not apt." Sure enough, "One day," Edison recalled with bitterness many years later, "I heard the teacher tell the visiting school inspector that I was *addled* and it would not be worthwhile keeping me in school any longer. I was so hurt by this last straw that I burst out crying and went home and told my mother about it.

"Mother love was aroused, mother pride wounded to the quick," Edison continued. This trauma was the beginning of the end of Thomas Edison's conventional schooling, a period that lasted a total of three months. His indignant mother "brought [him] back to the school and angrily told the teacher that he didn't know what he was talking about, that I had more brains than he himself." Mrs. Edison summarily pulled Al out of Mr. Crawford's domain, and commenced instructing her son at home: "She was determined," biographer Arthur Palmer enthused, "that no formalism would cramp his style, no fetters hobble the free rein, the full sweep of his imagination." As poetically enlightened as this may seem, teaching one's child in the home was considered an appropriate role for a self-respecting mother in the 1850s. We must also bear in mind that in the still-rustic Midwest of Thomas Edison's youth—where every hand that could be mustered was urgently needed to tend to chores and household duties, and to do his or her fair share—beginning at age ten, most children only attended the winter session of school, those few months when farm work tapered off. Only a small minority of teenagers attended secondary school of any kind in nineteenth-century America.

Al's hearing problems, Nancy's protectiveness, the absence of older siblings, and the relative geographical isolation of their house more than a mile from town all contributed to turning the boy inward, accelerating his aptitude for and inclination toward reading. Thomas Edison went on to become a voracious, even omnivorous, life-long reader. Living on his estate in Llewellyn Park, New Jersey, from the mid 1880s onward, he maintained a standing account at Brentano's bookstore in Manhattan, and was known to order literally hundreds of dollars' worth of books a month. If Edison came across a subject that interested him, he would request every available book and periodical on that subject for immediate delivery, without a moment's hesitation.

According to Edison, one of his earliest American literary heroes was Thomas Paine. In 1925, following a generous, pacesetting contribution of $300 to the Thomas Paine Memorial Library Building Fund, Edison wrote a laudatory introduction to a new biography of Paine, recalling that at the age of thirteen, he discovered the eighteenth-century political theorist's writings on Samuel's bookshelf. An autodidact himself, and follower of

Archimedes, Paine was a man who made independence into virtually a religious pursuit, who placed civil rights before natural rights. It is clear why Edison would have admired him, and his insistence that "under all discouragements, [man] pursues his object, and yields to nothing but impossibilities."

"Paine's works are a crystallization of human reasoning," Edison observed in 1925, but, of course, he had "always been much interested in Paine as an inventor." A less obvious but more intriguing source of the affinity Edison felt with his subject was the constant criticism Paine received for his libertarian ideas. "For writing his next great book, *Age of Reason*, Paine was burnt in effigy, and vilified outrageously," Edison tells us; but, identifying strongly, Edison concedes that it is the lot of "the world's great reformers, from Christ on down" to be thus treated, "crucified and burned at the stake . . . [as] victims of intolerance."

As Sam was rather limited in his literary tastes, we question whether the Paine volume belonged to father or mother. Childhood chums remember Nancy summoning her son indoors from play on a lovely summer's day to attend to his reading and writing lessons. Other people who knew Edison as a boy recalled that he also delved into David Hume's *History of England*, Edward Gibbon's *Decline and Fall of the Roman Empire*, and Barnas Sears's *History of the World*. To help meet Al's curiosity about the natural world, his mother gave him R. G. Parker's standard *A School Compendium of Natural and Experimental Philosophy*. This richly illustrated tome laid great emphasis on construction of simple experiments, demonstrating in detail a variety of batteries and electrical toys. Following the instructions, Al built two rudimentary machines to generate electricity, one by friction, the other by magnetic action. The Parker book also contained a detailed chart of "Morse's Telegraphic Alphabet," with the appropriate dots and dashes for each letter.

Early biographers note that as Al's physics and chemistry dabblings and concomitant hoarding of arcane, but (to him) necessary, materials such as mercury, feathers, sulfur, beeswax, alum, cornstalk pitch, and acids grew to an intolerable level, his mother banished him to the basement to set up shop. The 1940 MGM movie version of *Young Tom Edison*—starring Mickey Rooney as the spunky, intrepid lad with the jaunty cap and the devil-may-care, unfailingly sincere, indomitable attitude—reinforced this aspect of his upbringing. Mocked by schoolmates for his head-in-the-clouds clumsiness, banned from attending classes by a cantankerous, "old-maid" schoolmarm, intrepid Al labors over thick volumes, reading aloud to himself, leaning

over a wooden table in the gloomy cellar. He is surrounded by shelves laden
with bottles of all shapes and sizes, taking a break now and then to indulge
in his favorite treat, apple pie and milk, brought downstairs to him by his
loving mother. Nancy staunchly defends the lad as being different, "*not like
other children*," embracing her son warmly, assuring him, misty-eyed and resolute, of the "wider world out there beyond Port Huron," where, one day, he
will make his mark.

In the summer of 1976, a team of anthropologists and archeologists
from Oakland University in Rochester, Michigan, under the direction of
Professor Richard Stamps, began the excavation of the Edison homestead at
the corner of Thomas and Erie Streets in Port Huron. Fresh evidence unearthed at the site supports the existence of Thomas Edison's first "laboratory." A large distribution of bottle glass and ceramic fragments in this
heavily used vicinity lends credence to the assertion that Thomas Edison
worked here, "far from the madding crowd," among bushels of potatoes,
bins of carrots and onions, and crocks of peaches.

As Edison reached adolescence, the "miraculous decade" in the development
of the American railroad system was in full flower. On November 21, 1859,
the Grand Trunk Railroad of Canada, having reached Point Edward, Ontario, just across the St. Clair River from Fort Gratiot, loaded its trains onto
ferry boats and formally established a depot in Port Huron, Michigan, clos-

GRAND TRUNK RAILROAD DEPOT AT MT. CLEMENS, MICHIGAN

ing the crucial link to Detroit. The Chicago, Detroit, and Canada Grand Trunk Junction Railway was another step in the accelerated nationalization of the railroad as individual lines connected into a full-fledged network. The Lake Shore Railroad and the Michigan Southern Railroad had been completed five years before—through service from New York to Chicago was already possible. The commerce of North America was transformed as major north-south lines like the Grand Trunk intersected with east-west arteries, and the canals were all but forgotten.

In Port Huron the restored, brightly painted train station by the gently curving riverside tracks can still be seen today, just down the hill within walking distance of the Edison home, temptingly close at hand for young Al. At twelve years, he was impatient to move on, make his own way. Sam had gone to Milan to fetch his son after a recent trip to visit sister Marion and brother-in-law Homer. There were many boxes to ship home. Father and son were at the train station among their piles of goods. While waiting, never one to be idle, Al found a pot and brush, and marked the crates clearly for delivery to Port Huron. The station master, who knew Sam from the old days, said he did not have a man around the place who could do such a fine job, and offered Al, then and there, a spot for $30 a month plus room and board.

But the boy had set his sights on bigger opportunities. Nancy grudgingly agreed when Sam helped Al obtain the coveted position of "news butch" on the Grand Trunk, departing Port Huron daily at 7:00 a.m., arriving more than three hours later in Detroit for a stopover most of the day, returning to Port Huron at 9:00 p.m. Near this renovated Port Huron depot a heroic, bronze statue of Al has been positioned alongside the St. Clair's rushing current. Larger than life, the boy strides boldly forward, a confident smile on his face, scarf blowing in the breeze, a crate of merchandise for sale cradled under his right arm.

Al's wares, sold during the train ride from his well-stocked tray, extended beyond newspapers and magazines; candy, peanuts, cigars, postcards, guidebooks, jokebooks, dime novels (which Al read avidly before selling), fruit, sandwiches and other sundries were also dispensed. Robert Louis Stevenson, after a Western trip, listed the butches' diverse stock as including "soap, towels, tin washing basins, coffee pitchers, coffee, tea, sugar, and tinned eatables, mostly hash or beans or bacon."

Another touring Scotsman, the Reverend David Macrae, praised the butches' social savvy and sense of timing, ". . . by-and-by making appearances with a basket of apples, or nuts, or grapes. . . . Should he fail to tempt you with these, he returns with maple sugar, or figs, or candy," and then, finally, when your appetite for sweets has been satisfied, he "re-appears with an armful of books, magazines, or illustrated papers."

News butches were not always well-liked by their fellow crew members because, as Stevenson trenchantly pointed out, at times they "imitated the manners of the more vulgar drummers or travelling salesmen who were their steadiest customers." Some of the more daring butches built false bottoms into their carrying trays, and beneath them stashed lascivious books known as "Paris packages," sold for a dollar a copy, wrapped shut with the proviso that they could not be opened until the customer was off the train. Others had aggressive sales routines, such as this exchange illustrated, taken from the caption to a Thomas Nast cartoon: "*News butch*: Rock candy, rock candy, sir? *Passenger*: No, no, go away. I don't have any teeth. *News butch*: Gum drops, sir?"

Other than what Al Edison provided, there were few amenities on the bumpy, vibrating, grueling trip to Detroit. At speeds of up to twenty miles per hour, derailments were frequent and dangerous, since the wood-burning iron stoves at the center of each car, the only source of heat, could be upended. Meals were either cobbled together from the butch's box, or wolfed down hurriedly in railroad lunchrooms en route, since dining cars were not yet a feature. As automatic air brakes would not come into use until the 1870s, it was necessary for brakemen to run along the tops of the cars and twist handbrake wheels manually. Unlike hierarchical European cars, American railway cars were not divided into compartments. Modeled upon the design of the riverboats and canal packets they superseded, they were like large rooms, with a passage a foot and a half to two feet wide down the middle, seats arranged on either side of this aisle intended for two passengers each. The backs of the seats could be turned so that passengers could face in either direction. In certain cars, another British visitor noted in 1857, "a portion about seven feet long . . . is partitioned off, in which is a small room for the convenience of ladies nursing, and a watercloset."

The American train was a truly democratic, classless, and cheap conveyance, a veritable parable of American equality, with its spinning wheels, belches of smoke, mournful bell, piercing whistle, and surge of motion powerfully symbolic of the new industrial culture—and an irresistible microcosm of movement and the spirit of progress for Al Edison.

During the long layovers in Detroit, Al broadened his literary world considerably. He joined the Young Men's Society, so that he could make use of its large library. There, he discovered Isaac Newton's *Principia Mathemat-*

ica and Thomas Burton's *Anatomy of Melancholy*. From the editorial offices of the Detroit *Free Press*, one of the newspapers Al sold on the train, he was able to cadge discarded type, castoff composing sticks, ink, and paper. With a small, flatbed proof press (his mother's cheese press might have served as well), set up first in the baggage car of the train, and later in the basement of his home, Al launched the weekly Grand Trunk *Herald*, available by subscription for eight cents a month.

The Oakland University archeology team has discovered beakers containing vestigial stains of printer's ink, as well as 185 pieces of type, almost all of it in a room on the south side of the basement of the Edisons' Port Huron house. Over 30 percent of the type pieces unearthed were identical to the type used to print the text in the *Herald*. Other pieces of type precisely matched the letters in the weekly's headlines, advertisements, and train schedules.

Like the scores of subsequent ventures that would emanate from Edison's imagination over the next seven decades, this little newspaper had its roots in a commercial realization: that Al had a captive audience, and, as he staggered and swayed down the narrow aisle of the train anxious to earn enough so that he could set aside a dollar a day for his mother, he could purvey his own material just as easily as what was provided at commission. Although Thomas Edison has become renowned as a pioneering boy journalist, due in no small measure to his own justifiably enduring pride in the effort—the first issue of the *Herald*, not much larger than a sheet of typing paper, hung framed on the wall of Edison's second-floor den at Glenmont until the death of his second wife in 1947—he was in fact joining a movement characteristic of the time. As early as 1846 there were eight teenage pressmen in Boston alone, and an equal number in Worcester. Dozens of amateur newspapers produced by literary tyros across America during the years leading up to the Civil War reveal an unsuspected degree of complexity to adolescent life, and suggest an attempt to grow up faster than earlier generations. But the unique content of the Grand Trunk *Herald* and the distinction of having been published aboard a moving train set it apart from other representatives of the genre. Rather than predominantly excerpt from other periodicals, as most teenage editors (mostly boys) were wont to do, Al Edison understood his constituency and focused upon what he saw as a dearth of local news currently available along the route. At its peak, the paper had five hundred paid subscribers, and sold about two hundred copies per day on the train rides.

Only two issues of the *Herald* survive, out of the twenty-four that were published during its short life spanning the first six months of 1862. Together, they provide a precious and entertaining portrait of the ambitious young man with the byline of "A. Edison"—and the quirks of his orthography. As expected, "Local Intelegence [*sic*]" takes up a good deal of space in the *Herald*. Early on, Al was a keen observer of the human condition, not shy about expressing his opinion: "We [note the correct editorial usage] have rode with Mr. E. L. Northrop, one of [the Grand Trunk's] Engineers and we do not believe you could fall in with another Engineer, more careful, or attentive to his Engine, being the most steady driver that we have ever rode behind . . . always kind, and obligeing, and ever at his post."

Further personnel assessments consider "S. A. Frink, driver" to be "one of the oldest and most careful drivers known in the State." The editor was further "imformed [*sic*] that Mr. Eden is about to retire from the Grand Trunk Companys eating house at Point Edwards. . . . He has filled the place with the greatest credit to himself and the Company since the road opened," in the opinion of Editor Edison, "and we are shure [*sic*] that he retires with the well wishes of community at large."

The *Herald* also reveals the earliest documented example of Al's puritanical, work-ethic streak. "The more to do the more done," he opines. "We have observed along the line of railway at the different stations where there is only one Porter, such as at Utica, where he is fully engaged, from morning until late at night, that he has everything clean, and in first class order, even the platforms the snow does not lie for a week after it has fallen, but is swept off before it is almost down, [whereas] at other stations where there is [*sic*] two Porters things are visa a versa." He likewise casts a critical eye upon the loose drinking habits of the community: "The law requireing Saloons and Grog shops to close on Sunday is being enforced in Port Huron, a thing greatly needed, as they are a complete nuisance." Thievery has no place in Al's moral universe: "A gentleman . . . recently tried to swindle the Grand Trunk Railway company out of sixty seven dollars the price of a valise he claimed to have lost at Sarnia. . . . But by the indominatable [*sic*] perseverance and energy of Mr. W. Smith, detective of the company, the case was cleared up." Al goes on to say that he believes some form of retribution is appropriate, "We feel that the villian [*sic*] should have his name posted up in the verious [*sic*] R. R. in the country, and then he will be able to travel in his true colors."

After fulfilling the conventional responsibilities of the journal—printing assiduous timetables for the express and "mixed" trains in both directions, as well as for stage coaches and omnibuses scheduled to meet the arriving trains; advertisements, no doubt paid, for innkeepers and merchants in towns en route; birth announcements and lost and found notices; market prices for farm produce, dry goods, and services, and so on—Al found space for the occasional, if macabre, joke: "Let me collect myself, as the man said when he was blown up by a powder mill." The *Herald* was succeeded by equally short-lived gossip sheets called *Paul Pry* and *The Blowhard* of which, unfortunately, there are no known copies extant.

By his own account, Al also sold sweet corn, radishes, onions, parsnips, and beets from his father's acreage. He was able to hire two boys as helpers, initially on board the train, to hawk the papers, and then at a vegetable stand he opened in Port Huron. This establishment was also the outlet for goods that Al obtained at the Detroit market, shipped back freight-free on the Grand Trunk, and sold at a profit. Edison also recalled that he bought butter and blackberries wholesale from farmers along the train line, which he then resold to the wives of the engineers and trainmen.

The cluttered corner of the baggage car that served as Al's print shop, storage space, and mobile laboratory where he tried to conduct rudimentary experiments eventually became too crowded, which led to his commandeering a corner of his parents' basement. Edison told a tale rationalizing the inevitable move as the result of a chemistry concoction gone awry, during which he accidentally set a fire in his cramped space. The incensed conductor, a Scotsman named Alexander Stevenson, burst in and summarily evicted the boy from the train at Smith's Creek station, hurling Al and his paraphernalia onto the platform—"Off ye go, lock, stock, and ivry drap o' chimicals with ye. Ah must a' been daft when Ah let ye br-r-ring thim aboord!" At various times, Edison also blamed the evil Stevenson for causing his early deafness: One story accuses the irate conductor of boxing the boy's ears before throwing him from the baggage car, and the other states that Stevenson lifted Edison, laden with newspapers and unable to get a firm foothold, up into the car by his ears, at which point he "felt something snap inside [his] head."

It is far more romantic to collect apocryphal images of a plucky lad staying his course despite the persecutions of authoritarian, remote grownups than it is to concede to congenital physical problems. Regardless of the boundaries of truth and fable, Al's resourcefulness was unequivocally boundless.

CHAPTER

4

THOMAS EDISON AT FIFTEEN

AT FOURTEEN, HE PREFERRED TO BE called Tom. His coming of age was governed by the railroad, his lifeline, a gateway to the ever wider world of commerce and communication. And as the railroad spread in extent and influence, so did the telegraph in its wake. Both brought the frontier within reach. Understanding this symbiotic technological relationship helps us appreciate Tom's personal media shift, breaking away from static newspaper type and into "writing with lightning." No new technology up to that time ever had a more accelerated growth. By the time Edison was fifteen, telegraph wires reached all across America, from Washington, D.C., to San Francisco.

The telegraph was born a mere ten years before Thomas Edison. Invented by Samuel F. B. Morse in 1837, the first telegraph instrument resembled an electric switch, allowing current to pass for a limited time before being shut off with the touch of a finger. Messages were transmitted by such electric pulses passing over a single wire. Marks—dots and dashes—were made on a paper tape moving around a cylinder, varying in relation to the duration of the pulse.

The earliest telegraphic signals could only be transmitted about twenty miles, beyond which they were too weak to be recorded. Morse developed a relay—essentially an additional surge of current on the line—to repeat the original signal and send it further.

In 1833, Alexis de Tocqueville had been struck by the American outright affection for freedom of the press, aptly linking this obsession with the equally indigenous quality of self-reliance which "the inhabitant of the United States learns from birth." This "tumultuous agitation" of the spirit made the exemplary American restless, enterprising, enraptured by speed, and, above all else, an innovator.

The young Edison was a perfect exemplar of Ralph Waldo Emerson's "Representative Man," for apart from simply needing to make a living, he soon came to recognize the imperatives of commercial necessity even as he responded to a more profound creative impulse. His first patented invention was now a mere half-dozen years away, and his most important communications-related inventions in the future would consistently be refinements of telegraphic principles, structured to accelerate the transfer of information. The telegraph, derived from that special brand of American speed—the first message to outrun the messenger, as Marshall McLuhan observed—thus came into being as an expedient outgrowth of the newspaper.

The first practical application of the rudimentary magnetic telegraph was safety. With the unmoderated growth of rail traffic, trains dispatched by the schedule had a disturbing tendency to collide with each other. More lines were strung up from poles in the cleared space parallel to the tracks, and could be observed for breaks frequently by passing trainmen. One of Tom's first reponsibilities was to keep a sharp eye out for such problems on the line and to report them to the station agents along the way. It was the agent's responsibility, in turn, once the train pulled out, to alert the next stop on the line of its pending arrival. Under Western Union system guidelines, fully in place by the end of the Civil War, railroad messages pertaining to safe operation of the trains always took precedence over private messages, and the railroad depot itself became the community focal point for universal communication.

It cannot be emphasized too strongly that before the telegraph, communication was linked to established transportation means—infrastructure systems, roads, canals—and likewise to fixed conveyance on the page—newspapers, books, and personal correspondence—that was only as quick as the means by which the mail bag was carried. The telegraph transcended these limits, moving information electrically. The telegraph as a pivot-point marked the beginning of the end of nineteenth-century small worlds.

Short and long bursts of current seduced an ambitious, hard-of-hearing boy into a clattering embrace of dots and dashes he could understand: "When in a telegraph office," Edison recalled of those first enraptured encounters, "I could hear only the instrument directly on the table at which I sat, and unlike the other operators I was not bothered by the other instruments." Edison's pattern of downplaying his deafness began during this initiation time. What to others seemed a poignant tragedy—that after the age of twelve he "could not hear a bird sing"—was to the inventor, the fabricator, a fortuitous break. Ambient noise was immaterial: "Broadway is as quiet to me as a country village is to a person with normal hearing," he waxed poetically.

A canny analogy: Edison's idealized pastoral roots would stay with him forever, fixed in time like a sepia photograph. But his deafness evolved into more than a nostalgic echo simply because its onset coincided with his boyhood in a simpler era. Edison would always insist that being deaf set him apart from the masses of men, gave him an excuse to turn away from tiresome social involvements, making him a far more productive thinker. When he became a household name, Edison the capitalist, the "Wizard," would receive hundreds of letters from hearing-impaired people all over the world pleading with him to harness his formidable imagination to find a remedy for deafness or to invent an apparatus that would become the ultimate hearing aid, the miracle solution, the cure-all. These solicitations were routinely ignored, marked in blunt pencil "No ans." and passed along to his secretary. Edison refused to surrender what was essentially his passport to the inner world.

By the fall of 1862, in his words, Tom "commenced to neglect [his] regular business," subcontracting his "butchering" responsibilities to friends Jim Clancy and Tommy Southerland; he ceased publication of the Grand Trunk *Herald*, much preferring to spend his day loitering about the Mount Clemens depot on the Port Huron to Detroit line, where the stationmaster, J. U. MacKenzie, allowed him to listen in on the flow of telegraph transmissions. There was considerable commercial traffic through Mount Clemens, a constant shunting of freight cars to be attached to the daily mixed train.

MacKenzie's three-year-old son Jimmie was playing on the platform one morning as Tom and the stationmaster were talking shop. The story

goes, according to Edison—
and immortalized in the fabu-
lously dramatic etching in
Scribner's magazine, "Saving
the Child"—that the toddler
wandered blithely onto the
tracks as a freight car came
rolling toward him. The
brakeman,.braced on the roof
of the car, alarmed, called out,
unable to turn the brake wheel
quickly enough, and quick-
eyed Tom darted into the path

"SAVING THE CHILD"

of the oncoming car, dove onto the tracks, and grabbed the child on the fly,
shoving him to safety, neither of them badly hurt beyond gravel cuts and
bruises. The hysterical mother presses her hands to her brow, the brakeman
atop the freight strains at his "stemwinder"; and Tom, on the run, impelled
forward, holds the child by his arms and lifts him into the air, as the steel
wheels make their inexorable way.

In gratitude, MacKenzie offered to teach Tom telegraphy, since the lad
had the interest and the aptitude, and had already taught himself the Morse
code and built a rudimentary telegraph key. Speed was Tom's primary goal.
In three months of intensive lessons he mastered the skill, and landed a job
at Micah Walker's loosely named "jewelry" store in downtown Port Huron,
site of the Western Union telegraph office. It was an eclectic kind of place.
Walker sold clocks, watches, old spoons and forks, and did gold and silver
electroplating. He sold rifles, organs, dominoes, and chinaware. He sold
schoolbooks, stationery, and all manner of writing implements, day books,
blotters, and sheet music, as well as technical manuals and scientific magazines, which Tom had the liberty to peruse during his downtime. As Tom
read, he forced himself into the habit of grouping several words at once, always aiming toward greater efficiency. In the evening, after supper, he returned to the shop to practice receiving press reports in solitude, to satisfy
his hunger for information as well as to gain additional facility with longer,
more complex journalistic dispatches from the front.

The Civil War, nearing the end of its first year, raged far away from
young Tom. The North's incursions down the Tennessee, Cumberland, and
Mississippi rivers, culminating in the attack on Fort Henry under the lead-

ership of General Ulysses S. Grant; the violent battle of Shiloh in early spring, 1862, when 20,000 men were killed or wounded; the capture of New Orleans by Flag-Officer David Farragut, and his ensuing failure to take Vicksburg; Stonewall Jackson's march across the Blue Ridge Mountains and his Shenandoah Valley campaign to defend Richmond; the Confederate invasions of the Western territories; the capture of Harper's Ferry; the deluges at Corinth, Bull Run, and Antietam—to Tom Edison listening on the wire, these were remote rumblings.

The language of Edison's original, spontaneous, handwritten reminiscences about this apprenticeship phase of his life is noteworthy in its portrayal of a relentlessly driven, self-starting, and solitary being, with singular disregard for the constraints of a typical day: "I was small and industrious," he scrawls with matter-of-fact pride in himself as a boy. "I could fill the position all right. . . . Night jobs suited me as I could have the whole day to myself. . . . After working all day, I worked in the office nights as well. . . . I seldom reached home before 11:30 at night." To Edison in retrospect (he was sixty-one when he wrote these words), his adolescent life seems to have been an unbroken chain of tasks with the one purpose of self-betterment, far more competitive with himself than with the other guy.

"In the face of the eighty-year-old man," wrote Emil Ludwig, the German biographer of Napoleon, after a visit to Edison at his winter estate in Fort Myers, Florida, in March 1928, *I recognize the fifteen-year-old boy whose picture we, as children, saw when we heard that there was a land where a newsboy could become a great man* and that land was called America."

Tom outgrew Walker's shop, and took the 7:00 p.m. to 7:00 a.m. railway telegrapher's shift at Stratford Junction, Ontario, on the Grand Trunk Line, seventy-five miles north of Port Huron toward Toronto, his first job away from home—again despite Nancy's fears—as a "Knight of the Key." He had cultivated the talent of sleeping a minute or two here and there—sitting in his chair, lying on top of or under tables with the crook of his elbow for a pillow—for which he eventually became much renowned. To catch his forty (or twenty) winks, Tom made an arrangement with the night yardman to awaken him when his signal came over the wire. This deal failed when a freight train rushed through on Tom's watch and he was not able to detain it, and a collision with another train coming from the opposite direction was narrowly avoided. Summoned for a dressing-down before the district superintendent in Toronto, the stationmaster was "hauled over the coals" for permitting such a high degree of responsibility to such a young man.

Six months later, while Tom was back in Port Huron licking his wounds in the wake of this precipitous failure, his beloved sister Tannie died in childbirth at the age of thirty. In despair, he picked up his itinerant lifestyle again, and was off to Lenawee Junction, near Adrian, Michigan, fifty miles southwest of Detroit, again working as a night telegraph operator, this time for the Lake Shore and Michigan Southern Railroad.

He preferred the less favored night jobs because they provided him with "leisure" during the daylight hours and gave him readier opportunities for employment. Yet even following orders, Tom landed in difficulty. He was handed an urgent dispatch by his supervisor and advised that, if necessary, he should break in to whatever transmission was currently on the wire. Tom spent ten minutes struggling with the operator further down the line for transmission time. Unbeknownst to him, this anonymous agent happened to be the district superintendent, who appeared at the Adrian office infuriated. Tom defended himself, saying that he had been told to interrupt the current transmission, appealing to his boss for support, but the man summarily denied giving any such command. Tom was the scapegoat. "Their families were socially close & I became a wanderer," Edison wrote of the affair. "My faith in human nature got a slight jar."

In the early winter months of 1864, it was on to Fort Wayne, Indiana, for another half-year stint, a day job on the Pittsburgh, Fort Wayne and Chicago Railroad; and then to Indianapolis, working for Western Union. Tom found a room at a boardinghouse just four blocks away from the Union depot. Although still not hired for the coveted press wire job, Edison, living nearby, could come back after hours and sit next to the press operator, copying dispatches on his own time, until the early morning hours.

Finding it difficult to keep up with the more complicated press messages using conventional handwriting, Edison came up with an ingenious solution. He rigged up two Morse registers, the first of which recorded the dots and dashes of the incoming message in the form of indentations on a continuous strip of paper. He then ran this encoded strip through the second register, activating its sounder, but at a pace moderated to a frequency considerably slower than the original forty words per minute. With the help of a co-worker controlling the second machine, Edison was able to make virtually flawless copy. He had succeeded in adapting the machinery in tune with his particular sensory requirements.

It was next on to the big city of the Ohio River Valley, the state's pre-eminent metropolis, Cincinnati, proud home of Procter and Gamble, less proud of its image as the pork-producing capital of the midwest. Tom was still paying his dues in the telegraphers' caste system as a "plug," or second-class, night shift operator at the main Western Union office, on Third Street. He worked the wire that ran to Portsmouth, Ohio, and still kept up his practice by offering to substitute for any press wire operators who wanted some time off. His goal was "to become proficient in the very shortest time."

At the Bevis House in downtown Cincinnati, he roomed with two actors and with fellow-telegrapher Ezra Gilliland, whom he had first met in Adrian. The two would become friends and business associates for the next quarter-century. Edison regaled his comrades with tales of exploits from the "butch" days, describing himself as a "peanut boy" who had once taken over the train when the engineer and fireman fell asleep. The conductor signaled to go ahead, and Edison, unable to wake the two, pulled out of the siding, originally as a lark, but soon as serious business, and made it all the way to Detroit.

Gilliland having been apprenticed to a gunsmith, the two mechanically minded young men had much in common. Edison showed his friend a "little steam engine he had made out of brass tubing, with a very ingenious valve motion, and told of several other inventions he had gotten up." It was at this time that Edison began tinkering with old telegraph repeaters, obsessed with the idea—perhaps still smarting and frustrated from the Adrian fiasco—that it should be possible to send two messages over one wire simultaneously in opposite directions: "duplex" telegraphy. Edison disaggregated and reassembled these prototypical, modified machines, their parts connected with sealing wax and string, an old cigar box for a base.

Edison was becoming more collegial. While in Cincinnati, he joined with his co-workers to form the local chapter of the National Telegraphic Union, founded the year before as an attempt to consolidate this rather makeshift, exponentially growing, unruly profession of drifting young men, and give them a united arbitration voice. The Union also published and distributed its own journal, *The Telegrapher*. The prophetic motto on its masthead loftily, if rhetorically, asked, "Is it not a feat sublime?/Intellect hath conquered time." Thomas Edison became a persistent editorial voice in its pages.

Promoted finally to first-class operator, with a commensurate big salary boost to $125 a month, Tom felt footloose again in the face of actual stability. A friend in Memphis wired him with news of the possibilities there, "and as [he] wanted to see the country [he] accepted it." His brief stay on the Mississippi was occupied by deeper voyages into literature. Consistent with his admiration for the founding fathers and in true democratic spirit, Tom read a biography of Thomas Jefferson. He claimed that in November 1865 he manufactured a repeater that for the first time since the end of the Civil War brought New York City and Memphis back into telegraphic contact. For ambiguous reasons, Tom was once again dismissed.

His southern sojourn continued at the Associated Press Bureau in Louisville, Kentucky, a debris-strewn, ramshackle office crowded with minuscule desks, the plaster flaking from the ceiling, and a dysfunctional soot-jammed woodburning stove. The switchboard was woefully cramped and inadequate, its connections blackened and crystallized with age, and the planked floor of the battery room was corroded with nitric acid. It was a tumultuous place, operators barging in half-drunk at all hours of the day and night.

Back in the refuge of his seedy room, the teetotaling Tom wrote to his family, putting on a brave front, "I dont look much like a Boy Now . . . I have growed considerably . . . Spanish very good now," he continued, always tutoring himself in every available scrap of time. "Before I Come home I will be able to Speak Spanish & Read & write it as fast as any Spaniard. I want to get Some more Books."

There was a purpose, of course, to the Spanish lessons. In midsummer of 1866, Tom and some emancipated cohorts had read a notice in the Louisville *Journal* advertising a need for American telegraphers in Brazil. They decided the trip was worth the gamble, and their departure was even cited in the newspaper with the valedictory, "Success to the young adventurers!" Arriving in New Orleans by way of Nashville, Tom and his colleagues discovered race riots in the streets. In order to bring in troops to quell the disturbances, the militia had commandeered under martial law the steamer Tom was meant to take. He abandoned plans for spontaneous exile and returned to Port Huron for a brief stay before resuming his old Louisville job.

Edison's year on the Louisville Western Union press wire was a valuable one. Transcribing dispatches with increasing facility, to compensate for his admitted deficiencies in sending messages, he perfected the speedy, copperplate handwriting with abbreviations that would become his lifelong trademark. He frequented the editorial offices of the Louisville *Journal*, and, in the early hours of the morning, as soon as the paper had been put to bed, Edison would take home the daily dispatches and read them long in advance of the rest of the world.

His competitive spirit thrived at being on the leading edge of current events in the most immediate sense and also in the broader cultural arena. In later years, Edison was fond of telling the story of his purchase at an auction in Louisville of twenty unbound volumes of the *North American Review*. Founded in 1815, the *Review* managed to remain for 125 years the most eclectic yet mainstream American journal of the era, "the organ of the most cultivated and scholarly minds of the country," as its editorial credo proclaimed. "The subjects with which the *Review* will deal will be limited by no programme laid down in advance; whatever topics are at the time prominent in the public mind will be taken up and treated with thoroughness and vigor."

Economic theory, educational ideals, the uses of poetry, American self-identity and "expansionism," reconstruction and the role of the Negro, the "woman" question, the "Irish" question, the "Jewish" question, the "Indian" question, taxation, civilization and its emerging discontents—no subject of concern to Americans at this crucible time was too insignificant or too grandiose to be tackled in the forum of its pages by the best theorists of the generation—Ralph Waldo Emerson, Walt Whitman, William Dean Howells, Henry George, Andrew Carnegie.

Indeed, Thomas Edison, with his indiscriminately hungry mind, remained a regular reader of the *Review* during the next half-century, and went on to publish no less than six articles over his own byline in this prestigious journal, chronicling—and, as we shall see, defending—the progress of his most significant inventions between 1878 and 1902.

Edison purchased his precious twenty volumes for the bargain price of two dollars and had them delivered to the telegraph office. Walking home after work with the bundle of books on his shoulder through the dim, silent streets of Louisville to his solitary room over a saloon (and no doubt dreaming of the engrossing hours he would spend perusing these many pages) he was accosted by a policeman who—thinking the books had been stolen and receiving no response to his cries—fired a warning shot. The young man, being hard of hearing and deep in thought, had not heard a thing.

Fired for spilling sulfuric acid in the bureau's battery room, which ate through the floor and wreaked havoc in the manager's office below, Tom re-

traced another earlier trail and returned to Western Union's Cincinnati bureau, which had now moved from Fourth and Walnut to newer, larger quarters on Third Street. Friend Ezra Gilliland and other former colleagues of
Edison's recalled with amusement the many times Tom would manage to
get himself excused from work, pleading illness, and invariably "strike a bee
line" for the Mechanic's Library on Sixth and Vine, or the Cincinnati Free
Library. Edison's few surviving pocket notebooks bear witness to his assiduous research and to the wide spectrum of his interests: tables of conductivity, comparing the relative merits of copper and silver; sketches for self-
adjusting relays that through the use of more sensitized springs would help
make the process of receiving transmissions more efficient; and enhanced repeaters, to allow for longer-distance transmissions without having to transcribe and retransmit messages.

In the library stacks, Tom tracked down Dionysius Lardner's classic
work on the *Electric Telegraph*, as well as his *Handbook of Electricity, Magnetism
and Acoustics*. He read Richard Culley's *Handbook of Practical Telegraphy*,
Charles Walker's *Electric Telegraph Manipulation*, and Robert Sabine's *History
and Practice of the Electric Telegraph*.

But heading Thomas Edison's neatly jotted list of research needs was
the first volume of *Experimental Researches in Electricity and Magnetism* by
Michael Faraday, a pervasive influence who, like Thomas Paine, was a definitive role model for Edison as his working methodology coalesced. The com-
monalities between the son of a journeyman blacksmith born in the outskirts of London in the waning years of the eighteenth century and the
plucky Midwestern youth are striking. Swept up by the ethos of self-
improvement, Michael Faraday dropped out of school at thirteen to become
an errand boy for a bookbinder. On the job, Faraday made his way, volume
by volume, through the *Encyclopaedia Britannica*, and began keeping a "common place book" containing quotations and questions arising from his reading, wherein the boy wrote of his "desire to escape from trade, which I
thought vicious and selfish, and to enter into the services of Science."

At the Royal Institution of Great Britain on Albemarle Street in London, where he began as the assistant to its legendary director, Sir Humphrey
Davy (who warned the boy that science was "a harsh mistress"), Faraday
launched a fifty-year career as a self-styled "natural philosopher," essentially
establishing the interconnectedness of the worlds of analytical and organic
chemistry with electricity and magnetism—as Edison thus understood
them in the summer of 1867. Faraday discovered the principle of electro-

magnetic induction, demonstrating that an electric current could be produced by thrusting a magnet into a coil of wire and withdrawing it, thus giving birth to the dynamo and the transformer. In so doing, he broadened the domain and identity of electricity, envisioning not only an abstract power, but a commodity, and laid the foundations for the modern electrical industry. He showed that within the confines of his basement research laboratory, where he ostensibly focused upon the pursuit of knowledge for its own sake, theories could be formulated with commercial applications; research indeed led to development. Thomas Edison would soon make the identical transition from the workshop to the world at large.

Faraday the philosopher penetrated below the applications of electricity to its inherent qualities, regardless of the sources—voltaic, electrostatic, animal, thermal. All electricity, he declared, was constant in the universe and "identical in nature," as were by extension *all* the forces "by which we know matter." In his dogged pursuit of solutions to problems (even to the extent of fabricating his own tools), vacillation between humility and pride, belief in the holistic unity of the phenomenal world, extended work days, criticism of standardized education in deference to pragmatic learning, and knack for popularizing the most arcane knowledge, Michael Faraday emerges as Thomas Edison's scientific older brother.

By the time he turned twenty, Tom was back in Port Huron. All was not well at home. The family had been evicted by the army from the big house near Fort Gratiot; after being commandeered, it burned down. The Edisons were now on Cherry Street in more modest quarters, taking on boarders to help make ends meet. Nancy Edison was suffering, as she did more and more often in these years, from vaguely defined but chronic "nervous trouble." Sam's ongoing land speculation and contracting activities were failing. Tom, the tired and hungry prodigal son, became ill and was forced to spend the winter of 1867 in bed.

Then a letter arrived from Milton Adams, Tom's friend from Cincinnati, now working in Boston for the Franklin Telegraph Company. There were openings at the Western Union Boston bureau, and Milt could get Tom a job—if he came quickly.

CHAPTER

5

WITH SYMBOLIC APTNESS, EDISON'S BOSTON-BOUND GRAND
Trunk rail route took him back northeast through Canada, in the midst of a
raging blizzard, one more wintry pass traversing the rough-hewn terrain of
his forebears. Between Toronto and Montreal, the train was utterly snowed
under. Edison and his fellow passengers were stranded for four days in a
country inn, but the frontier was familiar to him. The Canadian North—
like the American West he would soon visit—was deeply rooted in Edison's
psyche. He had long since surrendered (with enough insight to describe
himself unambiguously at this time as "restless") to the urge for incessant
movement and expansion, his "energy . . . continually demanding a wider
field for its exercise," as the renowned historian Frederick Jackson Turner
wrote so tellingly of other pioneers.

"The appeal of the undiscovered is strong in America," Turner observed in his 1914 essay, *The West and American Ideals.* Edison had internal-
ized the notion of "frontier" to include wherever he had not been; it was a
frame of mind as well as a longing for a specific place. And, too, it was wherever he could go with the potential for change, as a native son of Ohio—
geographic middle kingdom between the populism of the prairies and the
capitalism of western Pennsylvania.

The farm was his past; the city was his future. Edison was part of a
mass American trend that between 1860 and 1890 saw twenty midwestern-

ers (despite the warnings of moral pitfalls in the great metropolis) move to a major urban area for every single city laborer who moved to the country.

Indeed, meeting George Milliken, the manager of the Western Union office in Boston, for his initial job interview, Edison in his draft memoirs self-consciously depicts himself as being poorly dressed and disheveled from the long trip, a naive, ingenuous, although genuinely decent, country hayseed standing embarrassed in the midst of professionally attired, efficient office workers. He could thus have been an incarnation of Ragged Dick, hero of a Horatio Alger, Jr., series, appearing on the popular literary scene that very same year.

The newsboys who were favorite subjects for Alger's success stories may have been homeless, and Thomas Edison in Boston was a cut above that line, but he subscribed to the same prevailing ethos: "Poverty is no bar to a man's advancement." Furthermore, Alger's heroic boys inherited a sense of virtue from their absent parents to help them weather the vicissitudes of being set adrift in the hard world. They depended upon this rural legacy, and upon benevolent merchants as surrogate fathers, to keep them afloat amidst corruption and help them gain their own wealth, which would be untainted.

Self-made success was not invented by Alger, but it had reached cult status in Victorian America by the time Tom Edison and "Ragged Dick" emerged together. This appealing character's quick-talking, beguiling ability to overcharge rich bankers for their shoeshines and "to obtain six jobs in an hour and a half," his carelessness with daily earnings, his unstoppable search for more and more customers, his willingness to sleep on whatever floor available and to cadge favors from friends and strangers alike cannot help but remind us of "Al" Edison, like his fictional counterpart, "a bright looking boy, with brown hair, a ruddy complexion, and dark-blue eyes, who looked, and was, frank and manly."

"I was a little feller to take care of myself, but," Dick continued with pardonable pride, "I did it." "What did you do?" his friend inquired. "Sometimes one thing, and sometimes another," said Dick. "I changed my business as I had to."

But the true self-made man, as exemplified by Alger's boys, was not *just* lucky or shrewd. He could not simply stumble upon success; he had to be deserving of good fortune and possess a measure of quasi-religious dignity. He had to be a common man as well as an adamant individual willing to choose dissent, and even rebellion, when warranted.

Installed on his favorite night shift as a receiver on the prestigious "Number One" press wire from New York City, Edison launched the compulsive, documentary habit which never left him as an inventor: assiduously recording each and every technological advancement he could articulate. He noted these discoveries either in articles or letters to the Editor in the *Telegrapher*, where his accomplishments would be seen by his colleagues in the trade— he wrote seven such pieces in the ensuing year—or in the *Journal of the Telegraph*, which was Western Union's corporate newsletter, or in his ever-present pocket notebooks. Ownership of ideas was Edison's most-valued currency, even if, as in the much-vaunted case of duplex telegraphy, they did not entirely originate with him.

Edison is justly known and revered for having registered more than one thousand patents in the course of a six-decade inventive odyssey. Patents were for concrete things. But there were thousands more inspirations of a less practical, more conceptual and speculative nature that sprang into his incessantly active mind and found their way to the pages of Edison's journals.

The practical nature of invention required that concepts be formally registered—the date, time, and place of their origin marked clearly—to protect the inventor against future encroachments. But the way Edison prefaced the ideas he transcribed reveals rather more specific intentions; beyond using the oft-repeated word, "record," he referred early on to the sketches he had drawn for various machines as "for myself exclusively, and not for any small-brained capitalist," self-consciously drawing the distinction between the value of innovation for its own sake and underlying market considerations. At other times, for example, in a notebook begun in July 1871, Edison reminded himself that any ideas he did "not see fit" to hand over to the Gold & Stock Company he would "reserve" for himself, as the creator. Furthermore, surveying his imaginative territory as far as the (inner) eye could see, Edison wrote that he was going to be tracking his ideas "previously formed, some of which have been tried, some that have been sketched and described, and some that have never been sketched, tried or described." Less than one short week later, having filled this notebook, he was into a fresh one, asserting that his discoveries and observations would be deployed in the event of "any contest or disputes regarding priority of ideas or invention." He was equally quick throughout his career to notice and contradict claims against him by others that he felt were unjustified, on principle, as if he were the vigilant, self-legislated guardian of innovation in its purest form.

Despite the intense activity in the telegraphic field during his twelve months in Boston, Edison revealed significantly wider commercial ambitions in the first patent issued to him, U. S. No. 90,646, for an electric vote recorder which would instantaneously note "Yes" or "No" through the same principle as a chemically recording telegraph. Current was directed via a switch to the Yes or No column, where each member's name was listed in metal type. Chemically prepared paper was laid over the charged type, and the clerk of the legislature would then pass a metal roller over the paper, pressing down upon the type. As the electric current passed through the paper, it would serve to dissolve the chemicals, leaving the impression of the names in each voting column. The instrument was meant to provide an efficient shortcut to the roll-call process, and for that very reason was never adapted by any legislature: truncating that honored give-and-take known as the filibuster would have had severe negative political implications. The lawyer who processed Edison's first patent application for him, Carroll D. Wright, later remembered his young client as "uncouth in manner, a chewer rather than a smoker of tobacco, but full of intelligence and ideas."

Undaunted by the vote recorder's failure, Edison ran a brief announcement in the *Telegrapher,* putting the world on notice that he had resigned from Western Union to devote all of his time "to bringing out his inventions." In so doing, he moved one decisive step further toward defining the course of the rest of his life. Heeding the credos of another lifelong mentor, Ralph Waldo Emerson, Thomas Edison, striking off on his own, was "making his circumstances"; he would "build therefore [his] own world." Emerson insisted that invention was "a spiritual act," and the inventor's daring mission was no less than to come face to face with the primal forces of industry, to understand and adapt them with intelligence to his own designs: "Shall he then renounce steam, fire and electricity, or shall he learn to deal with them? The rule for this whole class of agencies is—all *plus* is good; only put it in the right place." The inventor, as one of the representative men of his time, is compelled to learn "the last lesson in life . . . a voluntary obedience, a necessitated freedom."

And knowing Edison's erratic employment history, even this early in his career, we are not surprised that he craved emancipation from the cockroaches that invaded his desk at the telegraph office, from the foul, close air and the "cloakroom full of vermin playing leap frog and base ball on broadcloth coats and picknicking on the lunches of married men." It was no coincidence that consumption became the number one killer of nineteenth-century telegraphers.

Having rented space in a corner of Charles Williams's telegraphic instrument manufacturing company on Court Street, and later on Wilson Lane, near the Boston Stock Exchange, Edison began attracting his earliest financial backers, including Boston businessman E. Baker Welch.

"The greatest invention of the nineteenth century," the philosopher Alfred North Whitehead has written, "was the invention of the *method* of invention." Imagination, a gift Romantic poets understood, Emerson glorified, and Edison dearly valued, was the spark required to bridge the gap between the idea for the thing and any potential exploitation of it. Even though one could argue that Edison "failed" in his first officially documented attempt, the vote recorder, this failure, and countless others like it (and far more emotionally devastating) in the years to come, only served to push him forward with redoubled resolve. Taking invention in its vastest sense, viewing patent applications as the tip of the iceberg, it was the sheer delight in the mental *process*, and the ripples of ferment engendered thereby in his capacious brain and in many others beyond himself, that intrigued Edison.

He started a rudimentary stock-quotation service, with an initial twenty-five subscribers, including the Boston offices of Kidder, Peabody & Company. The resulting revenues allowed Edison to develop a primitive but efficient private-line alphabetical dial telegraph, upon which the operator spelled out words letter-by-letter instead of tapping them in code. These machines were restricted to use by Boston businesses—a lumber company, a sugar refinery—enabling Edison to maintain close control over his franchise; so close, in fact, that he used the roofs of neighboring houses to string his telegraph lines without even asking permission from the owners. To expand the network, Edison went out and solicited testimonial letters from his clients, attesting to "the simplicity and certainty" of his system.

Backed by the sum of $800, which would turn out to be his final loan from the trusting Mr. Welch—he had by turns warned his benefactor that he had "only enough money to last ten days," then reassured him with characteristic bravado that "I'll never say 'fail' . . . there is no use letting anything stand in the way of successfully getting this apparatus perfected"— Edison came up with a plan to set up an experimental duplex telegraphic line between Rochester and New York City. Edison hoped that the well-publicized completion of a four-hundred-mile transmission would open up markets all up and down the east coast and midwest. But one week of trials proved fruitless. "Alas!" he wrote sarcastically to his friend and shop-partner, Frank Hanaford, "that the bright vision of untold wealth should vanish like the fabric of a dream, that two ambitious mortals gifted by the genii of enterprise, and strength, should drink the bitter dregs of premature failure in their grand dreams for the advancement of science."

Despite his erratic track record, Edison was developing a reputation within the closely knit telegraphic fraternity. He was referred to in print and in correspondence as "a man of genius . . . a young man of the highest order of mechanical talent." Daring and resourcefulness in this still-growing field were valued in and of themselves, and were responsible for redesigning existing systems such as the double transmitter, the combination repeater, and the automatic telegraph to make them basically into clearer and faster machines. Edison's notoriety preceded him when he headed for New York City, leaving Frank Hanaford behind in Boston to manage their shop on Wilson Lane—which his colleague did for a few more months before dispiritedly returning to his old post at the Franklin Company. This time Edison was better-tempered, not as wet behind the ears, harboring no illusions, and no money either, but managing to hold his head high as "neither a dead beat or a selfish person."

"It is all I can do to keep the wolf from the door," he wrote Hanaford in July 1869, with his increasing flair for the melodramatic, "the Grey-eyed spectre of destiny has Been our guardian angel, for no matter what I may do I reap nothing but Trouble, and the blues." Desperation, or, perhaps, the illusion of it, motivated Edison to rescue himself from the wolf's jaws through a combination of mechanical facility and connections. Parlaying his special knowledge of financial reporting into a job offer, he took a fortuitous step, replacing Franklin Pope as superintendent of the Gold & Stock Reporting Telegraph Company, founded by Samuel Laws and situated in lower Manhattan. This was the big-league version of Edison's home-made Boston network. Price quotations for gold were sent out from Laws's transmitter to subscribing offices of bankers and brokers around New York.

This relatively stable state of affairs lasted only so long as the temporarily retired Pope was out of the city on impatient and purportedly restorative travels around the West. When he returned to New York, Pope invited the transient Edison to board at his home in Elizabeth, New Jersey, where Pope's mother tried with her massive meals to put some meat on Edison's 5-feet-9-inch, 135-pound frame. Camaraderie developed, and Edison then left Gold & Stock and joined forces with Pope, to start their own "establishment of a Bureau of Electrical and Telegraphic Engineering" to compete with their former employer.

Pope, Edison & Co. of Nos. 78 and 80 Broadway, would lease private lines to businesses for $25 per week, maintain and repair existing lines provided by other shops, install burglar and fire alarms, help draft patent applications for other entrepreneurs, and subcontract the expertise of the principals to purchase all manner of telegraphic supplies—cables, wire, insulators, as well as books.

However, by year's end of 1870, Edison, Pope, and their third partner, James Ashley, had assigned their patents for printing telegraphs to Gold & Stock; Marshall Lefferts, the new president of Gold & Stock, then commissioned Edison to design a new stock ticker exclusively for his company. Edison had also been approached by two well-connected and financially secure investors, Daniel Craig, former head of the Associated Press, in concert with George Harrington, former assistant secretary of the treasury, both of whom wanted to harness his talents to develop a speedier automatic telegraph. Thanks to Craig and Harrington's money, Edison was able to move to a new shop on Railroad Avenue in Newark, New Jersey, where he increasingly pursued his own path. Pope and Ashley responded by moving to dissolve the barely born partnership.

Edison's two high-powered backers wanted immediate results, so much so that the inventor, in the shop from seven o'clock every morning until two o'clock the following morning, had phrases written into his lease that he would be permitted to run machines at night and that he would not be limited in the amount of machinery he used. Even with his earliest underwriters, Thomas Edison set the tone and terms for the ways in which he needed to do business and conduct his work. He was a hands-on manager. He told Harrington on several occasions that he could not leave the shop to meet with him because he needed to "watch the men and give instructions." He set virtually intolerable deadlines, pinning down target dates for delivery of completed copying printers, transmitting and receiving motors, and punching machines, literally specifying the day of the week they would be available.

Most critically, Edison reminded his financial supporters that they were paying just as much for the experiments themselves as they were for successful results: "It is of no consequence whether [the punching machine] worked or not. It was an experiment as I told you once before," he scolded

Harrington, *"not made to show, but to satisfy me that I was all right."* Six days later, Edison repeated a similar lesson on the nature of scientific validation to "Friend Craig," blaming Harrington for paying an impromptu visit, distracting the engineers "by his fidiggity manner" and meddling in mechanical affairs about which he was woefully ignorant, thereby causing a setback in Edison's meticulously calibrated routine: "The Manufacture of Mechanism is a Slow operation being legitimate, and [like] all legitimate Businesses they are slow but sure, and *the slower the more sure.* . . . Mr. Harrington says that some of our experiments were useless. But," insinuated the sage and defiant Edison, "after he has had more experience in this business, he will find that *No experiments are useless."*

All this rapid-fire change in allegiance, with its stress and turmoil, may help explain why Thomas Edison at the tender age of twenty-three fancifully referred to his hair turning "damned near white. Man told me yesterday I was a walking churchyard."

More than fifty men labored on piecework for Edison in his larger shop on Ward Street, where he had moved the Newark Telegraph Works, designing and manufacturing perforators, transmitters, ink recorders, and typewriters specifically for automatic telegraphy. In a contemporary photograph seen in perspective from across the street, the workers (with a smattering of boys and women among them) are grouped on the sidewalk along the front of the four-story brick building. Peeking from a window on the top floor is a familiar figure in shirtsleeves—Edison as the self-styled foreman for both night and day shifts. "Sleep," he said, "was a scarce article" for him at the time. He commented wryly that he was too busy working when the picture was taken to come all the way downstairs and join the gang for the photo opportunity.

Aside from symbolically closing a circle by situating himself so near to the landing site of his original Tory ancestors in America, Edison the artisan was at home in Newark because it represented the flowering of a scale of work he understood well, work born at a simpler time, still steam-driven, still intimate enough for a man with a dominant sensibility to control. With Paterson, Newark was renowned as a center of finance in New Jersey, "the Birmingham of America," the state's workshop. Its inland waterways, most notably the Morris Canal and deep-water harbor on the mighty Passaic River; strategic proximity to New York City, a mere nine miles to the thriving port of New York; great railroad depot; and more than two hun-

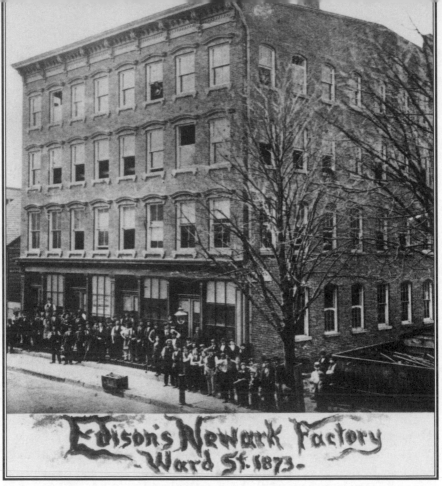

dred different industries, including all-important foundries—these attributes made Newark a natural resource, from which Edison was able to draw for his growing labor pool.

No sooner had Edison begun to hit his stride when word came from Port Huron, where he had not been for three long years, that his beloved mother had died on April 9, 1871. He went back home for the funeral. Nancy Edison was laid to rest at the old military cemetery on the grounds of Fort Gratiot. A mere three weeks later, widowed Samuel Edison took up with sixteen-year-old Mary Sharlow, who had been the family housekeeper. They remained together for more than two decades and had three daughters, euphoniously called Marietta, Maude, and Mabel. Local Port Huron historians insist that Sam and Mary were properly married, but it may have been a common-law arrangement. In any case, Sam refers to Mary as his "dear wife" in loving correspondence.

As seemingly intolerable as it was for Sam to be without a mate, so was it impossible for his son to be alone very much longer. The following fall,

MARY STILWELL EDISON

Thomas Edison found his own sixteen-year-old companion. She began as an entry-level employee, a telegraph tape perforator in his newly established subsidiary, the News Reporting Telegraph Company. Blonde, gray-eyed, full-figured, and sociable, Mary Stilwell was the second daughter and one of five children of Newark sawyer Nicholas Stilwell and his second wife, the former Margaret Crane of Elizabeth. Like the earliest "Edesons," three Stilwell brothers had come to the New World from Holland in the mid-1600s, originally by way of Lincolnshire, England, where they were born. The oldest brother, Nicholas—Mary's father's namesake—settled first in New Haven, Connecticut, and then moved on through Manhattan Island to the Gravesend district of Staten Island. The family lived there for another generation before migrating in a familiar pattern across to Whitehouse, New Jersey, then to New Brunswick, and finally Newark.

The courtship of the inventor and his teenage clerk lasted less than three months, and is virtually undocumented. It should come as no surprise that when Thomas Edison set his mind on something (or someone) he would countenance no obstruction in his urge to possess it (or her). On an unusually warm Christmas Day, Thomas Edison and Mary Stilwell were married in Newark by the Reverend W. S. Galloway. According to the writeup in the Newark *Daily Journal* of December 26, 1871, "the beautiful balmy weather of Sunday reminded all of a morning in May."

CHAPTER

6

ON THE MORNING OF THE GROUP-PORTRAIT SESSION HELD OUT-
side the Ward Street shop, there was one other hardworking young man
who declined to descend those four rickety flights of stairs from the research
laboratory and appear in the photograph. It was precisely in character for
Charles Batchelor to remain in the background, avoiding the hubbub and
the interruptions, seated at his familiar post on the workbench next to his
boss.

Born in Dalston, outside London, in 1845, the son of a mechanic,
Batchelor grew up and went to school in Manchester. He visited America
for the first time at the end of the Civil War for a brief stay to introduce a
cotton-processing machine. Returning to England, he went to work for the
J. P. Coates Company, whereupon he was chosen once again to be an erstwhile ambassador and sent to Newark to oversee the installation of machinery in the Clark Thread Mills on the banks of the Passaic, where Union army
uniforms were made during the Civil War. By late 1870, he had met Edison, was on the payroll of the American Telegraph Works, and had found a
house on Ward Street near the shop, where he lived with his wife, Rosanna,
and daughter, Emma.

The prototypical locomotive *John Bull* lumbered its way to Philadelphia, inspired by the *Puffing Billy* from the dark, Satanic mills of England's
industrial midlands in the early decades of the century. And ambitious rep-

resentatives of English technology, children of another revolution, crossed over, hoping for greater advancement in America. Batchelor's arrival was part of a rampant transatlantic knowledge transfer, propelled by the high regard in Europe for a uniquely American inventive tradition spawned in Benjamin Franklin's time.

Edison's senior by two years, Batchelor was a soft-spoken, modest, precise draftsman. His hand was as steady as his meticulously organized mind. His penmanship was copperplate. Every single entry in his Notion Books was dated and numbered. Every single news clipping in his scrapbooks was labeled with its source and wherever possible grouped according to subject matter; although they most often pertained to the increasingly voluminous media response to Thomas Edison's latest innovation, they also reflected deep research—especially during the fiercely competitive incandescent lamp period—into developments in other sectors and other countries.

With his slender build and prepossessing black beard, Batchelor resembled a portrait of D. H. Lawrence redrawn as a bookish, ascetic philosopher. He was a cultured fellow, fond of nearby New York's opera, theater, and fine restaurants; a conscientious reader; a devoted family man who enjoyed nothing more than spending Sundays off tramping through the woods collecting leaves for his daughter to press into her little homemade album, or taking her on a horse-drawn sleigh ride. An astute mutual friend presented Batchelor with a leatherbound set of Boswell's *Life of Samuel Johnson* for Christmas ten years into Batchelor's close association with Thomas Edison.

"Batch," as he was usually called, was Thomas Edison's first and most loyal long-term collaborator, sticking with "Mr. Edison" for twenty-five years, from the consolidating days in Newark when the marketplace conspired to propel Edison onward, straight through the heady corporate era, and on to the dusty stretches of ore mining, which practically ruined his health. But collaborator is perhaps not a wholly accurate way in which to refer to the long succession of male sidekicks and co-workers, self-styled "The Boys," no matter how old in fact they were, who inhabited the inner-shop sanctum and enjoyed "The Old Man's" tenuous confidence. (Yes, even those older than he was thus referred to Edison in his absence, in respect to the combination of his prematurely gray hair and precociously dogmatic, paternalistic manner.) Translator might be a better term, for it plays upon the idea of anonymously and successfully preserving the concept and voice of the original author.

In Newark, Edison's workshop research and development "methodology" crystallized—albeit unconsciously—and Batchelor was the first of several important colleagues to be present at the creation. The laboratory notebooks reveal that raw ideas for each future project sprang rough and fully formed from Edison's brain. Hurriedly scrawled, immediate phrases such as "I claim . . . I believe myself the first one to use this method . . . I have another device which . . . Another idea has just occurred to me . . . This is a novel affair . . . Try this . . . Test this . . . See if you can get this . . . " scramble across the pages, punctuated and often obliterated here and there by inkblots, acid-stains, and proud declarations by "E. & B." that the experiment just concluded at 11:00 a.m. had begun at 3:14 a.m., or assertions that "we worked all night" or "32 continuous hrs." or "60 hrs." or "six days this week" on one matter or another.

These frenzied, quasi-legible words sometimes read more like mysterious recipes with shifting ingredients than formal documentation: "Another thing we have noticed," Edison admits incoherently, for example, at the height of one frenzy, "is that we sometimes work a long time before [the machine] will work, and then it suddenly commences to work and it will work anyhow we fix it and we don't know how we have fixed it to make it work."

Such transcripts of the imagination connected in tumbling streams and envisioning new machines are offset by Batchelor's rounded, letter-by-letter inscriptions serving as restrained captions to byzantine, da Vincian pen-and-ink drawings shaded and laid out in classic perspective, populated with rotating gears, levers, springs, interlocking rods, circles upon circles, ratchets, cylinders, pivot-points, and labyrinthine wire strands leading along circuitous paths to relays and terminals, which appear to have been crafted with all the time in the world.

"We work all night experimenting," Batchelor wrote to his brother, Tom, "and sleep till noon in the day. We have got 54 different things on the carpet & some we have been on for 4 or 5 years. Edison is an indefatigable worker & there is no kind of failure however disastrous affects him."

Because Edison believed in the infinite perfectibility of his ideas, from one entry to the next, from one day to the next, or even later the same day, he might occasionally contradict himself, as a mercurial refinement of an earlier idea took precedence. It was Batchelor's responsibility during the years of their association to be certain that final articulated images for instruments were passed to the next step down the labor chain, to be actually assembled—though still under Edison's unwavering, proprietary eye. For while on the one hand he maintained a hierarchical structure in order to make what needed to be made and maintain his authority, Edison was "compelled," as he put it, to wander democratically and visibly up and down the aisles, ubiquitous, endlessly snooping, his sleeves rolled up and his unattended cigar ash dropping onto the shoulders of welders and die-cutters hunched over the scarred wooden tables.

Charles Batchelor astutely signed on with Edison at the most prolific moment in his telegraphic career, restlessly pursuing perfections upon his universal stock printer and universal private-line printer and continuing to crank them out at $125 apiece (although Edison felt that $200 was a fairer price for this "very cheap model which on the old plan of day work [instead of piece work] would cost double") for the insatiable Gold & Stock Company, which had now extended its interests overseas to England. Edison took his first trip abroad, to London, in the spring of 1873, to promote his automatic telegraph system for the British Post Office. There simply was no abatement in demand for telegraphic machines of all varieties: "My experience covers over 63 different printing telegraphs of my own," Edison wrote in his characteristically hyperbolic manner on April 21, 1873, to Marshall Lefferts, president of Gold & Stock, "and a constant work of 5 years 17 hours a day—in a diversity of mechanisms as complicated as 40 bushels of spiders webs."

Edison offered additional refinements for multiple telegraphy to Western Union. He worked to develop an "escapement" feature which would allow any single printer on the network to engage or be disengaged at will through rapidly breaking and then reconnecting electrical circuits. One surviving leather-bound notebook from February 1872 stamped ESCAPEMENTS in gold on the cover contains no less than 209 *different* drawings. Edison also continued to work upon the efficiency of the perforating and recording functions of the automatic telegraph: a single hole in the tape signaled a "dot," and three holes of equal size grouped closely together signaled a "dash"; his idea for a new perforator made use of electromagnetic rather than actual finger power. He was further fascinated by the potential for the telegraph to print ever faster, in actual roman letters: typed words. This necessitated type wheels (in essence resembling the IBM "Selectric" models) that could spin at lightning speed; a subsequent variation was based upon pens that formed letters in dot-matrix fashion. He also moved aggressively into the growing, popular sector of district telegraphy—"personal," domestic telegraph systems used in the home to send for messengers and place emergency calls to the police and fire departments (the precursor of "911"). Despite all this frenetic activity, Edison actually found the time to sit down and write a topic list of contents for a proposed book on telegraphy, summarizing his work from the beginning. In the ensuing months he went on to compose drafts for isolated chapters, but unfortunately, what could have been the definitive book on the subject (and, knowing Edison's instincts, a surefire bestseller) was never published.

However, other literary activities occupied Edison's imagination during the many nights he passed at the shop with Charles Batchelor and their business partner, Joseph Murray. Sardonic, oddball doggerel quatrains of multiple authorship in alternating handwriting styles (a poetry contest, perhaps?) illustrated with irreverent cartoons adorn their journals:

Who took me from my little cot
And put me on the cold cold pot
Whether I wanted to or not
My Mother

I tell you what an operator
Is a most mysterious crathur
With taste & elegance they dress
But how they do it I can't guess

If a man who "turnips" cries
Cries not when his father dies
'Tis a proof that he would rather
Have a turnip than his father

When you owe your Livery stable
And to pay you are not able
Obliged to walk off on your ear
I tell you it feels awful queer

How nice it is in Panic times
To have a pocket full of dimes
I used to know how 'twas myself
But now alas I'm out of pelf

Thomas Edison not only weathered the panic years of the 1870s, particularly 1873, but thrived during the most perilous moments, brought about by the overbuilding and subsequent stagnation of the very same rail-road industry that had knitted the nation together. Western railroads had grown the fastest, but their far-reaching infrastructure still suffered from underuse. The financial markets wavered in the aftermath of the Crédit Mo-bilier scandal, which brought down the mighty Union Pacific. Foreign in-vestors frightened by the specter of hemorrhaging profits shied away from further involvement in the extravagant, now tainted railroad business. In September, 1873, Jay Cooke and Company, the bank that had been under-writing the Northern Pacific Railroad, failed, and the New York Stock Ex-change shut down for the first time in its history. Five thousand commercial firms and fifty-seven stock exchange companies went under. Nearly 25 per-cent of New York City's workers were unemployed in the ensuing winter. Large, ingrained blue-collar union shops in the Northeast—mills and mines—were hardest hit, as tensions between distant management and op-pressed labor threatened by the influx of unskilled, cheaper workers were exacerbated, and there was widespread, violent rioting.

On the other hand, between 1870 and 1880, overall nonfarm (that is, manufacturing) gross product in America as a percentage of overall Gross National Product increased from 60 percent to 74 percent, the single most dramatic rate by far in the half-century up to 1920. As production ex-panded, prices dropped, and real income rose 60 percent during the same crucial decade. Small businessmen with a clearly defined commodity and market niche, Edison among them, buoyed by access to shorter-term, less onerous credit, were able to move quickly into the vacuum created by the failures of the corrupt behemoths. As Robert H. Wiebe writes, in his semi-nal and defining work on the period, *The Search for Order*, it was time for the "purity, honesty and frugality" of the individual entrepreneur to repudiate the "reign of licentious extravagance" of big government and bureaucracy which, after all, had gotten the country into such a quagmire.

The telegram from Edison in New York to Batchelor at Ward Street in Newark was terse and to the point: "Inform Wife will not be home tonight. Come over [to the City] with Pyrogallic solution for domestic [telegraph] before Eleven 'clock. Call at Western Union for me." The stage was set for another all-nighter; as he always had since the earliest days, Edison preferred using the Western Union wires in an improvised basement workshop he had set up there for transmission experiments between midnight and seven o'clock in the morning, because the lines were quieter. The inventor admitted that the marble tiled floor at Western Union was "a very hard kind of floor to sleep on," but it would serve.

There was serious work to be done, and Thomas Edison was systemically incapable of letting go of a procedure once it had started. In this instance, the long-awaited quadruplex line (sending *two* messages in each direction simultaneously) between New York and Boston was nearly ready, and therefore it was of little consequence that Mary Edison would spend one more night alone at their brownstone on Wright Street with her eighteen-month-old baby, Marion Estelle (nicknamed "Dot" by her father in homage to the telegraphic world she had helplessly entered). After all, he reasoned, Mary was well provided for, with a generous monthly clothing and amenities allowance, and domestic help.

Edison's deep-seated struggle between the demands of work and family began with the birth of his first child and became progressively more difficult and chronic as his life went on. It was a struggle far more complex than the normal requisites of a professional life balanced against the responsibilities traditionally incumbent upon a Victorian-era father. As we have seen, he was a compulsive worker, worrying projects like a stubborn terrier with a bone in his mouth. His tenacity when faced with a problem—admirable in and of itself—blurred into utter possession and an aching urge to have issues resolved, and was further compounded by his addiction to working on several matters simultaneously—the more balls in the air at any given time, the more Edison felt challenged, and, consequently, the longer juggling it took for any one of them to alight safely.

In one seemingly repentant journal entry for June 10, 1875, Edison "resolve[s] to work daytimes & stay home nights." But two months later, he notes that he "worked on the electromotograph all night." The twenty-four-hour clock meant nothing to him, and conventional sleep, in his middle years at least, was irrelevant. He was downright proud of the "one hundred nights" straight that he worked on the quadruplex at Western Union. Bathing more than once a week, eating three "balanced" meals a day, and changing his clothes or even taking off his shoes before clambering into bed (when he did actually manage to sleep in his own bed at home) were similarly incidental, more inconveniences than social conventions to be heeded.

Mary Stilwell Edison resists interpretation. For the first decade of marriage, until her own tragic circumstances break to the surface and she becomes a fully fledged casualty of her husband's accumulated neglect, she drifts across the scene as a childike cipher, insecure, withdrawn, dependent upon the company of her older sister, Alice, and parents living nearby, desperate to please her frenetic and illustrious husband. Mary is deprived of her primary source of authenticity because of Edison's inability and unwillingness to accept the conventions of hearth and home, where, under contemporary mores, as a dutiful wife and mother, she would be expected to hold sway.

The notorious Jay Gould, president of the Atlantic and Pacific Telegraph Company, had been flirting with Edison's interests ever since the inventor had begun quadruplex research and development for Western Union, Gould's fiercest competitor. At the very end of 1874 and into the new year of 1875—in the absence of a binding contractual deal between Edison and Western Union—Gould launched a preemptive strike, offering Edison a staggering $30,000 to purchase all rights to the quadruplex, which Edison dryly referred to as "somewhat more than I thought I could get." To Edison, during this financially constrained period, with more than one hundred twenty men under his employ, turnabout was fair play. Despite entreaties (and subsequent lawsuits) by his former patron, Western Union president William Orton, with a stroke of a pen the inventor transferred his loyalties, and was hired by Gould as "company electrician." Because of the many ambiguities present in these earliest years pertaining to claims of authorship and rights of ownership, patent litigation disputes over the quadruplex system dragged on for nearly four decades.

Characteristically Edison did not hoard his newly won cash. Instant liquidity inspired him to invest in the Port Huron street railway owned by his older brother William Pitt. He acquired new equipment for the Ward Street shop and for a separate laboratory in the same building, dedicated exclusively to experimentation. Within six months, Edison spun off from the manufacturing partnership with Murray, and hired his father and nephew "Little Charlie," William Pitt's son, barely into his teens, to come east and work for him. Emboldened by Gould's support, he drew up plans for an exhibit space at the forthcoming United States Centennial Exhibition in Philadelphia, set to open on May 10, 1876.

Undaunted and unwilling to concede all claims to Edison's skills, Orton was determined not to be forced entirely out of the picture. Acutely mindful of the ongoing work of Alexander Graham Bell and Elisha Gray in acoustic telegraphy, and of a possible revolution in communication—sound vibrations created by tuning forks sent over telegraph lines, which could lead to harmonic, or "telephonic" sound transmission, a virtual "speaking telegraph"—Orton quietly offered the resources of Western Union to support Edison's intensified efforts in this arena. The race was to the swift, and Orton knew all too well of Edison's speediness. Because he was hard of hearing, Edison focused his attentions upon trying different types of metal (brass, then iron coated with tin) to make the diaphragm that vibrated in resonance with the incoming signal that would create the clearest possible sound-bursts.

Parts of the metal machinery Edison used for his acoustic telegraphy experiments gave off sparks between them, even when they were not conducting currents; the accidental contact of a wire with the core of a magnet could also produce bright sparks. In a speculative frame of mind, aware that such an occurrence, if properly harnessed, could provide a quantum leap beyond telegraphy restricted to the confines of wire—as well as instant publicity—Edison therefore immediately seized upon this phenomenon. He defined it as not only new but *nonelectric*, more than mere induction—the creation of an electric current in a conductor moving across a magnetic field—and asserted that this phenomenon represented the unconfined, visual flight of energy through space.

His journals toward the end of November 1875 are taken over by breathless descriptions laying claim to the discovery of "*a true unknown force*," which seemed to increase in intensity in relation to the size of the "body of iron" involved. "Sparks can be drawn from iron bodies in a magnetic field," Edison wrote, intent upon staking out new territory, "& these sparks or force can be transmitted over metallic conductors & *do not follow the laws of either voltaic or static electricity*." Batchelor described an improvisational experiment as "very *paradoxical*" in which Edison used his body as a conductor for the force, and emerged unscathed. The two colleagues even built a "black box," or "Eatheroscope," within which one could observe the sparks flying. In an unsubstantiated reply to an inquiry from the telegraph engineer Charles Buell, Edison claimed to have "photographed" the force.

When Edison finally called representatives of the news media into the Newark laboratory to witness these experiments, the range of responses was broader than he had anticipated, and not generally favorable. There was rampant skepticism from many quarters: "Here we go again! is the familiar exclamation of the clown as he tumbles head first into the circus ring. Similarly Edison, the great professor of duplicity and quadruplicity, once more astonishes the editor of the New York *Tribune* . . . Not knowing its character, or exactly the quality of his new 'What is It?' does not prevent the professor from naming it, as its merchantable value largely depends upon a name."

"Very little is at present known by the experimenters themselves," added the New York *World* of December 1, 1875, preferring to downgrade the so-called discovery—"crude and but half formed"—as something that had existed but not been understood heretofore, a variation upon induction, insofar as "it does not require a circuit." The resourceful *World* reporter tracked down Mr. Orton of Western Union for a reaction to the news: " 'Oh, yes, I guess Mr. Edison has discovered an asteroid,' he replied pleasantly, and then turned his attention to business of greater importance." To the *Herald* on the following day, "this force or principle" was "the direct offspring of electricity and magnetism" and therefore related to the work of the German chemist Karl Reichenbach, who had defined this species of dynamic magnetism more spiritually as an "odic force" back in the 1840s.

The *Evening Post* went even further, accusing Edison of simply lifting from an issue of the English periodical *Electrical News* dated November 15, 1875, an account of the "weak electric sparks" noted therein by the German physicist Petrus Reiss, "almost identical with those of Mr. Edison. Neither does more than observe. No application is made of the force. But Mr. Reiss's announcement was made in Europe nearly a year ago [actually, nine months: February 1875]. . . . The coincidence is considered a remarkable one."

Even the *Scientific American*, usually a staunch Edison partisan, published an article excerpted from the *Journal of the Franklin Institute*, in which the author, Professor Edwin Houston, claimed bluntly that etheric force was no more than inductive electricity. Edison was moved to take the high road, huffily defending his findings as "presuming to believe in the capacity of Nature to supply a new form of energy." This was the crux of the disagreement, and it carried on well into the winter.

Help for the beleaguered Edison came in the person of George Miller Beard, M.D. (1839–1883), a New York physician, editor and founder of the *Archives of Electrology and Neurology*, and specialist and pioneer in the field of electro-therapeutics and nervous diseases. For more than a year prior to the "etheric force" fracas (thus named by Edison because of its tendency to diffuse itself in many different directions through matter), Beard had been following Edison's career and, through a mutual acquaintance, Jarvis Edson of the Domestic Telegraph Company, had already offered an "exchange of views" on broader, medical applications of electricity that would, he hoped, involve a meeting of the minds.

Beard's research over the decade before his initial contact with Edison had taken an intriguing path. His first published monograph, written when he was twenty-eight, was concerned with what he called "electrization," the actual application of electric current to solve eating and drinking disorders, and the uses of "electricity as a tonic." He explored electricity in treating diseases of the skin, whooping cough, and "debility," and the effect of "atmospheric electricity" on health and disease. He went on to study topics as abstrusely avant-garde as the psychology and pathology of delusions (Animal Magnetism, Clairvoyance, Spiritualism, and Mind Reading), and the relation of the aging process to work and creativity ("The Longevity of Brain Workers," demonstrated that "great men live longer than ordinary men"); and as fundamental as the pathologies of hay fever and writer's cramp.

Serious, mainstream scientific methodology, faith in repeated experimentation, and a more than passing interest in the spiritual realms were united by a common denominator that would grow in importance in Beard's work, and finally become his claim to fame: the nature of electrical forces within the human nervous system. This strong theme—as well as their similarly entrepreneurial temperaments; their subsequent but no less salient discoveries that they shared a common disability, both being hard of hearing; and their mutual abiding affection for lunches at Delmonico's restaurant—led him to Edison. "Please pay no attention to newspaper abuse," Beard wrote to Edison in early December 1875, as the etheric arguments spread thick and fast.

Although in later years Edison even transferred credit to Beard for inventing the term "etheric force," in fact the good doctor suggested the Greek *apolia* as a more accurate description, "given," he observed, "that want of polarity is the leading fact of it. Even if subsequent study should prove that the force is capable of polarization like light, yet practically and for practical purposes it is non-polar, and that fact distinguishes it from the ordinary forms of electricity."

This assertion became the informing theme of Beard's major twenty-eight-page, illustrated essay, "Experiments with the Alleged New Force," in the November number of the *Archives of Electrology and Neurology*. The article was subsequently published in the form of an extended letter to the editor of the New York *Daily Tribune*, on December 9, 1875, as "Mr. Edison's 'New Force': Result of Physiological and Other Experiments—Characteristics of the Alleged Force—The Apparatus Used," and later in its most definitive form in the January 22 *Scientific American*, "The Nature of the Newly-Discovered Force."

Beard, although as capable as Edison of grabbing headlines, entered the fray as a disinterested, third-party participant with credentials in a different, but related field, and he made it clear from the outset that his articles resulted from a series of experiments that he had conducted with Batchelor and Edison, as well as on his own. He laid the case out clearly: The force, generated in the laboratory by a large electromagnet, was conducted by the human body, but it produced no harmful or even discernible physical effects. The force did *not* require a circuit to be generated, and did *not* decompose iodide of potassium, a widely accepted litmus test at the time for the presence of "ordinary" electricity.

Rather, Beard wrote, having observed the comparative ease with which the force passed through nonconductors such as air, water, glass, rubber, and paraffin, "this is a radiant force, somewhere between light and heat on the one hand and magnetism and electricity on the other," a kind of electricity heretofore undefined, not subject to normal rules, in short, "something new to science." In no way, Beard insisted, should the fact that Edison had discovered this force by accident undermine its credibility; after all, "the element of chance has largely entered previous discoveries of a similar character."

Beard even attempted to console those who might be interested in attempting their own etheric experiments, only to find that no spark was produced, offering several reasons for failure, including insufficient battery power; an electromagnet that was too large; an inadequately darkened room; or inadvertent connection of the person doing the experiment with the conductor, thus "drawing off a part of the force before it reaches the dark box."

By the spring, Edison had dropped attempts to further validate etheric force, or to develop practical channels for it, although he remained in touch with his soulmate, George Beard, and their paths would cross soon again under more exciting circumstances. Edison later regretted the decision to move on so precipitously; but, being irrepressible, the *idea* of penetrating new territory always captivated him, and after the often premature announcement of innovation had been made, the struggle to sustain his crusade invariably bored him.

CHAPTER

7

IT WAS ONE OF RALPH WALDO EMERSON'S COUNTLESS APHO-
ristic, homespun but trenchant pronouncements that Thomas Edison loved
best: "If a man can write a better book, preach a better sermon, or make a
better mousetrap than his neighbor, though he build his house in the
woods, the world will make a beaten path to his door."

Toward the end of the century, William Graham Sumner revisited this
ideological dilemma and cast it in terms of industrialized America, observing, "A man who has a square mile to himself can easily do what he likes,
but a man who walks Broadway at noon or lives in a tenement house finds
his power to do as he likes limited by scores of considerations for the rights
and feelings of his fellowmen." As noted earlier, Alexis de Tocqueville, visiting these shores, had also perceived the essential tension in the American
psyche between aggressiveness and withdrawn individualism. It was the
struggle between "society and solitude" which Emerson debated in his mind
and enacted in his writings for his entire life, and into which Edison now,
on the cusp of celebrity, was about to enter. Even as Emerson guarded the
dialectic of what he called "armed neutrality," seeking always to condition,
day by day, the degree to which he engaged with the world at large, and to
which the world reciprocally was permitted to gain glimpses of his selfhood,
Edison—personifying the American experiment, commercial civilization in
full flower—nearing thirty years of age, likewise sought a level of socializa-

tion that would allow him to live and work comfortably and authentically.

According to Emerson's lights, Edison had already proven himself a genius by virtue of strength of will transcending mere talent, and had become the paradigmatic "new man" because he intuitively and directly grasped that, in America at least, progress and competition were identical. Edison was now on the threshold of the sacred temple wherein the "superior class" of American dwelled: "In every company one finds the best man," Emerson prophesied in 1848, "and if there be any question, it is decided the instant they enter into any practical enterprise. If the finders of glass, gunpowder, printing, electricity—if the healer of small-pox, the contriver of the safety-lamp, of the aqueduct, of the bridge, of the tunnel; if the finders of parallax, of new planets, of steam power for boat and carriage, the finder of sulphuric ether and the electric telegraph—if these men should keep their secrets, or only communicate them with each other, must not the whole race of mankind serve them as gods?"

Edison joked offhandedly that he could not keep up his rental payments in Newark, and that economic inconvenience led him to send his father, who was on the payroll, out into the surrounding New Jersey countryside to search for suitable real estate. It was typical of Edison to trivialize in retrospect such a pivotal move; yet, as we have seen, he had already set the design of separating his invention activities from his manufacturing efforts. He was always trying to find a better way to be a more creative and effective inventor, which required conditions of solitude for the purest informing moments. Then, by extension, he continually sought the most efficient (and controlling) method to connect ideas with their implementation, in conformity with his will and timetable, always keeping the men he trusted most closely at hand for when he needed them. In this manner, Edison was defining, disaggregating, and exploring the interaction of what we now refer to as R & D, research and development, a very modern pursuit.

Furthermore, with the birth of Thomas Alva Edison, Jr., in the new year (nicknamed "Dash" to complement his older sister's "Dot"), the Edisons' brownstone household was becoming insufferably constricted. And there was a deeper emotional motivation to move, to find a setting uniquely his own; we might imagine it as the "you can't go home again" syndrome. Edison alluded to his bitterness toward Port Huron in a series of brief, acerbic letters to his father around this period of flux, forcibly rejecting what might have been a proposed option of relocating back there to set up shop— "I do not think that any living human being will ever see me there again. . . .

I don't want you to stay in that god damn hole of a Port Huron which contains the most despicable remenants [sic] of the human race that can be found on the earth." Hot-headed ventilation? Perhaps, but also an essential developmental step for this young man on the way up and out. Port Huron was the place where Edison had been stigmatized and ridiculed as a child, the place where his beloved mother lay buried, where he had sunk countless thousands of dollars into helping his plodding brother William Pitt breathe life into a moribund street railroad company, the place where patriarch Sam and his child-bride had been the brunt of myopic, smalltown gossip.

At Menlo Park—a farming village twelve miles south of Newark and twenty-five miles out of New York City, strategically located on the Pennsylvania Railroad line between the Manhattan metropolis and Philadelphia—a dwelling and two tracts of land were purchased for $5,200 from a Mr. George Goodyear. It was a secluded rural site, yet with convenient access to industrial necessities. The cleared property on a hillside backed right up against virgin forest, and fronted acres of pastures and streams bounded by white rail fences meandering past ponds bordered by willows.

Alighting at the Menlo Park combination train station–telegraph office–post office on Lincoln Highway, parallel to the railroad line, you crossed the tracks past the conveniently situated Lighthouse Tavern, where you could moisten your throat parched from the dusty journey and indulge in a quick game of billiards, and then had to proceed about two hundred feet uphill on a boardwalk to reach the Edison home, a three-story white clapboard house with a windmill in back of it. Another hundred yards' trudge farther along unpaved Christie Street, at the very peak of the hill, was the laboratory building, constructed under the general supervision of Samuel Edison between January and March 1876, with sturdy hemlock planks purchased from the estimable firm of Hoover, Harris & Company of Philipsburg, Pennsylvania. This two-story frame structure, one hundred feet by thirty feet, graced with a commodious, arched porch and an outbuilding, most likely a carpentry shop, originally served as office, laboratory, and machine shop.

In the early days at Menlo Park, the second floor of the laboratory building was the focal point for the six-day, ten-hours-a-day work week. On the broad balcony looking southward over the cow pastures, Edison conducted his megaphone experiments. Just inside and to the right of the door leading to the balcony was Edison's small, modest, two-drawer work table, which he liked to say followed him from lab to lab, wherever he went. It was set squarely into a corner, facing the wall, isolated from the main flow of communal work yet in plain sight, a strong metaphor for Edison's collegial but distanced attitude toward his men, the spot where he settled down when not in motion around the shop. The walls of the laboratory were lined from floor to ceiling with more than 2,500 bottles of chemicals. Work stations with liquid batteries and telegraphic and telephonic equipment ran the entire length of the room. Downstairs, the machine shop with steam engine was originally housed at the rear of the laboratory building, while at the front were found earlier inventions and patent models which could be pillaged for spare parts as needed. There was also a hidden cabinet at floor level where, the story went, Edison would hide away for a nap when he did not want to be sought out.

Although Edison intended the laboratory to be literally off the beaten track, the inventor made certain at the same time that the broader public, especially the "scientific men," knew he was ensconced at Menlo Park for the long haul with the intention of pursuing serious work; as industrial historian David A. Hounshell has persuasively observed, Edison at this juncture in his career was in need of legitimizing his visionary pursuits in the aftermath of the etheric force debacle. And so, the move to Menlo Park was a deliberate image corrective as well as a domestic, commercial, and emotionally symbolic decision—at least at first.

There is a logic to the course of Edison's work during the three-year period beginning just before the Menlo Park shift. Although he liked to create the impression of being overtly scattershot, beneath the array of projects begun and never consummated and theories posited but never substantiated lay an inexorable progression driven by Edison's enduring captivation with varieties of communication media.

His first patent application in more than a year, filed on March 13, 1876, is a typical chapter in the story. It was for an "Improvement in Autographic Printing," more familiarly known as the electric pen, and Edison's (or more truly Batchelor's, who modestly and unfailingly employed the editorial "we") first experimental work in document copying and multiple duplication. Even as Edison progressed from duplex to quadruplex to sextuplex and then grandiosely to visions of octuplex telegraphy, he and Batchelor saw the niche for a technique that would allow businesses to reproduce many more copies from an original document than were currently

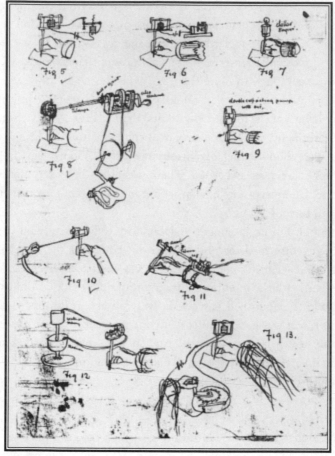

SKETCHES FOR THE ELECTRIC PEN

permitted by the cumbersome and messy papyrograph system.

The patent application relies upon significant analogies from embroidery and fresco painting in order to explain the principle underlying autographic printing—puncturing holes in the surface of a sheet of paper to make use of it as a stencil. The key to Edison's method was, once again, speed. You held the "pen" perpendicular to the surface of the paper, as you would any conventional writing implement, except this one contained a needle that "reciprocated [i.e., moved up and down] with great rapidity." As you "wrote" in the normal way, the needle made minuscule holes in the paper, spaced just the right distance apart so as not to tear the surface.

The movement of the needle was generated by an electromagnetic motor powered by a portable galvanic battery, opening and closing an electrical circuit, thereby spinning a revolving-armature flywheel. Once the sheet had been perforated, its tiny holes were filled using a felt-covered roller impregnated with semiliquid printer's ink thinned with castor oil; aniline colors could be created, "mixed with glycerine and molasses." Then you placed a clean sheet of paper underneath the stencil and worked the ink through with a clean roller. To expedite the printing process, the stencil and under-sheet were held securely in place on a frame by clamping springs. The frame was hinged along the top so that it could be lifted for the purpose of "inspecting the impression."

Once the holes in the stencil had been filled, subsequent impressions were made easily, and the ink could be replenished from time to time as required, In the early product development stages of the electric pen, Batchelor left Edison an autographic letter (signed "Yours in scientific love") in the lab, proudly noting that it was one of fifty copies taken from a single stencil.

Edison and Batchelor's advertising techniques were meant to underscore the universality of their creation and to portray the versatile electric pen, a bargain at $35.00 complete with supplies and book of instructions, as able to duplicate "Letters, Circulars, Price Lists, Market Quotations, Circular Letters, Pamphlets, Catalogues, Lawyers' Briefs, Contracts, Legal Documents, Manifests, Labels, Letter and Billheads, Maps, Architectural and Mechanical Outline Drawings, Music, Cypher, Codes, Press Reports, Financial Exhibits, Ruled Forms and [last but not least] Artistic Drawing."

Another advertisement was more romantically appealing and impressionistic. The machine was not illustrated; rather, an embracing couple were seen, surrounded by words floating halolike in the air: "Like Kissing— Every Succeeding Impression is as Good as the First—Endorsed By Every One Who Has Tried It!—Only A Gentle Pressure Used."

The electric pen sold well, through this lively and classically Edisonian combination of niche marketing and adept promotion. It was available by the new year 1876 all up and down the east coast, the midwest, and even as far away as British Columbia in Canada. As would soon become the practice with all his new ventures, Edison offered Charles Batchelor, as principal aide in the formulation of the pen, a straight percentage of profits in exchange for managing the day-to-day operations of the subsidiary—by March, they were selling more than one hundred fifty pens a month, and then went public; Batchelor, bitten by the American capitalistic spirit and a touch of spring fever, wrote from Menlo Park to his father in Britain that "the 'Electric Pen' [was] fairly out on royalty and in a very short time I shall have nothing whatever to do for it except to receive my share of Royalty."

But despite these and subsequent incentives—he would eventually retire a rich man—"Batch," ever the loyal soldier, was always careful to defer to Edison's interests when discussing any new invention the team were working on, cautioning his father, even though he was far away in Landudno, North Wales, that in all reports on their work, including casual conversation, he must be sure to give Edison due and *exclusive* credit.

Eventually, rights to the electric pen were sold to the Western Electric Company; then reverted back to Edison once again, and he finally sold the patent to A. B. Dick of Chicago, pioneer of the mimeograph.

From graphic replication of human writing, it is not a difficult step to acoustic replication of human speech. Edison was pushed to deeper experimentation by the announcement of Alexander Graham Bell's breakthrough on March 10, 1876—"This is a great day with me," Bell wrote, "I feel that I have at last struck the solution of a great problem—and the day is coming when telegraph wires will be laid on to houses just like water or gas—and friends converse with each other without leaving home." Where Bell subsequently stumbled, "disheartened," he confessed only a few weeks later, "at the intensity of the horizons opened out to me . . . like the first mariner in an unknown sea—uncertain which way to go," Edison seized the advantage. The most cumbersome problem with Bell's telephone—the liability at the crux of the thing—was that you had to shout into it, over and over again; and the same instrument, dependent upon a varying electromagnetic current, was used both for speaking and listening. Also, messages could only be sent for short distances. There had to be a better variable-resistance transmitter, or simply put, *there had to be a way to make the sound better.* If Edison could unlock that secret, he could successfully circumvent Bell's patent.

Underwritten by Western Union and belligerently scornful of Bell's achievement, Edison, with Batchelor at his side, pressed on intermittently with the "speaking telegraph" throughout the next two years (according to Walter L. Welch, Batchelor's biographer, for 188 days and 55 nights in 1877 alone) picking up where he had become sidetracked by etheric force pursuits, testing the electronic resistance properties of more than two thousand different chemical compounds under varying degrees of pressure, drawing upon the same acoustic principles used in his multiple telegraph. Edison's hearing deficit forced him when testing different materials' acoustic properties to follow the same bizarre technique he would use decades later when auditioning pianists for his phonograph records: clenching his teeth around a metal plate attached to the sounding apparatus, so that vibrations were conveyed through his resonating jawbone—meaning, in effect, that *he virtually heard through his teeth.* It was during this extended period that Edison also first seized upon the possibility—a glimmer of the phonograph to come—that impulses coming over the telephone might be recorded and then reproduced.

Bell's telephonic transmitter depended upon the extreme sensitivity of a vibrating diaphragm of hammered-thin metal. The movement in and out of the membrane, determined by the sound waves from the voice of the speaker, was translated into varying degrees of current sent out over the line to an electromagnetic receiver. The receiver contained an identical diaphragm that vibrated in tune with the input tones. As a necessary refinement on the transmission end, Edison interposed two carbon buttons derived from humble lampblack, one in direct contact with the iron diaphragm, and the other in loose contact with the first button. This additional component proved to be dramatically more responsive to the pressure of spoken voice sound waves, which moved the buttons concomitantly closer together. Electrical resistance, conditioned by the minute changes in distance of the buttons from each other, varied in response to the sounds, and thus made possible not only greater volume, but more shading and distinction between the vowel and sibilant characteristics of words.

The manufacture of carbon "plumbago" buttons for Western Union became one of the first fully developed cottage industries at Menlo Park. Out behind the laboratory building and to the west, along the fence that ran around the entire compound, stood the Carbon Shed, reconstructed today at Henry Ford's Greenfield Village in Dearborn, Michigan, just as it once was in the 1870s. About a dozen kerosene lamps would be kept burning there all night long. Periodically, the night watchman scraped the soot deposits off the glass lamp hoods, weighed the carbon carefully, rolled it until worm-shaped and of a pastelike consistency, and then sliced it into segments each three hundred milligrams thick. The trick was to catch the soot at just the right moment, because if it was allowed to accumulate and cook too long in the lamps, it would become crisp, therefore too brittle and resistant, unsuitable for use in the telephone.

Critics of Edison's "cut and try" methodology—the inventor's global conviction from his earliest days that the only way to be successful was to understand the inherent physical and chemical properties of every substance known to man, and, in later life, even substances unknown—fail to acknowledge the viability of this kind of approach when it is applied to a dispassionate and strategic patent campaign. Refining the telephone, as with other technological innovations, Edison intervened, not as originator but rather as finessing perfector, with the knowledge that a chain in the interacting components of a machine was only as strong as its weakest link. The classic Edisonian maxim, "Genius is ninety-nine percent perspiration and one percent inspiration," is simply a rationalization for the approach that worked best for him, bringing to bear the broadest conceivable range of options into focus upon the narrowest possible range of a challenge.

A small and intrepid band of disciples labored at the Menlo Park microcosm during its first few years, no more than a dozen men at any given time. The innermost circle came out with Edison and Batchelor from the Ward Street shop. Charles Stilwell, Edison's brother-in-law, was among them, joined soon after by Edison's nephew Charles, William Pitt's son from Port Huron. Complementing Batchelor the strategist was John Kruesi, the adept Swiss machinist who had come to America in 1870 after stints in Zurich, where he had been apprenticed to a clockmaker, and Paris; he had then worked for the Singer Sewing Machine Company in Elizabeth before joining the Edison cadre. Like Batchelor, Kruesi would go on to become a twenty-year man with the organization, and he appears to have been another kindred spirit, able to decipher Edison's raw scrawls that captured rough drawings left on his desk (often written so quickly that all the words were connected), to "Please start on this right away," or to "Make this,"—and then fabricating them before dawn from wood and brass and tin.

This was the formative period leading up to the phonograph, when everything would become wholly metamorphosed, including perceptions of the place as articulated by the men who worked there, by the news media, and by the culture at large. This was also the most informal period, a time when men who worked for the love of it as much as for the money could also feel free to "lean and loafe" as brothers, in Walt Whitman's vision, before the sacred precincts of Menlo Park became permeable to sophisticated society. And it is not a contradiction, but, rather, an acceptable and characteristic American practice that this pastoral place possessed an intensely utilitarian purpose. Here was an industrial farm, a place of machines in the garden of New Jersey, governed by natural, irregular work rhythms. Skilled immigrants like John Kruesi and Charles Batchelor rose up out of an artisan tradition at a refreshing time before labor hierarchies and the gospel of efficiency took hold in the American factory. Thomas Edison revered this tradition and fought to preserve it. In later years, he freely dispensed advice to the ardent young men who came to him seeking the gospel, saying, "Don't go to college. Get into a shop and work out your own salvation," echoing Benjamin Franklin's declaration to the American worker to "Keep your shop and your shop will keep you." This was the symbolic tone for Menlo Park which led Edison, if not consciously then culturally, to first find a job in his early youth, then transform it into a calling, then bring others with him into an unadorned wilderness.

In the communal scheme of Menlo Park, under Thomas Edison's all-seeing, all-knowing eye, it made sense that Kruesi and Batchelor and their families would share a house a stone's throw from him, across Christie Street. Requiring his tightly knit work force likewise to be in close proximity, Edison imported another distant relative, the recently widowed Sarah Jordan, at the time living in Newark with her daughter, Ida, offering the women the opportunity to establish a boarding house on Christie Street midway between the Edison home and the laboratory. Here, single men and occasional transients were accommodated in the six upstairs bedrooms of the twelve-room duplex house. As traffic in and out of the compound intensified, Mrs. Jordan hung out a Lunch Room shingle, charging thirty cents a meal to all comers, and set her establishment up to cater to the local men who did not go home to the village for their repast, as well as to serve visiting dignitaries and journalists—when the public spotlight inevitably found its target.

CHAPTER

8

THE SPLENDID LEGENDS OF THE EXPLOSIVE BIRTH OF THE phonograph, which we shall recount later in this chapter, do not do justice to Edison's methodically developmental manner of conducting business. Rather than springing fully formed from the workshop of his imagination, the phonograph, in its first incarnation, had a gradual gestation, culminating in Edison's first broadly published essay, a virtual manifesto of progress.

Certain of its roots in the underlying principle that sounds—or, more pertinently, voices—could be recorded and saved can be tracked back to French inventor Leon Scott's "phonautograph" of 1837. A pig's bristle fixed to a diaphragm of stretched pigskin replicated patterns of sound vibration onto a sheet of paper coated with lampblack rotating on a cylinder. Thus, a visual imprint of sound's "sinuous continuity" was created.

And in 1863, F. B. Fenby, of Worcester, Massachusetts, received a patent for a device called The Electro Magnet Phonograph, which made a visual recording of telegraphic dots and dashes.

We have earlier noted Edison's idea during his working telegrapher days of recording incoming Morse code messages, essentially "holding them up" so that he could receive them more accurately, and then replay them at comfortable speeds. Early in 1877, during his telephone experiments, Edison applied for a patent formally to enhance the automatic telegraph as an extension of this initial concept, by "indenting upon a sheet of paper [actually, embossing upon a disc revolving in spiral pattern on a platen] the characters received from a distant station, and use such sheet to transmit the same message, thus providing an automatic device for transmitting the same message more than once from one station to different stations; and for transmitting it automatically where it has to pass through several offices."

This kind of synergistic product development strategy illustrates yet again that Edison did not indulge in closed-end thinking. He was constantly making synaptic leaps from the heart of one idea to the inception of another. Francis Jehl, who arrived on Edison's doorstep in Menlo Park as a lad of eighteen in 1879, took a job as a laboratory assistant, and remained a colleague for half a century, has left a delightful, loving, if somewhat uneven and conjectural memoir, *Menlo Park Reminiscences,* in which he refers to this device: "One day," he writes, "I happened to be standing looking at this instrument, when Mr. Edison came up. He saw my interest in it and told me a few things about it, one of which I have never forgotten. *'That machine,' said he, 'was the father of the phonograph.'* "

During the spring of 1877, the local press slowly began to notice Edison as he continued to work on a "Telephonic apparatus." Spurred on by tantalizing reports about Professor Bell's telephone, Edison presented journalists with an adversarial persona that would become increasingly familiar in the years ahead, that of a young man "so often scoffed at that it has no other effect upon him than to stimulate him to increased study and labor." While Professor Bell, with the "singular and inexplicable peculiarity" of his startling invention, was allied with the "witches of Salem," Mr. Edison meanwhile tooled with a *music*-transmitting apparatus that incorporated components very closely related to his telegraphic machine. The Motograph, as it was called, depended once more upon the principle that a vibrating metal diaphragm could be used to activate and then interrupt an electrical current through resonating contact levers. The central principle persisted: mechanical vibrations to electrical impulses at one end, and a reversal of the dynamic at the other end, to result in *enhanced communication*.

Edison and Batchelor moved back and forth between the telegraphic and the telephonic on into the springtime. Their fame (or notoriety) had reached such distant shores that they were visited by the esteemed Sir William H. Preece, Engineer-in-Chief and Electrician to the Royal Post Office in London, on his first trip to America. Preece looked in on the drawn-out litigation over the quadruplex, partly out of curiosity, to compare American and British legal systems, and also to get a peek at Thomas Edison, on the stand for eight days. "He is known here as the Professor of Duplicity," Preece confided to his diary on April 30. "[Western Union president] William Orton told me in England of Edison, 'that young man has a vacuum where his conscience ought to be.'"

Regardless of such negative public perceptions and not dissuaded by Orton's attitude (in itself duplicitous, since Edison was still on retainer with his company), Preece, after spending the evening attending a lecture of "telephonic talk" by the hyperbolically celebrated Professor Bell at Chickering Music Hall ("The press gives a great deal of prominence to the gathering of armies on the Danube and in Armenia, but it is not unlikely that the score of distinguished gentlemen who witnessed the first operation of the speaking telephone in New York last night assisted at an event the importance of which is greater than any twenty of the greatest wars recorded in history."), made the trip out to the wilds of Menlo Park with great excitement. "Another blazing day which I spent . . . with Edison—*an ingenious electrician*—experimenting and examining apparatus. He gave me for dinner *raw ham*! tea and iced water!! [No mention is made of Mrs. Edison. Perhaps she sent the meal for the busy men up to the lab so they would not have to take a break in their discussions.] It is nearly thirty miles off [from New York City]. The railways here have no fences and they go bang through the streets of the towns. The whistles have the most horrid howls—more like an elephant's trumpet than anything else. The stations have no names and there are no porters about. Everyone has to look out for oneself. We nearly missed our station and as it was had to jump out while the train was moving. At level crossings the only notice put up is—'Look out for the locomotive.'" It was a rough and tumble journey, but Preece was seduced by the countryside at Menlo Park, which reminded him of England, "chestnuts and lilacs in full bloom and nature clothed in brilliant green. I saw a few gorgeous butterflies flitting about and red and blue backed birds."

Edison and Batchelor demonstrated for the enraptured Preece their pressure relays and tuning forks for splitting telegraphic line signals; at night, after Preece departed—he had a dinner date back in New York City with Professor Bell; the Englishman reveled in shuttling back and forth between these two star rivals—they continued to work on a diaphragm speaker: such were the obsessions of the season. And so on into the summer months, when they came to terms with the fact that for the actual human voice to be recorded, preserved, and played back, it would be possible to eliminate the electrical mode and simply rely upon the mechanical: "Just

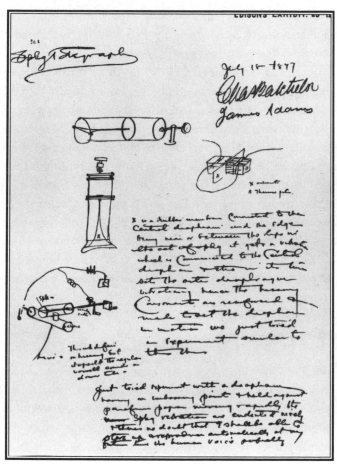

EDISON'S FIRST KNOWN NOTATION ON THE PHONOGRAPH
("SPEAKING TELEGRAPH")

tried experiment," Edison wrote on a laboratory sheet witnessed by Charles Batchelor and James Adams and dated July 18, 1877, "with a diaphragm having an embossing point and held against paraffin paper moving rapidly. The spg. [speaking] vibrations are indented nicely and there is no doubt that I shall be able to store up & reproduce at any future time the human voice perfectly." This brief entry is vintage Edison: it does not mention anyone else's name; it is definitively and prematurely conclusive; and it self-confidently corners the market on the future, all in two sentences. A few days later, a slight but significant change was added—instead of using flat, unmodified waxed paper, a ridged tape was employed, like a stock ticker-tape, and a stylus connected to a telephone receiver diaphragm was applied directly to the ridge on the tape, creating "voice-impressions." We know from subsequent, clearly dated drawings that this approach predominated into September and was still in effect in November.

In August and September, following this "direct sound" break-through, the actual word "phonograph" begins to appear in Edison and Batchelor's laboratory notebooks. It may be that rudimentary models were fashioned by this time. Edison was concentrating upon the interaction of the stylus needle with various forms of receiving media, since ordinary or even chemically coated paper left a voice tracing, but could not then be em-ployed to play back the tracing with any dependable coherence. At the same time, however, in conformity with Edison's typical work structure, "speaker telephone" work prevailed in parallel fashion, Batchelor noting in his metic-ulous journals that he was repeatedly commuting to Western Union head-quarters in Manhattan to show his corporate underwriters how the machines were coming along. By early fall, word began to leak out that something new was brewing at Menlo Park. Batchelor noted with characteristic under-statement in his diary on September 2, 1877, "A *Times* reporter called [in the English sense, meaning, to visit] at night to see the Speaking Tele-graph."

It is not altogether clear, however, what this reporter would have actu-ally seen. According to a chronological study made by Edward Jay Pershey of Edison's drawings during this intensive period, a wide assortment of pro-totypical sketches were done at Menlo Park. Some were "thought experi-ment" machines captioned with the word "phonograph," combining tele-graphic and telephonic images. Other drawings were for machines to be constructed as models. Still others concentrated on the particulars of the sty-lus and its mechanical movement. Pershey further notes that the frequency of the work, including notes and sketches cosigned and annotated by other workers, picked up during October and into November, which is signifi-cant, because we know from Batchelor's diary entries that at this time Edi-son was away for at least ten days, fetching his wife, who had gone "out West" for a rest cure. The phonograph, like all Edison inventions, was the labor of many hands.

And even though the phonograph was not in fact ready, the accumu-lated enthusiasm of one of these hands, Edward H. Johnson, could no longer be contained. On November 17, perhaps at Edison's behest, he wrote to the *Scientific American*, disclosing that a "telephone repeater," in essence only an idea, even if on the brink of actualization, had been developed at Menlo

Park. A WONDERFUL INVENTION—SPEECH CAPABLE OF INDEFINITE REPETI-
TION FROM AUTOMATIC RECORDS—blared the exultant editorial that very
day. It was a fait accompli—Science *can* be sensational, after all. Perhaps letter writing will become obsolete? Perhaps we will no longer have to journey
to the opera house to hear the great divas of our day? Perhaps congressmen
will be able to deliver speeches while simultaneously "enjoying a cigar in
the corridor or drinking iced tea in the restaurant"?

As yet unprotected by a registered patent, Edison's phonograph had
now been announced for all the world to admire. The *Scientific American* article refers several times to "an indented strip of paper," and to "vibrating
plates" (i.e., diaphragms), and the accompanying drawings show the
schematic results of a stylus applied to ridged paper, creating a jagged edge,
so it is clear that while Edison had the recording dynamic figured out, he
still had not hit upon the playback method which would give a purer sound.

The twelve-day period following Johnson's premature announcement
must have been frenetic. On November 29, 1877, Edison made a hasty
sketch of what would become the first phonograph, showing the rudimentary elements of the machine: a spiral-grooved, solid brass cylinder three
and one-half inches in diameter was mounted on a feed-screw operated with

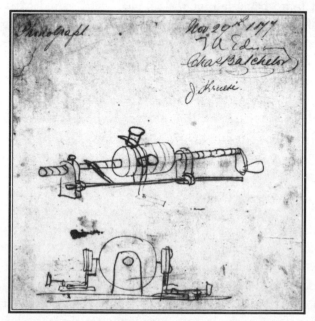

FIRST REPRESENTATIONAL SKETCH OF THE PHONOGRAPH

a hand crank. Like a barbecued pig roasting slowly on a spit, the cylinder turned and moved laterally from right to left, wrapped with a sheet of thin tinfoil—this was the new recording medium!—pressed firmly into its grooves. Riding on the surface of the foil was a stylus connected to the center of a diaphragm, activated by the human voice spoken (actually, yelled) into a funnel-like mouthpiece made from a telephone transmitter.

How did Edison extrapolate in such a short span of time from paper to foil, solving the main technological problem stalling the completion of the phonograph? No written evidence remains, but there is one specific graphic analogy between the penultimate and final recording techniques. At some point between the middle and the end of November, Edison abandoned the linear method described in Johnson's article, in which a strip of treated paper was fed beneath the stylus and then collected in a spool, and moved to the design that he instructed John Kruesi to follow, in which the recording medium rigid

THE TINFOIL PHONOGRAPH

enough to enable clear playback, remained fixed beneath the stylus. When unwrapped from the surface of the spiraled metal cylinder, the foil was about the size and shape of a rectangular sheet covered with linear indentations—what Edison described prophetically as "stereoscopic views of the voice . . . a matrix of the words and voice."

Operating the phonograph successfully required excellent eye-hand coordination and a stout pair of lungs. To begin with, the foil had to be wrapped around the cylinder with just the right amount of tension so that it would sink the proper distance into the grooves under pressure from the stylus, yet not tear in the process. Then, the feed screw had to be cranked at a consistent pace while one had at most ten seconds to blare a message into the funnel. This explains why "Mary had a little lamb, its fleece was white as snow, and everywhere that Mary went, the lamb was sure to go," aside from its mainstream American qualities, was purportedly Edison's first trial statement into his new phonograph. Once the message was engraved as a series of "bumps" in the foil surface, the stylus and funnel were pulled back away from the cylinder, which could then be "rewound" to its original position. Then, a corresponding "playback" spring-held stylus mounted on the opposite side of the machine was gently applied to the "recording," and, if you were skillful enough to recrank the machine at the same rate at which you had recorded your message, you—and your admiring audience—would hear it replicated.

A thicket of sanctioned folk tales has the ingenuous John Kruesi simply carrying out the Old Man's instructions—Edison thrusts a crude sketch into his hands and says bluntly "The machine must talk"—that November night and into the following week without a clue as to what would result, as if the men had not spent the past ten months in manic pursuit of the idea. There were stories of skeptical bets with cigars and dollars wagered, knowing laughter, and winks; Edison had finally gone off the deep end.

And when "Mary had a little lamb" went into the mouthpiece and then came back in Edison's reedy tone, "honest John Kruesi" exclaimed, in true peasant character, *Mein Gott in Himmel!*" and Edison conceded that he "was never so taken aback in all my life. . . . I was always afraid of anything that worked the first time." Ninety-nine percent of his perspiration seemed to have evaporated, and only one percent inspiration remained to tell the story.

Surely it was serendipitous, yet another quirk of American intellectual history, that *Popular Science Monthly* in November 1877, carried an article called "The Fixation of Belief," by a young, neurasthenic philosopher living in Hoboken named Charles Sanders Peirce, an essay that William James hailed as "the birth certificate of pragmatism." The piece was a hymn to empirical knowledge and an outright debunking of metaphysics; as an acknowledged snapshot of the inception of a mode of thought, it helps further locate the eminently pragmatic Edison in his time and place.

"The genius of a man's logical method should be loved and reverenced as his bride," Peirce wrote, " . . . she is the one that he has chosen, and in doing so he only honors her more." Drawing a parallel with Antoine Lavoisier, Peirce recalls that the great French chemist's "way was to carry his mind into his laboratory, and to make of his alembics and cucurbits instruments of thought, giving a new conception of reasoning as something which was to be done with one's eyes open, by manipulating real things instead of words and fancies." Thus did Edison, with endless drawings and working models and incessant collaborative attempts, dwell extremely in the world of experience. And equally thus did he have the alchemical capacity to exploit that hard experience.

According to Edison's self-aggrandizing account, on the morning of December 7, 1877, following the theatrical demonstration at the lab, after which everyone stayed up all night playing with the thing, he took the newly minted phonograph into New York City to make it perform for Alfred Beach and the other editors of *Scientific American*. Such a huge crowd assembled around Beach's desk, Edison recalled, that the floor was in danger of collapsing. No mention was made of Charles Batchelor's accompanying him, even though Batchelor raved in a letter dated December 6 to the Chicago electrical supplies manufacturer George Bliss that, "You probably remember when you were down here about Edison's idea of recording the human voice and afterwards reproducing it. Well, we have done it and have today shown it in New York to the *Scientific American* people who are now sketching the apparatus for a future issue. . . . The Phonograph is going to be a magnificent success!!!!!! [*sic*]." On the following day, once the story had been told to *Scientific American*, Batchelor self-effacingly wrote back home to the editor of the *English Mechanic* in Covent Garden that "Mr. Thos. A. Edison of New York, the well-known electrician has just developed a method of recording and reproducing the human voice."

Regardless of whether or not a structural disaster almost occurred at the renowned magazine that day, Edison and Batchelor accomplished their purpose: " . . . the machine inquired as to our health," Mr. Beach marveled in his December 22 column, "asked us how we liked the phonograph, informed us that *it* was very well, and bid us a cordial good-night. . . . Here is a little affair of a few pieces of metal, set up roughly on an iron stand about a foot square, that talks in such a way that . . . there can be no doubt but that the inflections are those of nothing else than the human voice."

Edison filed his patent application for the "phonograph or speaking machine" two days later. He explicitly sought to make the distinction between his machine, which could both reproduce sound and render it "audible at a future time," and the telephone, which was limited only to conveying sound. Alexander Graham Bell expressed deep regret to his father-in-law Gardiner Hubbard that he had "let this invention slip through my fingers," considering how many of the same principles he had been exploring so deeply for the past many years. In principle, Bell had discussed the idea that phonautographic tracing should indeed be possible—but he had never pushed the concept far enough. Robert Bruce, Bell's biographer, points out that Bell kept these morose speculations private while congratulating Edison in public for his "most ingenious instrument."

The other fascinating insight Edison revealed in his patent for the phonograph was the observation that diaphragm recordings of the human voice were visually "separate and distinct," as opposed to overlapping or "superimposed." This propensity to be recorded sequentially made it logically follow that sounds could be replayed replicatively. Thus, Edison envisioned that the foil, or as he called it, the "indented," or "yielding" material, could conceivably be detached from the machine and "stereotyped by means of a plaster of paris process." With these two observations, Edison paved the way for the basic principles in his motion pictures, the phenomenon of persistence of vision: separate, sequential photographic frames projected at forty or more per second.

He did not rest there. Edison's improvisational mind spun out many more applications for the phonograph than could ever possibly be contained on paper, and he loaded as many of them as he could into a *North American Review* essay, "The Phonograph and Its Future," written that winter and published in the May-June 1878 issue. Because the phonograph could capture heretofore "fugitive" sound waves and retain them, Edison described the possibilities for the machine as "illimitable"— letter writing and other forms of dictation; reading books for the blind; teaching elocution; singing children to sleep; preserving "sayings, voices and last words of great men"; publishing a new form of book, with forty thousand words engraved upon a metal plate ten inches square; music boxes and talking dolls; clocks that could tell you the hour of the day, "call you to lunch, send your lover home at ten, etc."; and, last but certainly not least, (italics Edison's) *"perfect the telephone."*

Thomas Alva Edison on the crest of his most publicly celebrated invention was approaching his thirty-first birthday. An unprepossessing figure physically, a touch below medium height at about five feet eight inches, he weighed in the 155-pound range, and even though he guarded his diet closely, a few too many ravenous snacks of pie and coffee could send him toward portliness. His soft, straight hair was either dark brown lightly touched with gray, or prematurely gray flecked with a few stray patches of brown, depending upon the light, but usually unruly and wandering down over his forehead, so that he was constantly brushing it back with an impatient sweep of his acid-stained hand. Edison's complexion was pale, even sickly at times, as befitting a man who spent most of his waking hours indoors, sleeping irregularly. No wrinkles disturbed the broad and smooth, rounded brow above the prominent nose, where one might discern a charcoal smudge during especially busy laboratory sessions. The light eyes, with hooded lids, spaced far apart, went from gray to blue, moistening to a sheen when he was excited. He was rigorously beardless, except for a brief period of less than a year when he affected a scrawny mustache, or when he simply forgot to shave and allowed a stubble to peek through. He was known to have a big, winning smile, more like a grin, showing his tobacco- and tea-discolored teeth—and a ready laugh, more so if the joke were at his expense.

Edison tended to walk slowly and with a slight stoop, even at this young an age. On first meeting, he spoke quietly, at times even shyly; with familiarity, once in the laboratory and honed in on the day's labor, he became more animated, but always with a slight tilt of the head and a fixed gaze, as his hearing impairment forced him to listen with an added measure of attentiveness.

In his attire, Edison did not stand on ceremony or go out of his way to tidy himself up for visitors. Most often you got the hayseed look, a man sporting a slouch hat or a workman's cap perched squarely on the thatch of hair, with a nondescript frock coat, dark vest, and collarless shirt, and absolutely no jewelry—no ring, no studs, not even a watch fob.

Reflecting back over half a century to his years with Edison, Francis Jehl recalled Menlo Park as an Arcadian place where "sentiments were magnified almost to the verge of hallucination," where the laboratory "lighted up with a bewitching glimmer—like an alchemist's den."

Was the presiding genius over Menlo Park simply an aw-shucks, salt of the earth, plainspoken, grownup schoolboy—or was he a latter-day incarnation of Nathaniel Hawthorne, as a *Scribner's* writer fancifully decided, reborn as a sorcerer "whose personages are chemicals," believing, like that virtuoso spinner of weird tales, that "the fantasies of one day are the deepest realities of a future one"?

CHAPTER

9

AS THE PHONOGRAPH BEWITCHED THE POPULAR IMAGINATION the hordes descended upon Menlo Park. Newspaper reporters and miscellaneous dignitaries by the carload crossed the Rahway River and swampy meadows on the Pennsylvania Railroad, and, up its south branch to the Raritan Valley, a few miles from Perth Amboy in Middlesex County, signaled for the diminutive station stop—rumor was they were going to rename it Edisonville. Occasionally, one might make a courtesy call at the Victorian homestead, a light-colored frame house with dark-red trim, wide and shallow with cross-gables on the roof and a porch shadowed by a distinctive latticework pattern.

The retiring Mrs. Edison did not always come downstairs. Self-conscious and shy by nature, incessantly lonely for her husband who was never there for dinner—or any other meal for that matter—Mary Edison, pregnant with her third child, had no choice but to keep close to home. Little Tom, Jr., was often sick and needed constant care. He was no playmate for his big sister, the vivacious, quick-minded Marion. Long blond hair streaming, dressed in lilac silk, she was a familiar figure around Menlo Park, skipping merrily along the boardwalk leading up to the laboratory, always on the lookout for the black snakes she feared, carrying Father's lunch to him in a basket—sandwiches, tea, and a huge piece of pie. After morning tutoring sessions with her governess, Marion loved to make quick visits to

the laboratory, that exotic playland. If she rested her elbows on Papa's knee and gazed up into his open face with just the right plaintive expression, he might set down his clay pipe, dig deep into his pocket, and come out with a dime for her to buy candy. Marion avoided the strange, bearded men who labored by Edison's side and called her "Dot," and looked as if they were hiding behind masks. Visitors were told by the Negro housekeeper to proceed to the laboratory.

It was "Professor" Thomas Edison, "the Wizard," "the New Jersey Columbus," "the Napoleon of Invention" the throngs were coming to see at firsthand in his hilltop hideaway, "Monte Cristo's cave." In a private moment of Thoreauvian inspiration, Edison protested to a friend that "the reporters who come down here have already unstrung my nerves [so much] that I think of taking to the woods." However, the public Edison possessed more than a touch of the P. T. Barnum to offset whatever reluctance he may have felt deep within—most of his visitors were there by invitation.

Adding to the Oz-like drama was the topography of the countryside, a low rise in the first hill after the station preventing the visitor from noticing the architecture of the laboratory from the train depot. The farmhouse-tabernacle-schoolhouse-frame tenement-experimental *atelier* (one wag likened it to a "country shoe factory") was festooned with telegraph wires and lightning rods, with two brick chimneys rising from one side. It came slowly into view only after you had passed the Edison home. Mounted on a piazza in front of the laboratory porch was a powerful telescope through which, on a clear day, you could actually see the cables of the great East River bridge under construction, "rising to the naked eye like a swarthy giant on the misty horizon." At the back of the building, next to a pond stocked with bullheads to indulge Edison's penchant for fishing, a twisted apple tree stood, enhancing the rural air, surrounded by old barrels, rickety machinery, and heaps of discarded junk.

On the height of the bluff, in the country peace, crickets and tree-toads chirped, bees hummed, breezes blew, buttercups wavered. You drew nearer to the white building and noticed peeling paint as you mounted the sagging stairs to the balcony-shaded verandah, with a sensation more akin to entering a meeting house or town hall rather than a laboratory.

Mr. Edison was not immediately available. Just to the right inside the huge double-doors was a reception area and drafting alcove with glass-fronted cabinets containing Edison instruments, telegraphic keys, Morse sounders, and galvanometers. Open shelves lined with books—*Chambers's Encyclopaedia, Watt's Dictionary of the Sciences,* James Dwight Dana's *Manual of Geology*—gave continued evidence of Edison's obsession with surrounding himself with all conceivable information at his fingertips. From dead straight ahead, at the other end of the first floor, past a carpentry room and through an archway, came the sounds of metalwork, steam hissing from a ten-horsepower Baxter engine, lathes whirring, drills chattering, hammers, files, saws, and forges.

A flight of uncarpeted wooden steps led up to the second floor, and thus visitors were directed by Mr. Griffin or Mr. Johnson. Benches and tables were crowded to distraction with batteries, microscopes, magnifying glasses, crucibles and retorts, "a wilderness of wires, jars of vitriol, strips of tin foil, old clay pipes, copies of the great daily newspapers, and sundry bits of machinery of unknown power." Sunlight passing through twenty windows onto chemical bottles arrayed floor to ceiling as if in an apothecary shop gone mad—again, the "one of each" mentality—gave a rainbow cast to the air of the laboratory's unbroken space, dominated at the far end by a massive pipe organ, the gift of its manufacturer, Hilbourne Roosevelt, Theodore's cousin. Around this impressive, sonorous instrument Edison (who was fond of pounding out improvised musical compositions of his own devising) and his men would gather in the evenings to play and sing selections from Moody and Sankey's popular *Tune Book* as well as less tasteful ditties in the later hours.

If Edison was otherwise engaged, which he invariably made it his practice to be, he would not notice or hear the approaching visitors, and so their first impression would be of a man with a diffidently distracted air, an "abstracted look," caught in the midst of a dreamy thought, an artless fellow unused to society, "riding rough shod over the ordinary mannerisms of life," who cared not for material things. Once brought around to the situation at hand, he would be earnest, courteous, and affable, eager to please—and in control. Many remarked on the inventor's brightly boyish, "provincial" manner and quick blush as he cupped his hand to his ear while courteously attending to the interviewer's questions.

Edison drew his visitors' attention to the phonograph, on a table at the center of the room—often referring to it as his "baby," patting the machine affectionately and confessing that he "expected it to grow up to be a big feller and support me in my old age," before settling down to perform a wide variety of theatrical demonstrations. In addition to the much-vaunted "Mary Had a Little Lamb," played forwards and then backwards to the great amusement of assembled admirers, and the occasional doggerel renditions of "Rub a dub dub, three men in a tub," "Three Blind Mice," and "Jack and Jill Went Up the Hill," he would recite the first four lines from the "Now is the winter of our discontent" speech of *Richard III*, the story going that as a journeyman telegrapher in Cincinnati, Edison learned it while rooming with two actors he met there—and a variant on this tale being that Edison actually *performed* the role on stage. Other favorite pieces included excerpts from the currently popular songs "John Brown's Body," "Old Uncle Ned," "La Grande Duchesse," and "The Wandering Refugee," and snippets from such middlebrow poetic chestnuts as Grey's *"Elegy,"* Poe's *"Annabel Lee,"* and Edison's lifelong favorite, *"Evangeline,"* by Longfellow.

Edison also liked to recite passages from the Bible and then replay the cylinder, interrupting himself by barking "Oh, shut up!" "Dry up!" or "Help! Police! Police!" "My little instrument," he confided one day to the phonograph, duly transcribed by a correspondent from the Rochester *Daily Union and Advertiser*, "if I had had you at the Centennial Exhibition [in 1876] wouldn't we have astonished the foreigners? I rather think so. Wouldn't the Chinese have opened their almond eyes when they heard you talking to them in broken China? Heigho! Ah, ha-a-a-a!"

But all was not fun and games on a visit to Thomas Edison's laboratory. Indeed, the phonograph was exploited quite cannily as Edison's opening act, a teasing appetizer, at which everyone could take a turn, after which he guided the guests on a tour of his other inventions and works in progress strategically situated around the room. He also spoke at great length about the further commercial and cultural applications for the instrument as set forth in his *North American Review* article—artistic and literary preservation and educational instruction of course high on the list. By now the first few publicly available tinfoil phonographs were coming out of the Menlo Park machine shop at a price of $100 apiece, and others were being made for leasing by Sigmund Bergmann's satellite workshop in lower Manhattan. Edison hinted to one and all that by Christmas of 1878 many novelties and toys derived from the phonograph could be widely available if the right backers were found. And another phonograph was meanwhile in development—he called it a "plate machine"—employing a flat disc instead of a cylinder, with a funnel-like attachment that would be employed to greatly amplify the playback of recorded sounds. A prototype model was available for inspection, as was, on a nearby stand, the electric pen. Here, Edison paused and wrote samples—his name and the date of the visit—for visitors to take away as souvenirs. The electromotograph (essentially a telegraphic instrument in which the sounder was operated nonmagnetically), harmonic engine, galvanometer, carbon microphone, and musical telephone were also near at hand, as was the pièce de résistance, unveiled in early summer, out on the terrace—the new ear trumpet, or telescopophone: three huge funnels mounted on a tripod, the central one equipped with a mouthpiece for amplified transmission, and the two outside ones with rubber tubing that plugged into each ear. "I stood here the other day," said Edison, turning his eager gray eyes upon the news-hungry correspondents from the Daily Graphic, "and I put one of these to my ear. I heard a child cry; I heard the telegraph instrument tick down at the station, 1,000 feet away; and I heard one curious, softly grinding sound that I could not identify till I followed up and found it to be a cow biting off and chewing grass, two-fifths of a mile distant."

Whereupon Mr. Batchelor, accompanied by Uriah Painter, the Philadelphia Inquirer correspondent, set off down the hill into the pasture 600 feet distant, ear-trumpets in hand. The observers on the second-floor terrace raised their opera glasses and confirmed that the megaphone-like devices were at the ready. "Go a mile further off," said Edison, in a normal voice, and, lo and behold, away the two men strolled out of sight, pausing to wave a white handkerchief, barely visible through the trees.

To promote the phonograph and develop a constituency for its hoped-for commercial success, Edison waged an all-out promotional campaign: in person at Menlo Park; through theatrical demonstrations before huge audiences by agents of the new Edison Speaking Phonograph Company in New York City and subsequently around the country; in print, through his serious essays and occasional letters to the editors; and, as we have seen, by cultivating writers in the popular press. However, the publicity spread much more quickly than he had anticipated, and not all of it added up to momentum in the marketplace. For as many business and artistic benefits as Edison enumerated, there inevitably arose in response an equal number of satiric ones—a smiling, half-asleep wife surreptitiously playing the phonograph crying out alarms of "Murder! Police! Fire!" in the middle of the night, awakening her terrified husband; a disgruntled merchant putting off the aggressive salesman peddling his wares, gesturing instead to the phonograph stationed by his desk, and saying, "Oh, I'm busy! Tell your story to that machine and I'll grind it out tonight"; a church with a low budget setting up a line of "Tenor, Bass, Baritone, Contralto, and Soprano" phonographs to substitute for the Sunday morning choir; the newly built Statue of Liberty announcing "Welcome to our shores" to arriving immigrants.

In one particularly delightful *Punch* cartoon, the mistress of the house, just before her gala dinner party, instructs (and cautions) the butler on the musical program for the evening, all to be piped in by phonograph: "Now recollect, Robert, at a quarter to nine turn on *Voi Che Sapete* from Covent Garden; at ten, let in the stringed quartette from St. James's Hall; and at eleven turn the last quartette from *Rigoletto* full on.—*But mind you close one tap before opening the other!*"

Aside from its sheer novelty and amusement, qualities of perennial appeal to the American sensibility, among the many reasons the phonograph attracted such attention was its mystical capacity—repeatedly hyped by Edison—not only to *capture* the human voice, but to *retain* it for eternity, for the future benefit of ancestors as yet unborn. Regrets were also often voiced of how unfortunate it was that this magical machine did not exist in earlier times, so that even today we would have been able to attend to past "voices of the great."

In this manner, the phonograph at its birth and flourishing exercised its seductive appeal upon Helena Petrovna Blavatsky, living in New York City, where she had founded the Theosophical Society in 1875. Born in 1831 at Dnepropetrovsk in Ukraine, raised by her maternal grandparents, and endowed from childhood with psychic powers, Mme. Blavatsky had literally spent a quarter of a century—after an unconsummated, teenage marriage of only a few months to a man old enough to be her father—in search of enlightenment, criss-crossing the world through Eastern Europe, England, Canada, South America, the West Indies, Ceylon, India, Tibet (for three years), Japan, Burma, Java, Italy, Greece, Egypt, and Paris, from which, in July 1873, she was suddenly told by spiritual voices to embark for New York on one day's notice. A prepossessingly tall and broad woman with a Gertrude Stein–like bearing and limpid gaze, a propensity for rollicking laughter and a weakness for chain-smoking hand-rolled cigarettes, given to enveloping herself within richly brocaded velvet dresses, Mme. Blavatsky may best be considered the first modern occultist, an ambassador of her *mahatmas*, her Spirit Guides, the "Himalayan Sages," bringing tenets of Eastern religion and astral philosophy into American popular society. Her first major work, *Isis Unveiled*, a 1,200-page, two-volume epic published in the fall of 1877, was an immediate bestseller; the first edition sold out within ten days. While not an outright stemming of Darwinism's rising tide, it nevertheless attempted to counterbalance theories of evolution that emphasized, within the confines of exact science, the purely materialistic and *physical* nature of man—stressing instead the equally important and intangible role of consciousness and the enduring qualities of the *soul*, which is immortal; we must learn to dwell contemporaneously on this earth, she wrote, "with those whom we have lost in material form." Theosophy (a neo-Platonic term from the Greek, meaning "Divine Wisdom") was founded as a spiritual corrective to the deterministic themes of *The Origin of Species*; *Isis*, in contrast, insisted upon the existence of "unexplained laws" within the natural world. Furthermore, magic, equally a science in Mme. Blavatsky's value-system, had to be considered a valid art, and as she tellingly expressed it, "The cornerstone of Magic is an intimate practical knowledge of magnetism and electricity, their qualities, correlations and potencies."

At the peak of the phonograph hoopla, sensing her affinities with the Wizard of Menlo Park, Mme. Blavatsky sent Edison a gift copy of *Isis Unveiled* along with application forms for membership in the Theosophical Society, which already claimed among its diverse American adherents Abner Doubleday and William James. Edison immediately signed the papers, returning them to Mme. Blavatsky's colleague, Colonel Henry Steel Olcott, cofounder of the Society—a lapsed farmer, agricultural journalist, Francophone, customs lawyer, and native of Orange, New Jersey. Edison's cordial note read, "Please say to Madame Blavatsky that I have received her very curious work and I thank her for the same. I SHALL READ BETWEEN THE LINES!" (A decade later, however, perhaps in reaction to a report published in England by the Society for Psychical Research asserting that Mme. Blavatsky was a fraud, Edison vehemently denied articles appearing in the press quoting Colonel Olcott as stating that the famous inventor was a card-carrying, dues-paying Theosophist. Edison insisted that Olcott was mistaken even after the Colonel wrote to Edison that the signed membership form as well as a second letter from Edison acknowledging receipt of a Diploma of Fellowship were enshrined in the Society's international headquarters in Madras, India.)

Toward mid December 1878, four days before leaving America permanently for India, Mme. Blavatsky, who had become a naturalized citizen because, she said, "America is the only land of *true* freedom in the whole world," and Colonel Olcott arranged a trip to Menlo Park, as two members of the still ongoing celebrity parade to New Jersey. When Mme. Blavatsky fell ill with discomfort from an emergency tooth extraction, and much to her chagrin was unable to make the journey, the Colonel went alone. He approached Edison not only as a fellow Theosophist, but also as honorary secretary to a Citizens' National Committee created to act as a liaison with the French government for the purpose of planning a major international industrial exposition in Paris for the coming year. Olcott hoped that the inventor would want to participate, which he most assuredly did.

Having accomplished their business, the two men strayed off naturally enough into a conversation about "occult forces," a field in which Edison had already done some exploring. Edison regaled (and intrigued) Olcott by telling him that he had attempted through the dynamism of will power (conducted via rubber tubes extending from his forehead) to move a pendulum suspended on the wall of his laboratory. Olcott narrowed the discussion of mental energy in general into the directly psychological realm: How did Edison arrive at his creative ideas? The inventor replied that "often, perhaps while walking on Broadway with an acquaintance, and talking about quite other matters, amid the din and roar of the street, the thought would suddenly flash in his mind that such a desired thing might be accomplished in a certain way. He would hasten home, set to work on the idea, and not give it up until he had either succeeded or found the thing impracticable."

Olcott returned to Mme. Blavatsky's salon on Eighth Avenue and 47th Street with a "100-pound phonograph," purchased for delivery to the Bombay branch of the Society, whereupon she immediately and successfully recorded and played back herself whistling. Invited to a farewell reception for Olcott and Mme. Blavatsky—which, according to Olcott's diaries, turned into a "voice-receiver" party, with all the guests taking turns setting down messages—Edison demurred, but did send E. H. Johnson in his stead.

Over the years up to and even beyond the flap over Edison's affiliation

with the Theosophical movement, Mme. Blavatsky, who was extremely prolific—her *Collected Works* take up more than ten thousand pages in fourteen
volumes—published several articles discussing in partisan terms her ideas
about his work; even though they never actually met, she was a deep admirer with a clear sense of the affinities between what Edison was after and
what Theosophy valued: "Had our Brother Theosophist, Thomas Alva Edison, the inventor of the telephone and the phonograph, lived in the days of
Galileo," she wrote in her essay, "Magic," published in the *Dekkan Star*,
Poona, India, in March 1879, "he would have surely expiated on the rack or
at the stake his sin of having found the means to fix on a soft surface of
metal, and preserve for long years the sounds of the human voice; for his talent would have been pronounced the gift of Hell . . . *Divine Wisdom* has been
discovered by Mr. Edison . . . in the eternity of sound."

In a March 1890 essay, "The Cycle Moveth," she approvingly cited
Edison's Monadic conception of matter, his intimation that there was a
supreme unity manifesting itself within all particulate life, as expressed
during an interview with G. Parsons Lathrop in a recent issue of *Harper's*: "I
do not believe that matter is inert, acted upon by an outside force. To me,"
Edison said, "it seems that every atom is possessed by a certain amount of
primitive intelligence; look at the thousand ways in which atoms of hydrogen combine with those of other elements. . . . Do you mean to say they do
this without intelligence?"

One month later, incensed to discover that "Brother Edison" was
ridiculed and demeaned in the *Review of Reviews* as a "dreamer" for these
Harper's remarks, Mme. Blavatsky leaped to his defense yet again as a holistic thinker, a man with the spiritual wherewithal to accept the feasibility of
a universal reality. In a provocative, indignant piece called "Kosmic Mind,"
she took Edison's side: "Would to goodness the men of science exercised
their 'scientific imagination' a little more and their dogmatic and cold nega-
tions a little less. Dreams differ," she wrote. "In that strange state of being
which, as Byron has it, puts us in a position 'with seal'd eyes to see,' one
often perceives more real facts than when awake."

On the second tier of Edison's twelve-thousand-volume, gymnasium-
sized library in his West Orange laboratory complex there are several
shelves devoted to books about psychic power and reincarnation, including
an inscribed copy of Mme. Blavatsky's *The Key to Theosophy* (1889) alongside
Emanuel Swedenborg's *Heaven and Hell*. In keeping with his expansive
and recurrent style of thinking, Edison took as his province not only

the whole knowable world, but, by extension, as much of the spiritual world as anyone could humanly assimilate. And despite the fact that this persistent inclination to explore the unquantifiable aspects of reality undercut his scholarly credibility, Edison visited and revisited these mystical spheres periodically throughout his life, up until the utter end.

Much to Edison's delight, the phonograph did indeed attract the attention of the "scientific men" as strongly as it had the general public, and he was invited to demonstrate it at the spring meeting of the National Academy of Sciences in Washington, D.C. Created during the Civil War by the federal government to serve as an advisory body on inventions, scientific proposals, and military technology, with a broad mandate "to investigate, examine, experiment, and report upon any question of science and art," the Academy was now an honorary society. At the invitation of his friend, George Frederick Barker, professor of physics at the University of Pennsylvania and editor of the *Journal of the Franklin Institute*, Edison had spoken four years previously at the Academy, showing his electromotograph.

Accompanied by Charles Batchelor, Edison arrived in Washington on the morning of April 18, and the men were first taken by Uriah Painter to the studio of photographer Mathew Brady, where a group portrait with phonograph was made—Edison looking (for him) quite well turned-out in a spanking-new plaid suit. He stares fixedly at the camera with his right hand proprietarily on the crank of the phonograph, to which a flywheel had recently been added to aid in turn-

ing the cylinder. Before going on to a reception for congressmen and members of the diplomatic corps at the home of Gail Hamilton, a Washington socialite, they stopped by at the Smithsonian Institution to be greeted by its Director, Joseph Henry, who was also president of the Academy of Sciences. Departing from this meeting, Edison was waylaid by a writer from the Washington *Post*

PORTRAIT OF EDISON BY MATHEW BRADY

who accosted him on the street: "Good morning, Mr. Edison," said the reporter, "handsome grounds, these." "Yes," Edison replied. "What an immense stretch of telegraph wire without a support," he said, pointing to the network of wires running high in the air from the Smithsonian building, "I have been wondering where they reach, but when I went over to that tree just now I could see no more than I do here." "Naturally enough," the reporter commented parenthetically, "the inventor has no eyes for trees or flowers where telegraph or telephone wires are concerned."

The afternoon session of the Academy began promptly at 4:00 p.m. in the gothic, red-brick Smithsonian Building with a series of papers on such topics as "The Effective Force of Molecular Action," "Remarks on the Value of the Results Obtained for the Solar Parallax from the English Telescope Observations," and "The Vertebrate Fauna of the Permian Period of the United States." By the time George Barker stepped to the podium to introduce the great inventor, Thomas Edison, as "a man of deeds, not words," the lecture hall had become so crowded that its doors were removed from their hinges to allow for the overflow, which included women and children, extending into the adjoining dining room. Edison, true to Barker's characterization, did not speak, chronic deafness causing him more and more to shy away from large groups because of problems he had attending properly to more than one person at a time. Instead, he sat at a desk, "twisting a small rubber band nervously in his fingers," while Charles Batchelor did the honors, singing, shouting, "crowing," and crooning (once again) "Old Uncle Ned." Several different models of the carbon telephone were also placed on display, and a successful long-distance conversation was conducted by members of the audience with persons in Philadelphia.

At the reception following the demonstration, Edison discussed the phonograph discs currently in development at Menlo Park, made by stretching tin foil over a rim of straw board soaked in alum, "like the head of a drum," he said. He announced that he was exploring ways to send the discs through the mail without injuring them, so that phonographic correspondence could be carried forth.

On into the evening members of the Academy visited the United States Observatory to view the heavens through the giant telescope. Edison identified strongly with Galileo and the historic controversy greeting his seemingly impractical seventeenth-century invention—and indeed, much like the phonograph, the telescope had initially been trivialized as a popular toy. Edison and Batchelor stopped by at the Observatory on the late side, because they had first made another demonstration at the Washington bureau of the Philadelphia *Inquirer* as a favor to Uriah Painter. Finally, at about eleven that night, the two men were summoned to the White House by President Rutherford B. Hayes for a private phonograph session. They ended up staying past three in the morning, because the President, in his enthusiasm, awakened his wife so that she, too, could experience this marvelous machine.

Several times during the spring, Edison considered taking a summer break from the frenetic, hothouse atmosphere of Menlo Park. He alluded to a European trip. He spoke longingly of retreating to a hotel in the White Mountains of New Hampshire. He wanted to disappear from view into the wilds of northwestern Pennsylvania for a fishing jaunt. At the insistent request of the Rev. John Heyl Vincent of Plainfield, New Jersey, cofounder with Akron industrialist Lewis Miller of the new Chautauqua Institution in upstate New York, he made a commitment the week before leaving for Washington to spend three weeks at Chautauqua during August, ostensibly to demonstrate the phonograph, but lured equally by the promise of quiet evenings in a meditative woodland setting overlooking the tranquil lake, watching white sails flit by through stately pines.

How Edison finally decided to spend his summer may best be considered a busman's holiday. At the Washington Observatory gathering, seized by the romance of astronomy fever (or, as Edison put it so engagingly, "peeking at the almighty through a Keyhole"), George Barker invited Edison to join an expeditionary team planning to head out west to Rawlins, Wyoming, to observe the July 29 total eclipse of the sun. Edison readily accepted, because in addition to adventurous escape, the trip would provide him with the chance to test an as yet unnamed and even unbuilt new invention that had been on his mind under various formats for more than a decade.

Within the preceding year, he had been in enthusiastic correspondence with Henry Draper, a medical doctor, professor of natural science at New York University, telescope designer—and one of the prime movers behind the Rawlins trip—regarding Draper's pathfinding research in stellar spectroscopy. Along similar lines and contemporaneously, Edison had also been writing to Samuel Pierpont Langley, director of the Allegheny Observatory, who expressed the need for a supersensitive instrument that would exceed the limitations of the current thermopile in its capacity to detect the sun's infrared, or, as it was called in the parlance of the day, ultrared (long wave) radiations.

Edison had already shown that, when properly prepared, compressed carbon proved to be particularly sensitive to variable electric currents in the telephone. During sound experimentation, Edison discovered that these very same carbon buttons *distorted* sound transmission when exposed to radiant heat. This characteristic formed the crux for Edison's "tasimeter," with which he claimed he would be able to quantify the heat given off by the sun's corona during the coming eclipse. Made of brass and small enough to be hand-held, it worked upon the basic principles governing a thermometer. Hard rubber (called vulcanite) was the expansible material upon which heat was focused via a funnel not unlike the one used in Edison's phonograph. As the carbon button changed its size (and thereby its resistance) minutely in relation to the heat bearing down upon it through the vulcanite, this shift was calibrated by a finely threaded screw. The specific correlation between the degree of heat applied and size compression affected the current displayed on the dial of a mirror-galvanometer powered by a wet cell battery; the adapted carbon microphone was now employed to determine shifts in temperature, rather than sound.

By June, excited by Edison's reports on his progress in developing the tasimeter, Professor Langley—who, after all, had proposed the initial challenge—offered to test the instrument in his laboratory at Allegheny. Out of his ingrained sense of competition, Edison never sent one, despite Langley's repeated requests, even months after the eclipse experiments had been completed. In an essay on the tasimeter that he may or may not have written himself for publication in the *Proceedings* of the American Association for the Advancement of Science's August 1878 meeting in St. Louis, Edison claimed misleadingly that he really had not finished building the instrument to exact specifications until two days before he left on the journey west.

" 'Yes,' the inventor said, as he stood on the platform of the Pennsylvania Railroad depot on Saturday evening [July 13], about to start with the Draper observing party, 'it will measure any degree of heat that can be measured. If the sun's corona has any heat of its own or possesses any heat-reflecting power, the tasimeter will measure it accurately.' "

The line of the central eclipse, 116 miles wide, would start in northeastern Asia, cross the Bering Strait into Alaska, pass through "British America" (as the Canadian territory was then called) near its Pacific coastline, and then hit the northern boundary of the United States, running southeasterly through Wyoming, Colorado, Indian Territory, and Texas, passing into the Gulf of Mexico between Galveston and New Orleans.

Several scientific parties set forth westward from early to mid-July, variously headed by J. Norman Lockyer, the British editor and founder of *Nature* magazine; Professor Simon Newcomb, director of the government's Nautical Almanac Department; Professor Arthur Wright, of Yale; Professor Charles Young, of Princeton, a frequent visitor to Menlo Park; the afore-mentioned Professor Langley, whose goal was Pikes Peak; members of the Cambridge Observatory of Massachusetts; the Chicago Astronomical Soci-ety—and even a group of three priests from the Sistine Order of George-town, in Washington, D.C.

But the most heralded was the Edison-Barker-Draper party, because they brought along their own private journalist, Edison's crony, Marshall Fox, of the New York *Herald*. On the train journey to Rawlins, a frontier town merely nine years old with a population of eight hundred, Edison was hailed at every stop by local telegraphers, thrilled that a member of their fraternity who had ascended to undreamed-of heights was to be among them. Half a century later, one of those men, who in the interim had gone up through the ranks to become senior vice-president of the Union Pacific Railroad, recalled Edison's momentous arrival at Rawlins. Knowing that the Wizard still relished a good old practical joke, the telegraph agent rigged up a stuffed antelope (Edison later said it was a jackrabbit) about one hundred yards out in the sagebrush. At daybreak, the townsmen directed Edison's attention to the proliferation of game animals in the area, handed him a rifle, and invited him to take pot shots. After about a dozen blasts, he realized the trick. Edison on his part relished the prank played on Marshall Fox, whose vintage, reconditioned Springfield army musket was so bent out of shape that when he began trying to pick off some tin cans lying about the station yard, the first recoil knocked him to the ground and he had to be taken to the camp hospital.

Edison's claims for the tasimeter's acuity were borne out at first hand by the highly respected Mr. Lockyer of *Nature*, who joined up with the Rawlins group. Two nights before the eclipse was due to take place, Edison tested the instrument on the bright star Arcturus: "For its extreme delicacy I can personally vouch," Mr. Lockyer wrote in a dispatch to England. "The instrument, however, is so young, that doubtless there are many pitfalls to be discovered. Mr. Edison, however, is no unwary experimenter." Subse-quent astronomical studies [see Notes] based upon calculating the location and intensity of Arcturus on July 27, 1878, gauged against the known properties of vulcanite, the type of telescope Edison used (a four-inch Dolland achromatic refractor provided to him by Barker on the site), and the galvanometer figures noted in Lockyer's assiduous report confirm that the tasimeter was actually capable of registering a temperature change of 10^{-6} degrees Fahrenheit.

The experience of the eclipse was not quite as perfect, however. Suspense built as it rained torrentially for three hours the day before the event, pushing masses of dark clouds across the plains, and even though the sky cleared on July 29, it was still blustery. Marshall Fox noted with perhaps a touch of fancy that Edison took shelter in the doorway of a dilapidated henhouse and was forced to set up his instruments while dirt and debris swirled on all sides. (At the very least, it was a makeshift but firm wooden shed in a small yard protected by a six-foot-high wooden fence.) The duration, or *totality* of the complete eclipse, when the entire disc of the sun was obscured by the moon, was three minutes. Struggling with his machine during the final seconds of the final minute, Edison was able to focus the brilliant corona light into the tasimeter funnel. The galvanometer went off the scale, revealing an intensity of heat about fifteen times as great as the Arcturus measurement.

In stunning parallel to the path of development of the phonograph, Edison immediately began improvisationally, irrepressibly, to concoct further applications for his tasimeter: to detect invisible stars and create infrared maps of the heavens; to measure wind velocity; to weigh minutely sized objects; to take body temperature as an indicator of impending illness; on shipboard, to sense the presence of icebergs; as an "odorometer," to sense the presence of perfume; replacing the vulcanite rod with gelatin to measure moisture instead of temperature. Alas, intriguing as these outlandish ideas were, the fundamental machine was too fickle, delicate, and overly sensitive to touch and vibration to serve any broader utilitarian purpose.

The eclipse over, their mission fulfilled, Edison, Barker, and Draper continued westward. Roughing it, sleeping under the stars, they went deer and antelope hunting by bronco pony in Ute country; visited Virginia City, Nevada, where Edison went down into a silver mine; and traveled to Yosemite, riding to the top of Glacier Point, and on to Mariposa and San Francisco.

The main subject of conversation among the trio of friends on the long train journey homeward in late August concerned other as yet unexplored and unexploited landscapes—electricity and the transmission of electric power.

CHAPTER

10

THE LONG AND COMPLEX HISTORY OF EDISON'S WORK ON THE electric light, hardly Archimedes' "Eureka!" by any stretch of the imagination, is best characterized by the ingenuously aphoristic statement of Karl Friedrich Gauss (1777–1855)—the German mathematician who developed, among other innovations, the fundamental theorem of algebra, the prime number theorem, non-Euclidian geometry, and the normal probability graph—"I have the result, but I do not yet know how to get it."

Edison was hardly the first scientist in the nineteenth century to experiment with the uses of incandescent light. But before coming to the final, simple elegance of the archetypal bulb, we must acknowledge the crucial strain in his thinking that gave rise to it, a theme sounded in the last paragraphs of Edison's *North American Review* essay on the phonograph. Here, as he does so often when describing the immediate benefit of a new invention, he lifts off from discussing the inherent attributes of the machine itself to consider its influence within a larger canvas, "the telegraph company of the future—and that no distant one," which, Edison envisions, "will be simply an organization having *a huge system* of wires, central and subcentral stations."

Asked more than three decades later by a correspondent to settle a debate about which was his greatest invention, recognizing that the phonograph was a very wonderful thing, but that he had "also given us other things," Edison in response scrawled hugely across the bottom of the letter, *"Incandescent Electric Lighting and Power System."*

One consistent sign of Edison's genius—a trait that often tended to outrun his ability to deliver the goods, but endears him nonetheless—was his inclination to think globally long before achieving success locally. Concocting vast concepts at the outset of a new idea—the inspiration—protected him from becoming bogged down in the inevitable, much less enjoyable minutiae that he knew would follow—the perspiration. "All parts of the *system* must be constructed with reference to all other parts," he wrote of the electric light endeavor as he viewed it, "since, in one sense, all the parts form one machine." This kind of refreshingly modern rhetoric represents organic thinking at its best: the magnificent end justifies the means. Translated onto the hierarchy of Menlo Park in 1878, this mindset allowed Edison to maintain his vision and then distribute it as necessary into the welcoming arms of Batchelor, Kruesi, et al.

And holding to that vision—or, rather, trying to preserve the operating conditions that would allow him to continue to flourish—seemed to be of even greater importance to Edison after he returned from his western trip. Close friends understood the perambulations of Edison's psyche well and accepted his addiction to the new. Grosvenor Lowrey, general counsel for Western Union, and now Edison's legal and financial adviser, told Edison that potential income from the syndication of electric power would liberate Edison from market preoccupations and permanently endow the Menlo Park laboratory, so that he could devote all of his time to pure invention. It was Lowrey who became the prime mover behind the capitalization (at $300,000) and incorporation of the Edison Electric Light Company, a multifaceted industrial entity set up "to own, manufacture, operate and license the use of various apparatus used in producing light, heat, or power by electricity," drawing in the clout of Drexel, Morgan and Company and other powerful stockholders to corner the market in the early fall of 1878, before Edison had even embarked upon any formal research whatsoever.

Thus, a number of influential circumstances—typically, no single one of them sufficient in and of itself—converged to cause Edison to turn his energies toward electric lighting. An oft cited but undoubtedly epiphanic moment occurred while visiting mining sites in the Sierra Nevada and Rocky Mountains during his western tour. Overlooking the mighty Platte River, witnessing the laborious process of drilling ore by hand, Edison turned to his colleague, Professor Barker (or "Barky," as he was fond of calling him) and "exclaimed abstractedly, 'Why cannot the power of yonder river be transmitted to these men by electricity?' " Barker was the pragmatic man of science, and his encouragement took a different course than Lowrey's. His dialogue with Edison on the state of the electrical arts and the niche tantalizingly available to Edison (indeed, electricity had never been conveyed over a wide area, nor had efficient dynamos yet been produced that could generate it cheaply) carried on all the way back to New Jersey, and helps explain why all further work on the development of the phonograph was summarily suspended. The "little feller" would have to wait another decade before growing up: "It is taking care of itself," Edison told a visitor, "comatose for the time being. It is a child and will grow to be a man yet; but I have a bigger thing in hand and must finish it to the temporary neglect of all other phones and graphs."

"The first step is an intuition, and comes with a burst," Edison confided to an associate, Theodore Puskas, trying to find the right words to define his excitement. "It has been just so in all of my inventions." The inherent sense that he could make a difference came to Edison when he viewed himself as part of a great tradition spanning eight decades, during which more than a score of scientists—beginning with Faraday's mentor, Sir Humphry Davy, and his trials to sustain the glow of a platinum filament—had attempted to make a sustained, viable incandescent electric light. Farmer, de Moleyns, Starr, Shepard, Jobard, Goebel, Lodyguine, Swan, Sawyer, Man, Maxim, Lane-Fox—entering this international fray to compete with his forebears in technological history, both illustrious and lesser known, coupled with the chance to make his mark for posterity, had been powerful incentives for Edison in the past, and were now in play again.

Competition with the current status quo was an equally driving force. Arc lighting was very much in vogue: two pencil-like carbon rods, separated by an insulating layer of gypsum, which glowed when lit by a flame and, like a candle, burned down after an hour or two. Developed by Elihu Thomson, Edward Weston, and Charles Brush in America, greatly popularized by the Russian inventor Paul Jablochkoff, and displayed dramatically to much acclaim on the Avenue de l'Opéra during the recent Paris Exposition, the problem with arc lighting was that it was an outdoor phenomenon, unadjustable and hissing, throwing off embers, brilliant as a Fourth of July sparkler. These "lamps that outshine Canopus" were bright as daylight, too intense for the quieter needs of domesticity, where manufactured illuminating gas prevailed. But gas was costly; it could explode; it silently poisoned the ground, and water wells and cisterns, leaking out from the mains through which it invisibly traveled; it blackened walls, and discolored paintings; it devoured oxygen, made you dizzy and gave you headaches; it smelled of sulfur and ammonia. Gas did, however, move through a vestigial power system, ready to be usurped, ripe for Edison's head-on challenge.

Deeply tanned and sporting a new black sombrero, Edison had hardly settled back in at Menlo Park when Professor Barker, anxious to feed the fuel for his friend's brainstorm, prevailed upon him to take a field trip in early September to the Ansonia Brass and Clock Works in Connecticut, to view the pioneering electrical work being done there by its proprietor, William Wallace. The party, in addition to Edison and Barker, and reflecting its importance, consisted of Charles Batchelor and his brother James, visiting from England; Professor Charles Chandler, of Columbia University; and Charles Davis, a telegraphy expert on the Menlo Park staff.

For the illustrious visitors, Wallace demonstrated a generator-dynamo of his own manufacture, which converted mechanical energy into electrical energy, dividing the power it had produced and distributing it among a small series of eight arc lights, complete with primitive circuit breaker. For Edison this was a catalytic episode, to see at first hand electricity provided by a mode other than batteries; Wallace's machine was a visual metaphor from which Edison immediately leapt—"When ten lights have been produced by a single electric machine, it has been thought to be a great triumph of scientific skill. With the process I have just discovered, I can produce a thousand—aye ten thousand—from one machine. Indeed, the number may be said to be infinite."

In the laboratory Edison was focusing upon finding a durable filament as the primary, key building block to his lighting system, applying his telegraphic knowledge of serial circuitry to regulate the current properly so that the filament would not melt down. The filament wire, heated at first in the open air, then inside a bell jar, was attached to a copper wire that led from a current source. However, in the public eye Edison was already three years ahead of these simple, early trials, coopting the field with the promise that, "in a few weeks" the entire lower part of New York City would be ablaze with electric lights, radiating outward from a central dynamo on Nassau Street uptown to the Cooper Institute, down to the Battery and across both rivers. This prototype power station would pave the way for twenty such installations in different parts of the metropolis. Insulated wires would be laid under the ground in precisely the same manner as gas lines; gas burners and chandeliers could be easily adapted to the new electric light. The same wire bringing light would also bring power and heat. All you would have to do is turn a screw to bring in bright light or soft. And the whole thing would come in at 5 percent of the cost of gas.

In the laboratory, Edison had clearly articulated (one might say, invented) the *idea* of "infinite subdivision" of the light, but he had not by any means accomplished the method for doing so in *practice*. He knew, nevertheless, that his fundamental goal was to find a way to control current output over a vast network of lights. Meanwhile, as the gas companies fretted that they would all go belly-up, Edison's verbal hype continued, with the flow of predictions and guarantees unabated, but the free and easy access that had become the hallmark of Menlo Park suddenly ended. "When do you expect to have the invention completed, Mr. Edison?" asked the reporter from the New York *Herald*. "The substance of it is all right now," he hedged, "but there are the usual little details that must be attended to before it goes to the public. For instance, we have got to devise some arrangement for registering—a sort of meter; and again there are several different forms that we are experimenting on now in order to select the best." "Are the lights to be all of the same degree of brilliancy?" his interlocutor pressed on. "All the same," was the terse reply. "Have you run across any serious difficulties in it as yet?" "Well, no," replied the inventor, and we can almost hear the pregnant pause, "and that's what worries me, for in the telephone I found out about a thousand obstacles, and so in the quadruplex. I worked on both for over two years before I overcame them."

"Of all the things that we have discovered this is about the simplest," continued the equivocating "Napoleon of invention," "and the public will say so when it is explained. We have got it pretty well advanced now, but there are some experiments I still have on my mind. You see, it's got to be so fixed that it can't get out of order. Suppose where one light only is employed it got out of order once a year, where two were used it would get out of order twice a year, and where a thousand were used you can see that there would be much trouble in looking after them. Therefore when the light leaves the laboratory I want it to be in such shape that it cannot get out of order at all, except, of course, by some accident."

Edison was cautious regarding the release of specifics about his lighting system before he had registered any patents because he feared appropriation of his ideas by foreign interests: "The electric light is the light of the future—and it will be my light," he reiterated, with his usual competitive edge, "unless some other fellow gets up a better one." Besides increasingly impatient backers and a cadre of resentful journalists, besides fearing "the superfluous gabble of an imprudent press" who for the first time were confronted with "Positively No Admittance" signs when they alighted from the train, Edison was subject to other emotional pressures. His wife Mary was coming to full term, and in the anxious, hot, summertime weeks of her husband's absence had become nearly immobilized, imprisoned in her house, gaining an unconscionable amount of weight from overeating sweets. For several days, Edison himself was confined to bed with what his doctors were instructed to describe as "a severe case of neuralgia." Edison blamed the nervous condition on excessive exposure to electricity, which he said had burned the side of his face and caused his eyes to water and become tired so that he was constantly dabbing at them with his pocket handkerchief. On the morning of October 26, William Leslie Edison finally entered the world, all twelve pounds of him. That same day, Edison emerged from seclusion and was back in the laboratory.

Urged on by Lowrey, who told his friend euphemistically that he needed "to get the jug by the handle with a reasonable probability of carrying it safely

EDISON AND HIS CREW ON THE LABORATORY STEPS AT MENLO PARK (BATCHELOR
BEHIND EDISON'S LEFT SHOULDER; UPTON TWO MEN TO BATCHELOR'S LEFT)

to the well and bringing it back full," Edison agreed he needed more expertise and depth in the Menlo Park ranks if he was to bring the bulb research successfully to fruition. Through Lowrey's influence, help fortunately arrived in the person of Francis R. Upton, two years Edison's senior, Boston-born graduate of Bowdoin College and Princeton, expert in calculus, tempered by a year of postgraduate study at the University of Berlin with Hermann von Helmholtz. The latter must have commended the young man especially to Edison, because that great German scientist was a firm believer in the perception and quantification of physiological and inorganic forces by the senses. Von Helmholtz spent more than twenty years exploring the nature of muscle contraction, the velocity of nerve impulses, optics, and the resonance of hearing. By the time Upton came under his influence, von Helmholtz was conducting classes on mathematical ways to analyze electrodynamics.

Edison hired Upton with the oblique promise that perhaps at some future date he might be assigned to take charge of a telephone system in another city. It was a struggle for the diligent, admittedly absentminded Francis, engaged to be married to his beloved Lizzie, and dreaming of a sedentary existence, to enter the world of commerce so quickly and on such an unexpectedly heady level, especially as he had hoped to return to Germany and "the quiet life." But within the year, this self-effacing and loyal soldier—like Batchelor before him professing absolute faith in Edison's ability to achieve ultimate success, affectionately nicknamed "Culture" by his boss because of his introspective, learned mien, piano-playing talent, and impeccable educational credentials—had obtained a 5 percent interest in all profits accrued from the electric light, a lucrative deal, considering that the value of the stock offered by the Edison Electric Light Company eventually soared from $100 to $3,500 a share. Upton had promised his mother that working for Edison would help him "learn how to earn money," and it seemed that he had succeeded. "Everyone must work and it is not always the most agreeable thing in the world," he confessed grudgingly to her.

Upton's very first assignment for the company was to conduct an extensive literature search through existing domestic and foreign patents for all the information he could find about arc and incandescent lamps. The editorial comments interspersed throughout Upton's notebooks make it clear that his mission was to validate Edison's methodology by testing it against what had been ascertained by other scientists over the decades: "I never saw any appliance in the least like Mr. Edison's," he wrote, "although one has often been wished for—to produce a constant temperature with the galvanic current." Recurrent themes appear in Upton's research, that would inform Edison's course of action with respect to selecting components for the electric lamp; these included the use of platinum for the primary material of the filament, because it had a resistance about seven times greater than copper, could be heated to 5,000°, and still not burn out; the "closed vessel," as Upton called it, or vacuum inside the bulb which was necessary to keep a filament from deteriorating while it glowed with incandescence; the dynamoelectric source (as opposed to batteries) required to provide consistent power to the lamps; and the regulator, acting like a circuit breaker, so that the lamps in a series would remain engaged at equal, constant luminosity even if one of them should fail.

Francis Upton's arrival marked the transition from initial freneticism to a more reasoned approach at Menlo Park, not only because of the thoroughly grounded aspect of his research but also because his austere sensibility had a salutary influence on Edison. Upton was a student of the human condition who empathized with his new boss, and Edison soon trusted Upton so well that he charged him with drafting various speeches and articles under the Edison byline, "not liking the trouble of writing," and confident of Upton's ability to comprehend and represent his ideas. Upton was inspired by Edison's efficient use of time, by his ability to delegate and inspire simultaneously, his discipline of mind, and facility to "save energy when one point [goal] is reached, cutting even the thought," abandoning one line of enterprise temporarily, parenthetically delving into another area of research, and then picking up at the same point later on; marshaling "crudities," raw ideas, and refining them through his power to draw analogies; and by his healthy balancing act, to be "sanguine and confident" and yet doubting at the same moment, taking nothing for granted. Upton was never bored while working for Edison, a man "never in a rut because he was never satisfied" until he had what he wanted.

And Upton was especially impressed with the capacity Edison had to motivate others to reproduce thoughts into things. The new machine shop which had risen up behind the laboratory building, designed and constructed by the respected company of Babcock and Wilcox, reached its apotheosis under the stewardship of John Kruesi, not merely as a factory but as a testing ground for Edison's ideas. There, the huge C. H. Brown steam-powered engine with its twelve-foot-high flywheel drove the electric dynamos that powered the bulbs for a few hours at a time so that incandescence could be tested under extended conditions more closely approximating the final configuration of the lighting system as a whole. What actually happened to the physical structure of the platinum wire as it was subjected to intense heat, and what other metallic candidates for the filament could be ruled out? This kind of inquiry was Batchelor's forte. He loved to perform basic trials over and over and over again. Edison himself refused to be worn down by this process, and in fact it fed directly, even happily, into his compulsive nature.

The necessity to provide a vacuum within the bulb derived from the physical fact that, when the platinum filament was heated, it gave off gases causing it to expand, crack, fuse, and lose effective staying power. If the air was drawn out of the bulb with the use of a mercury pump—eventually down to a level of one-millionth of an atmosphere—and the filament was coated with magnesium oxide as an insulator, its light was found to be brighter. If these lamps were also of high resistance, that is, if they used a platinum filament that was spiraled and very long and thin, with a high melting point, they could be effectively employed in the vast system Edison had in mind.

By early spring, Edison had used a Gramme generator to power no less than eighteen lamps alight in a series, each equal in intensity to sixteen candles, but without the bothersome heat caused by gas. He was already fantasizing (although he had no idea what it would cost) about "lighting up" the town of Metuchen, about two miles distant, running the current over poles set along the country lanes so that "all kinds and classes of people may have an opportunity to judge of it. . . . It is more important to me," declared the Whitmanesque Edison, "that the servant in the kitchen should express her mind on the lamp, with regard simply to its lighting power, than that her mistress should wonder over it as a novelty without comparing it properly to gas."

Francis Upton, now firmly ensconced in the inner circle, took a steamer up the Hudson River with Edison and Batchelor to attend the 28th Meeting of the American Association for the Advancement of Science in Saratoga, "a fine chance for [him] to see the various scientific men of the country," including Alexander Graham Bell, who was to be in the audience. Professor Barker invited Upton to read the first full-length paper he had written for Edison, "On the Phenomena of Heating Metals in Vacuo by Means of an Electric Current," essentially a report on the progress of the

Menlo Park "insomnia squad" thus far. It announced Edison's findings that wires with smaller diameters were conversely much more difficult to melt— a low radiating surface permitted a higher temperature and thus a brighter light. The essay also reported that a better glow was obtained if the wire was brought to incandescence slowly through preheating.

Only temporarily buoyed by these important observations, Francis Upton was in a disconsolate mood (unusual for him) when he told his father in the hazy Indian summer that "The electric light goes on very slowly, I hope towards ultimate perfection. . . . Yet if it does not succeed I shall be contented with the experience. . . . We have not as yet what we want, but we have as good if not better than anyone else in the world." Had nearly a year passed so quickly, since Edison prematurely announced the existence of the light with such fanfare? He seemed to be a long way from illuminating the great city across the Hudson, let alone a neighboring town, or even his own village, or his own home for that matter. Edison and his men were stalled with a bulb that functioned perfectly well in the laboratory on a small scale for a limited period of time with the minimal current required— a prototype, newborn and naked, not ready to emerge into the wider world.

In the early fall of 1879, anxious to break the lull, Edison made the pivotal decision to abandon platinum, returning to his old and faithful ally, carbon, as the crux of the bulb. His own version of the reason for the substitution would have us believe once more that a serendipitous, mystical thought flitted through his mind as he was sitting one evening at his favorite corner table in the second floor laboratory mulling over the impasse he had reached, "abstractedly rolling between his fingers a piece of compressed lamp black mixed with tar for use in his telephone." He remained thus for some time, his thoughts wandering, as, like Rumpelstiltskin, his fingers spun the piece of tarred lamp black back and forth, back and forth across the scarred table top. Suddenly, he snapped to attention, in the present moment again, and *voila!* the alchemist looked down and saw the threadlike shape before him, revealed as the new filament.

Byron M. Vanderbilt, in his study, *Thomas Edison, Chemist*, believes that the most important factor in Edison's turning away from platinum was simply coming to terms with its scarcity and cost. Little commercial application was seen at the time for platinum, and Edison was not able to convince any corporate interests out west to engage in widespread mining for the native ores. Professor Barker also had recently advised Edison that it would be risky to expect large-scale access to platinum to the extent that would certainly be needed once the bulbs went into mass production. Finding a cheaper material for the filament would drive down the ultimate cost of the bulb and make it more acceptable to the consumer.

And in the most thorough recent account, *Edison's Electric Light: Biography of an Invention*, scholars Robert Friedel, Paul Israel, and Bernard S. Finn propose that the reason for the shift to carbon was securely expedient. During the summer of 1879, they point out, most of the manufac-

EDISON CARBONIZING PAPER FOR ELECTRIC LAMP FILAMENT

turing energies of the Menlo Park laboratory were directed toward providing carbon buttons for the chalk-drum telephone receiver being heavily promoted in England. Lampblack was ubiquitous on the premises, and its further application made sense—especially to Charles Batchelor, who made several entries in his laboratory notebooks about the viability of burnt lampblack shaped into carbon spirals. This insight alone leads us back to questioning Edison's version, inventor as dreamer, because we know how adept he was at appropriating the insights of his lieutenants—and how willing in turn they were to trade these contributory ideas in exchange for a piece of the action. And finally, we must remember that Edison claimed that as early as two years previously he had been toying with carbonized paper as a potential filament.

Whatever the reasons for the change, the shift to carbon was the deciding factor in regaining the ever seductive trail of progress that led to ultimate success. Any number of materials can be carbonized, of course, and the determined Batchelor tried many, including celluloid, cedar, coconut hair, fish line, and cotton soaked in boiling tar, before coming to rest, appropriately, on Clark's cotton thread, the product of the very same English com-

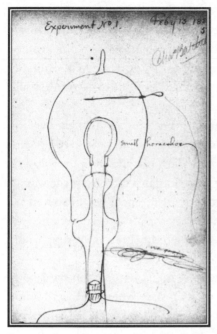

Experiment No. 1.   Feby 13 188[?]

Small horseshoe

EARLY SKETCH FOR ELECTRIC LAMP

pany in Newark where he had first made his start in America before coming to work for Edison. In the wee hours of October 22, a bulb with a fragile filament of carbonized thread glowed—in the immortal words of the ever-vigilant Marshall Fox, who broke the story in the New York *Herald*—"like the mellow sunset of an Italian autumn . . . a little globe of sunshine, a veritable Aladdin's lamp"—for 13 1/2 hours at a level of light equal to thirty candles. It was an all-time record, broken just a few days later by a bulb containing a horseshoe-shaped filament which prevailed for more than a hundred hours.

Events moved rapidly as Edison first set out to assuage the fears of his impatient financial supporters, who had been hovering in the wings for a long time, held back from deserting him only through the good graces and diplomatic assurances of Grosvenor Lowrey. Francis Upton's home and the main laboratory were selected for a post-Christmas exclusive demonstration. Forty lamps were kept alive from six until after ten in the evening for members of the board of the Edison Electric Light Company. Upton was feeling flush again, even though as the latest breakthrough took hold he still believed privately that Edison's "valuations are on his hopes more than on his realities." His reservations had not deterred him from buying a brand-new piano which was installed in the house he had acquired on the corner of Frederick and Monmouth streets in Menlo Park, surrounded by a neat picket fence, just behind Batchelor's place; and he and his proud bride then went over to New York to hire their first maidservant. The crowning glory was the installation of six bulbs in "Culture's" very own parlor.

Loops of charred bristol board encased in globes skillfully fashioned by Edison's virtuoso German glassblower (and occasional zither strummer) Ludwig Boehm glowed for many hours over the holiday season at Menlo Park as a new, larger generator, dubbed a "Faradic Machine" in honor of

Edison's paragon, was installed in the machine shop. The old Wallace machines had long since been discarded because their iron cores became overheated and wasted energy. In their place was a monstrosity nicknamed "Long-Legged" (later modestly changed to "Long-Waisted") Mary Ann, because each of its twin field magnets towered almost five feet high and weighed more than five hundred pounds.

Crowds of sightseers began to invade the compound in the waning days of December. Even extra trains put on the line could not keep up with the flow. The little depot was filled with the curious, and the narrow boardwalk leading up to the laboratory past Mrs. Jordan's establishment was almost impassable at times. Valuable laboratory equipment was broken in the crush, requests of "do not touch" ignored. Lamps had been mounted on wooden poles outdoors, and their eerie orange light threw quivering, firefly shadows over the snow-dusted ground.

Most evenings, over the telephone line rigged up between the laboratory and Edison's home down the hill, mother Mary would call, her hungry brood whining in the background, and ask Griffin or Batchelor—since Edison never came on the phone himself—to curtail her husband's sermonizing and hustle him off to dinner. Edison would typically relay the response back to her that he would be along imminently, then forget about it for the second and third times until Mary gave up and went ahead as usual without him.

The mood in the laboratory that winter was light and festive. Over midnight supper of smoked herring on hard crackers washed down with cold water, the men gathered around the organ even went so far as to improvise an irreverent tribute to their boss, sung to the tune of Gilbert & Sullivan's "I am the captain of the *Pinafore*" from *H.M.S. Pinafore*:

> *I am the Wizard of the electric light*
> *And a wide-awake Wizard, too.*
> *I see you're rather bright and appreciate the might*
> *Of what I daily do.*
> *Quadruplex telegraph or funny phonograph,*
> *It's all the same to me;*
> *With ideas I evolve and problems that I solve*
> *I'm never never stumped, you see.*
> *What, never?*

No, never!
What, never?
Well, hardly ever!

There was more to come. Marshall Fox promised from his bully pulpit that by the New Year "The Wizard of Menlo Park, after 'nourishing a youth sublime / With the fairy tales of science and the long results of time' will take his countrymen into his secret [and] present them with the 'lucky horseshoe,' the magician's wand, which is destined to solve more problems on heaven and earth than are dreamt of in the current philosophies."

CHAPTER

11

ONCE UPON A TIME AN ARCADIAN HAVEN, MENLO PARK NOW became an electric showplace. Against the black, brushstroke profiles of bare trees, slightly tilted white pine timber lampposts spaced fifty feet apart and crowned by helmet-shaped glass bulbs marched aglow across white outlying fields. Sightseeing throngs mingled and tramped through the rank and file of lamps looking for Professor Edison. In those few weeks of sweet fame, he could be distracted from the laboratory, emerging coatless into the cold air and greeting visitors in a blue flannel suit with a white silk handkerchief knotted hastily around his neck as an impromptu scarf.

But the glow soon faded with the return to the tough truths of daily routine. Piles of mail poured in, and Edison insisted upon spending the first two hours of his day sitting with Griffin and sorting through the strange amalgam of job applications from the younger generation of scientific men as well as amateurs anxious to serve with the Wizard. Hanging like a threat over his head and growing more intrusive with every passing day was the promise to illuminate New York, and in self-defense and with his ever-present sense of history, Edison referred his critics to fellow-innovators of the past: "There is a wide difference between completing an invention and putting the manufactured article on the market. The public, especially the public of journalism, stubbornly refuse to recognize this difference. It was years after photography was invented before the first photograph was taken;

EDISON AND THE CURIOUS PUBLIC AT MENLO PARK

years after the steamboat and telegraph were invented before they were actually set going," he declared. "George Stephenson built his first locomotive
ten years before he made his *Rocket* run from Liverpool to Manchester, and
he and Robert Fulton were called fool, charlatan, fraud, lunatic—everything
they could think of. They demanded that he should 'hurry up,' or acknowledge himself a humbug."

The present lighting challenge was direct and huge: to extrapolate from the laboratory setting to the actual urban site. This was a challenge primarily of scale, and Edison had rightly begun the task by adding more and more lights to the Menlo Park network—he wanted six hundred "set and ready," which meant in turn that a smoother system for mass-producing the bulbs had to be in place—pushing them to burn for longer and longer hours, since operating efficiency pointed the way to economic success. To the southeast of the laboratory complex, across the railroad tracks, the old electric pen factory would be refurbished by the summer, to serve in its new identity as the headquarters of the Edison Electric Lamp Company. Preliminary market-research field surveys were conducted by Edison's men among gas customers in the lower Manhattan district he had first targeted for his central station, to determine the level of pricing for his utility that would be necessary in order to be competitive. At the same time, Upton, Charles Clarke, and Samuel Mott priced out the relative costs of gas jets vs. electric lamps based upon average daily use.

While homes of the rich and famous, banks, and newspaper offices of New York City waited to become the grand testing ground for what eventually would indeed be, to cite Reyner Banham's dramatic evaluation, "the greatest environmental revolution in human history since the domestication of fire," the image of Edison in the zeitgeist of the new decade hovered between the poles of legitimate entrepreneur whose time had come, and unfulfilled prophet.

Both aspects of this image were encouraged by Edison's persistent knack of perceiving gaps in the material culture that he believed should be filled, by him. His obsession with electromagnets had led to the current search for bigger and better dynamos to power his electric lighting network, and, further, to two related fields of inquiry, although not immediately apparent as such.

The first was ore-milling. At this period, the early 1880s, the long-dominant iron ore industry on the eastern seaboard was faced with serious competition from the newer midwestern and western mines. Edison's idea, first seized upon in impulsive notebook sketches, bypassed the popular Bessemer process, first introduced in 1856, during which air was blown through molten pig iron. The Bessemer process required that the reducing gases be raised to extremely high temperatures and then forced through ore so that impurities were oxidized—stoking the furnace was costly and cumbersome, but finding higher grades of iron was difficult. Rather, Edison proposed a cheaper and faster route: dropping crushed magnetite sea-sands or "tailings" (as the discarded byproducts of mining were called) in a thin stream through a small opening in a funnel downward past the face of a magnet. The magnetite would naturally be drawn away from the rest of the material and into a separate hopper. After a model machine had been tested in Menlo Park, a concentration unit was established at Quonocontaug Beach in Rhode Island, with the capacity to deliver up to two hundred tons of magnetic iron ore per month. However, only one major corporate customer could be convinced to participate, and within three years the ore-milling operation had to be closed. Edison was down, but not out, and he would return to the field with a vengeance.

The second theme playing below the bulb project was the concept of an electric railroad. Old maps of Menlo Park reveal an experimental track running along Middlesex Avenue perpendicular to Christie Street, next to the dynamo-laden machine shop. Here, again, Edison was not the first to propose the idea—Werner von Siemens had demonstrated his model at the Berlin Industrial Exhibition the year before—but he was the first in America. The train-boy from Milan was never happier than when at the wheel (more like a lever) of his toylike twenty-five-horsepower locomotive, pulling a swaying, open carload like a surrey with the fringe on top loaded with colleagues and reporters at speeds up to forty miles per hour along a

THE ELECTRIC RAILROAD WITH BATCHELOR AT THE THROTTLE

three-quarters-of-a-mile track through the open meadow. Improved traction was key. In tests on a section of the railroad with a gradient of one foot in every hundred feet, the engine showed no signs of strain. Eventually the goal was to run a shuttle service between Perth Amboy and Rahway, wherever a horse could draw a wagon, and then, who could say?

Even while embroiled in the thick of worrisome labors, Edison was never too preoccupied with research to engage in image correction. He was quick to respond to one letter that emerged from the daily bushel basket, sent by a young freelance writer named John Michels, who had been at various points in his career an editorial writer for the New York *Times* and a contributor to *Popular Science Monthly*. Michels had met Edison in Menlo Park during the height of the phonograph and tasimeter hype two years earlier, and now was drawn back to the inventor by the most recent debates surrounding the bulb, especially its apparently imminent introduction in New York. Knowing of Edison's close acquaintance with J. Norman Lockyer, British founder of *Nature* and fellow eclipse-watcher, and of Edison's interest in seeing the launch of a similar journal on these shores, the resourceful Michels proposed starting a magazine—even offering sample names such as *Nature (American)* and *Intellectual Observer*—with Edison as publisher and himself as editor. Michels elegantly pointed to the increasing focus among the American public upon scientific matters, as well as to the uncrowded field in the genre at the time: "While Literature proper, and Art, both ornamental and useful, nay, almost every distinctive social and economic interest in the United States, have their several organs for the interchange of views or the diffusion of information, Science still remains without any weekly journal exclusively devoted to the chronicling of its progress."

Edison seized the opportunity and agreed to employ Michels on a trial basis at a salary of thirty dollars per week for a year and a half, set him up with a room and an assistant on the fifth floor of 229 Broadway, and agreed to pay all the bills—with the warning that in due course the journal would have to become at least marginally profitable. And thus was born *Science: A Weekly Journal of Scientific Progress,* direct precursor of the present magazine of the same name, still in existence more than a century later. Its mission, as announced by Michels in his "Salutatory" column, was "to afford scientific workers in the United States the opportunity of promptly recording the fruits of their researches . . . presenting immediate information of scientific events."

Although Thomas Edison kept his role as publisher of *Science* an anonymous one, his name nowhere to be found on the masthead, the editorial content of the journal was subject to his special brand of scrutiny, and he insisted upon reviewing all advance proofs prior to publication. A survey of the first numbers of the magazine reveals an occasionally prideful and at other times self-righteous defense of Edison's work, even within the larger context of general news and views. In the very first issue of July 3, 1880, for example, we find a piece by Edison's faithful amanuensis, Francis P. [*sic*] Upton, Esq., on "Electricity as Power," assuring the readership that "Mr. Edison feels very confident of success," compared to the work of his perennial rival Mr. Werner von Siemens of Berlin, in his initiative to create an electric railroad; at Menlo Park, where "an experimental track of one-half mile in length" has been laid, "a speed of twenty to thirty miles an hour has easily been reached." Further on, the diligent reader is reminded that "the steadiness of the incandescent light over that of the arc has long been understood."

Later that month, John Michels' indignant editorial voice resurfaces to quell "all meretricious arguments and scientific hair-splitting" rumors insinuating that "Mr. Edison has thrown up his electric light researches." On the contrary, "Taking the view that it is a waste of time to argue theoretically, on that which can be demonstrated practically, Edison, through all this wrangle has been silent, but not idle; while others *talked,* he has *worked,* and in a few short weeks all will be ready. . . . The public is becoming nauseated with the wearying cry of '*non possimus.*' "

And so the rhetoric went throughout the inaugural year of *Science*: Be advised, faithful readers in the scientific community, Alexander Graham Bell has *not* solved the problem of transmuting light pulsations into electrical ones; Edison's tasimeter, lest we forget, had accomplished this some years ago, through his exploitation of the electrical conductivity of carbon. And of course we know that there still remain "an enormous mass of details" to be mastered by Edison before the electrical lamp can work on a practical scale, but "his sincerity" in all things places the Wizard of Menlo Park "above reproach" and we must be patient with him at this critical time. Additionally, the editor says, admonishing his English brethren, you must be more hospitable to Mr. Edison's ideas across the seas; you are to be chastised for omitting Edison's instruments from the just-published *Gordon's Illustrated Catalogue;* and how dare you insinuate, Prof. J. W. Urquhart, C. E., in your standard reference work, *The Electric Light, Its Production and Use,* that "there is little probability that this lamp (the horse-shoe carbon) will prove constant . . . [and it is apparent] that Mr. Edison will abandon this imperfect burner"?

Despite its partisan banner-waving, it was not much longer before Edison became impatient with the shaky bottom line of *Science*. He felt that it was not living up to the premise of its masthead; economically, its progress was negligible, and he was too busy to pay attention anymore. Or perhaps it had simply exhausted its public relations purpose for him. Michels's lease ran out and he asked, to no avail, if the magazine's offices could be taken in at Edison's New York headquarters on Fifth Avenue. Edison told Michels that he must find another backer. Ironically, it was Alexander Graham Bell who in the end, impressed with the journal, came to the rescue, took it up in January 1883, after a one-year hiatus, and went on to publish it for a decade.

Meanwhile, Edison remained on the defensive in other arenas. Allen Rice Thorndike, the distinguished editor of the *North American Review,* had also recently visited Menlo Park, and was anxious for Edison to contribute another essay to his influential periodical. Thorndike even offered to send an editorial assistant to take dictation from Edison that would then be polished into final form. Edison handed the task over to Upton, and "The Success of the Electric Light" was written over Edison's byline during the summer of 1880 and published that fall. Once again, the "tardiness" of the implementation of the bulb was rationalized at great length. Edison (through Upton) even went so far as to acknowledge that the system had been declared a failure and a "hoax" in some quarters. In measured tones, Upton laid out the further perfections that had taken place in the lamp, especially the recent replacement of the cardboard horseshoe filament with, of all things, bamboo—tested up to a spectacular and durable life of nine hundred continuous hours of incandescence. Toward that end, during the summer of 1880, one of Edison's men, William H. Moore, was dispatched to China and Japan in search of the best form of bamboo. Arriving in Yokohama, he met with no lesser dignitaries than the prime minister and the minister of foreign affairs, who advised him that Kyoto bamboo, specifically from the small, outlying areas of Yawata and Saga, possessed the qualities Edison needed—mature, eight to ten years old, growing in iron-rich, naturally fertilized soil under temperate climatic conditions, producing a sticky, flexible stalk, used for centuries in arrows, tea ceremony utensils, and fans. Edison, finicky as always, insisted that the filament be derived from the central, vertical sections of the tree no closer than one yard from the ground. He wanted segments of fourteen to sixteen inches in length cut from the tough, outer shell of the stalk, harvested during the traditional middle of autumn season only, trimmed into one-centimeter strips, bundled into packages of 100 strips each and shipped to the Menlo Park Lamp Factory.

Upton also revealed the pervasive extent to which Edison had never stopped thinking *systemically,* already planning ahead to replicate but vary the forthcoming New York experiment with lighting plants that would be operable within single buildings virtually anywhere that a 120-horse-power dynamo (the long-legged variety) could be placed. These spin-off plants—which did not have to be built for the express purpose of housing electrical machines, but rather, could be fashioned from existing ware-houses and other general-use structures—would be managed by local companies under royalty-franchise deals with the Edison Electric Light Company.

As Upton insightfully observed, Edison was not only struggling to main-tain the credibility of his lamp by catching up to a public image of his own design, but he was also operating within a rapidly expanding international context of intense technological rivalry, especially from, as might be ex-pected, England, seedbed of the industrial revolution. In the horse race (at times more like a steeplechase) to bring the bulb to market, one contender more than any other was right behind Edison. But even though at signifi-cant times he moved into the lead as the acknowledged innovator, the intel-ligent but retiring Joseph Wilson Swan eventually capitulated to Edison's will.

Born the son of an anchorsmith at Pallion Hall, Sunderland, in En-gland's northeast county Durham, two decades before Edison, young Joseph grew up an inquisitive and mischievous lad, enthralled by the Tyneside shipbuilding and commerce at his doorstep. At school he became interested in science, and at fourteen was apprenticed to a local chemist and druggist. Present at the dawn of photography, Swan has been credited with the per-fection of the halftone process. But he is principally remembered for his pi-oneering work in the development of the incandescent lamp. As early as 1860, he had settled upon the classic format of the inverted bell jar resting upon a brass tray covering a carbon arc heated by a primary battery. In 1877, one year *before* Edison stated that he had begun work on carbonized filaments, Swan was working in his laboratory on the evacuation of the bulb in order to achieve incandescence. And in February 1879, eight months *before* Edison and Batchelor achieved their breakthrough, Swan demonstrated his carbon lamp at a meeting of the Literary and Philosophical Society in Newcastle. He began the lecture by reading aloud to the accompaniment of much derisive laughter from the audience Edison's earliest "sensational" public prediction that "illumination by carburetted hydrogen gas will be discarded." Swan was not quite so certain of this eventuality. His general approach was more moderate. He had already set up a small lighting system in the library, picture gallery, and dining room of "Cragside," the mansion of Sir William Armstrong, a merchant in Rothbury, and Swan believed that the wave of the future in England would be conciliatory coexistence with gas.

Powered by a six-horsepower turbine, a Siemens dynamo fed Sir William's forty-five platinum-filament lamps in frosted globes. Observers marveled at the pure and steady way in which Millais' painting *Chill October* was shown to such desirable effect in the gentleman's residence.

On September 24, 1880, just before his second and more elaborate lecture before the Newcastle group, Swan sat down and scrawled a letter in rapid longhand to Edison in which he laid claim to having been on the electric lighting scene far before his more celebrated American counterpart: "I think I am in advance of you in several points especially in the making of the carbons. . . . I can easily convince you if necessary that I have been working long on this subject and that carbonised cardboard was a material that I have for years been experimenting with and was actually working at the very time you announced your use of it. . . . I therefore had the mortification one fine morning," Swan continued, "of finding you on my track and in several particulars ahead of me—but now I think I have shot ahead of you." Found among Swan's papers after his death, the letter had never been sent.

The satiric Gilbert and Sullivan parody (of "When I First Put This Uniform On" from *Patience*), "T. A. E.'s Lament," which appeared in a London newspaper, rubbed salt deeper into Edison's imagined wounds:

> *I said when I turned the lamp on,*
> *It's plain to the veriest dunce,*
> *That all gas shareholders*
> *Will shrug up their shoulders,*
> *And sell off their gas shares at once.*

I thought I was first in the Field,
And this lamp a good income would yield;
But my hopes are all blighted,
Completely benighted,
By a Briton whose plans were concealed,
Which I never counted upon,
When I first turned this little lamp on.

Who was first? Ultimately, this was not the relevant question, as dispassionate observers in the French journal, *La Lumière Electrique* of December 1880, noted when they drew the fine nuance. Remarking upon the most recent salvo fired against Edison, the editors noted that "la lampe de M. Swan paraît avoir la *priorité* sur la lampe Edison, mais rien n'indique jusqu'à ici sa *supériorité.*" ("Swan's lamp may have come *before* Edison's, but nothing presently shows that it is *superior.*")

Naturally, Edison had known quite well of Swan's early researches. He also understood that Swan was a respected native son. And as he demonstrated time and time again during his long career, Edison believed that the race was not to the swift, but rather to the man with bravado, endurance, and money—none of which Swan possessed. Nevertheless, Edison had to protect himself. Tiresome and draining litigation began in England, the solicitors of the Edison Electric Light Company, Messrs. Waterhouse and Winterbotham, advising the Swan United Electric Light Company that the Swan incandescent lamp was an infringement of the Edison patent. The Swan Company directors, not spoiling for a fight, and in complete concert with the shareholders, mirrored the realistic sentiments of their guiding spirit in speculating that a merger with American colleagues must surely be the solution, and within three years, The Ediswan Company was created, "Makers of the Original, The Most Economical, The Most Reliable, The Best of all Incandescent Electric Lamps." Securely ensconced at 36 and 37 Queen Street in the heart of the City of London, it held a firm monopoly on the British market until 1893.

A few blocks away, at 6 Lombard Street, much the same conflict embroiled Edison's telephone interests in England during this period of flux. Under the supervision of Edison's agent, Colonel George E. Gouraud, resident director in London of the Mercantile Trust Company of New York, and with the added engineering expertise of Edward H. Johnson, also sent over by Edison, the Edison Telephone Company of London had set up an administrative office and was operating the first exchange outside of the United States, at 11 Queen Victoria Street, serving twenty-four subscribers, and with a waiting list by May 1880 of more than four hundred additional businesses and individuals. Edison squared off head to head with the Bell interests in negotiation with the British Post Office for the right of exclusivity to the phone lines, Edison's side vaunting the famous carbon-button transmitter, and subsequently the chalk-drum receiver, and Bell's people promoting the superiority of their receiver.

The postmaster general held that person-to-person communication through conversations transmitted over the telephone must be considered messages. As such, like a telegram, transmitting of telephonic talk was still the exclusive privilege of the Crown, as defined by the Telegraph Act of 1869; Edison was therefore profiting financially from revenues the Postmaster General insisted were legally the government's. Edison argued in turn that the telephone had not been invented, let alone conceived, in 1869, and thus the prohibition was specious. To which Her Majesty's attorney general replied that the definition of telegraph *subsumed* all telephone communication, since both were sent "by wire," electrically. And so it went.

In the spotlight of controversy, Johnson found himself in the unaccustomed position of demonstrating the superiority of Edison's instrument to Bell's. "Now," he said to the latest round of London correspondents, "I will ring the bell in the central station." He touched a little knob, and instantly came a ring. "That's the answer," he explained; and then, placing his mouth at the telephone, he called, "Hallo, hallo! (a contraction, we presume, of "Hail, ho, ho!"). A reply, "Yes, sir. What can we do for you?" "Well, put me on to the store room." Even against the rumble and roar of trains, horse-drawn carriages, and vehicular traffic that bustling spring day in London, the voice by telephone from the suburb of Norwood, fifteen miles away, came back loud and clear, and would do so between nine in the morning and six in the evening every day, except Sunday, when the telephone rested.

If Mr. Johnson was too busy to conduct a tour of the telephone exchange premises crowded into the basement at Queen Victoria Street, callers at the Edison offices would be ably indoctrinated by a gangly, twenty-three-year-old battery room clerk recently emigrated from Dublin named George Bernard Shaw, an aspiring writer who published music criticism under an assumed name. He "took boyish delight in the half-concealed incredulity of [the] visitors, and their obvious uncertainty, when the demonstration was over, as to whether they ought to tip [him] or not."

Shaw reveled in a brief association with the energetic, rowdy Americans, who revered Edison "as the greatest man of all time in every possible department of science, art, and philosophy, and execrated Mr. Graham Bell, the inventor of the rival telephone, as his Satanic adversary."

As with Swan, a settlement was eventually reached with Bell, to the benefit of both parties, resulting in the United Telephone Company of Great Britain, the government realizing that "the Defendant Company [are] proprietors of a most beautiful invention, and it has been used entirely to the public advantage."

Previous biographies of Thomas Edison—and they are legion—rightfully glorify his accomplishments and prodigious inventive output. However, they tend to position him as an abstraction, strangely apart from the ebb and flow of powerful intellectual currents, and because of his larger-than-life stature, oddly disconnected from prevailing trends. Actually, he was a creature very much in and of his time. Fellow New Jerseyan, Paterson-born William Graham Sumner, whose wisdom we have invoked previously, himself the "inventor" of sociology, might have been painting an Edison portrait when he wrote, "Nature contains certain materials which are capable of satisfying human needs, but those materials must, with rare and mean exceptions, be won by labor, and must be fitted to human use by more labor." Sumner saw industrial organizations as the new institutional heart of American society.

The operant word here for our purposes is *"won."* It is obvious now that well before the smooth-cheeked age of thirty-three, by which time he had already made his indelible impression upon the culture at large, Edison lived and breathed competition (to Sumner, as omnipresent as the law of gravity). If Thomas Edison had not existed, Charles Darwin would have had to invent him. If in the world of nature, struggle was the law, and only adaptation could guarantee survival, in Edison's world of creative commerce, the unrelenting war was fueled by Grosvenor Lowrey's and Henry Villard's capital, and at this juncture, Edison had not yet rebelled against his monied supporters. His jungle was New York City, looming ahead as the inevitable arena, constantly on his mind, mentioned virtually daily in conversation.

Adding depth to the argument, at the same time that Francis Upton in the *North American Review* was defending Edison's tenacity in pursuit of the best bulb, William James held forth in the October 1880 *Atlantic Monthly* with his observations on the "remarkable parallels" between social and zoological evolution in the essay, "Great Men, Great Thoughts, and the Environment." Evolutionary interpretations of history, so much in vogue, had no basis in reality, James insisted—unless they accepted the importance of "individual initiative." Once more, we cannot help thinking of Thomas Edison when we consider James's conception of genius: "To be fertile in hypothesis is the first requisite, and to be willing to throw them away the moment experience contradicts them is the next." Great epochs are thus created by a "community" of great men who can, by their assembled energies, cause *"the mass of the nation {to} glow incandescent."*

Such a glow had reached across the ocean to the great actress Sarah Bernhardt, arriving in New York on board the ship *L'Amérique* for her first American tour. Driving the critics into a frenzy with her passionate performance as Marguerite in Alexandre Dumas *fils' La Dame aux Camélias,* within hours of her final matinee, escorted by her manager, Edward Jarrett, and Robert Cutting, an Edison Electric Light Company trustee acting as an interpreter, she was on the train to Menlo Park to see the "Yankee Cagliostro" in person enthroned in his "temple of light." Tall, thin, dressed in black from head to toe, her wavy golden tresses tumbling out from beneath a hooded cloak, she arrived like an apparition in the late November evening. Bernhardt later recalled her determination to "vanquish the delightful, if bashful savant," whose imperious presence reminded her of that of Napoleon. Edison, withdrawn at first, took her on a guided tour of the laboratory—and revealed the secrets of the light. But he saved the best and most appropriate creation for last, singing the tried and true favorite, "John Brown's Body" into the foil phonograph. Sarah reciprocated with lines from Racine's *Phèdre* and then went into a selection from *Hernani,* by one of Edison's favorite authors, Victor Hugo; she would be opening in that play the following week up in Boston. Touched by her choice but not to be outdone, Edison had the last word with his own special version of—what else?— "Yankee Doodle."

The interminable show went on into another act soon after, with a visit by the entire Board of Aldermen of New York City, needing to be sold on the prospect of Edison's digging up their downtown, a privilege that they felt should be assessed at the outrageous fee, to Edison, of ten cents per linear foot of wire. Lowrey and the other Electric Company directors, anxious to avoid a bureaucratic logjam, sought to break it with ample bumpers of wine and champagne served in the laboratory.

"Napoleon" or not, it was clear that the New York campaign could no longer be waged from twenty-five miles distant. In the new year, with their application still on the table before the aldermen, the Edison forces made their move. The newly reconfigured "Edison Electric Illuminating Company of New York" took out a four-year lease on a four-story double brownstone, the former Bishop mansion at 65 Fifth Avenue. Here the engineers, draftsmen, electricians, house installers, street installers, and statistical and law departments would be housed. A golden plaque was hung proudly by the side of the front door, the name Edison emblazoned at the top of the front of the building in block letters on a pale green ground. Fabrication of the lamps, now exceeding one thousand per day, would continue at the Menlo Park factory under Upton's supervision for the foreseeable future. Sigmund Bergmann, one of the old Newark alumni, upgraded his manufacturing shop at 104-108 Wooster Street so that it would have the capability to make electroliers, cut-outs, switches, sockets, and other components of the electric lighting system. In the summer of 1881, Bergmann and Company, three hundred strong, moved to larger quarters on the northwest corner of 17th Street and Avenue B, and Edison established his personal laboratory on the top floor, complete with Roosevelt's pipe organ carted in from New Jersey. The former site of the Aetna Iron Works at 104 Goerck Street, just north of Grand Street in lower Manhattan, became the new home of The Edison Machine Works, essentially the transplanted Menlo Park machine shop. And the Edison Electric Tube Company was incorporated at 65 Washington Street, in January 1881, with a capital stock of $25,000 for the manufacture and sale of insulated tubes for conducting electricity, with shops first in Manhattan and then in Brooklyn.

The move to New York marked a shift in Edison's administrative structure, and therefore led to changes in the trusted inner circle. Charles Batchelor was sent off to Paris as technological ambassador, to begin the lengthy and laborious process of establishing Edison lighting interests on the European continent, and to prepare for the huge Edison Company presence at the International Electricity Exhibition set for that coming summer. Stockton L. Griffin stepped down as Edison's secretary, and was replaced by another Englishman, twenty-one-year-old Samuel Insull, who had been working in London for Gouraud and Johnson. Skilled at shorthand, Insull began his career at *Vanity Fair* magazine and had become intoxicated with the romantic myth of Edison's rise to fame. Johnson, recently returned to New York, sent for Insull when the job opportunity arose out of the new hub of activity at 65 Fifth Avenue. Embarking from the S.S. *City of Chester* on February 28, 1881, young "Sammy" Insull (his Edisonian nickname) was immediately taken to meet the great man in his office in the rear room on the ground floor.

After the intense buildup and all he had read about Thomas Edison in the pulp press, Insull was not impressed at first glance. Here was a "casually" dressed man in a "rather seedy" black coat, still sporting a broad-brimmed western sombrero, the familiar white handkerchief wound around his neck and falling over onto a worn-out, graying white shirt. Not standing on ceremony, Edison started right in, and talked straight through from eight o'clock that evening until five o'clock the following morning. Even though Edison had barely established his base in New York, he was already well into an imperialistic plan of action. The subject of the conversation with Insull was Europe, and how to raise capital from the progress of the telephone in London to help support the new "shops" now geared up to feed into the city's as yet nonexistent electrical system. Insull finally got to bed in the room he would share on the third floor with chief engineer Charles Clarke as dawn was breaking on what had now turned into his second day on American soil.

Batchelor's mission in Paris, with the assistance of Theodore Puskas, Otto Moses, and Joshua Bailey, was to create an installation for the International Electrical Exhibition in the Palais de l'Industrie on the Champs-Elysées that would establish the definitive dominance of Edison's incandescent lighting system over those of his chief rivals, the American Hiram Maxim, and the two Britishers Joseph Swan and George Lane-Fox, not to mention the more than fifty other versions of arc and incandescent lighting that would be on view over four months. Long before the floor plan of the Exhibition was finalized, the Edison men prevailed upon Gaston Tissandier, one of the principal organizers of the show, who was drafting the catalogue and guide, to give their company two enormous rooms with greater immediate visibility than any of the other exhibitors—on the first floor of the Palais, immediately next to the main staircase. They also approached Theodore du Moncel, the editor in chief and Directeur Scientifique of the influential French periodical *La lumière électrique,* who had been publishing articles critical of Edison for some time, and convinced him eventually to reverse his opinions. He had seen the crowds pressing night after night into the Edison exhibit; he had seen the soft glow of the heavenly globes. In a long, apologetic lead essay appearing in the fall of 1881, du Moncel went back over his transgressions committed against Edison and publicly regretted his past skepticism. He had not given these new inventions a fair chance to prove themselves. Yes, he had once, along with many other self-styled "savants," been guilty of doubting Monsieur Edison. It was much to his favor and a tribute to his strength of character that Edison had not backed down, but rather fought on despite the inordinately acerbic rhetoric thrown his way. He cautioned against focusing exclusively on the bulb, and, marvelling at Edison's whole system, rhapsodically and exhaustively reviewed all of its components, with charts and gravure illustrations.

The Exhibition opened with great fanfare at 10:45 a.m. on August 10. The band of the Garde Republicaine struck up the *Marseillaise* as the president of France, his wife, and assorted dignitaries arrived. The American ambassador held court at the entrance to the ornately—one might say baroquely—decorated Edison rooms. Crystal chandeliers augmented the smaller lights adorning the walls, illuminating paintings and tapestries commissioned expressly for the show at a cost exceeding 300,000 francs. "One would have thought," wrote one observer, gazing at a wall-hanging "representing children at play, some culling flowers, others cutting thorns, that the scene were real and was bathed in the warm glow of the autumnal sun."

The general atmosphere was of an elaborate, festive retrospective survey of Edison's entire output. Arrayed on low-level tables, some objects in vitrines or under glass domes and others accessible to the touch, were stock tickers, phonographs, telegraph transmitters, etheric force black boxes, telephones, small generators, and even a plan for the Pearl Street Station in New York, so visitors could have a preview of that exciting project. The Edison forces were also able to commandeer the great central staircase itself and to festoon it with seven hundred lamps, some of them formed into illuminated bunches of grapes held by classical bronze statues posed on the landings, as well as an attention-grabbing bulb constellation in the form of a giant E dominating the entire scene, topped with an illuminated Baccarat crown weighing two tons.

Puskas and Bailey hired the writer Camille Flammarion at a salary of one thousand francs per month for the duration of the Exhibition, and placed major articles lauding Edison in *Le Monde illustré, L'Illustration,* and *Le Figaro.* It was an intense, costly, but ultimately pivotal public relations campaign, because when the exhibition concluded, and five Gold Medals and a Diploma of Honor had been awarded it, Edison's system emerged as indisputably preeminent in Europe, and Batchelor was given his most prestigious offer ever, to provide the lighting for the Great Foyer of the Paris Opéra. This coup was followed soon after by the placement of three Edison chandeliers in the foyer of La Scala in Milan, which was not to be outdone.

In the aftermath (and afterglow) of the great Exposition, two companies were established with headquarters in Paris. The Compagnie Continentale Edison, at 27, Chaussée d'Antin, controlled Edison electric light patents in Austria, Belgium, Denmark, France, Germany, Hungary, Italy, Russia, and Spain. The Société Electrique Edison marketed isolated lighting plants in these countries under license from the Compagnie. These organizations advanced their activities by reporting to interested, potential investors on progress back in New York City, where two buildings had been purchased at 255 and 257 Pearl Street, to house the immense dynamos that would power the square-mile system south of Fulton Street; between six and seven miles of underground cables had finally been laid down, half the amount that would eventually be needed; and a functional meter had been perfected. The idea was to build a case for the commencement of similar incursions in Europe under the Edison aegis. Muncipalities could order the entire Edison network, from 30-ton, 125-horsepower dynamos to individual bulbs, and import them.

"My job here is no fool of a job," Charles Batchelor wearily reported home to his boss, "what with lamps, dynamos, chandeliers, and all the extras. I am just in up to my neck; then I have so much outside work of such a responsible nature and involving so much money that I wear a hat about three sizes larger than when I left New York."

At the same time, across the Channel in London, Edward Johnson had returned—along with engineer William Hammer from the Menlo Park gang—to familiar haunts, taking on *his* new assignment, the temporary electrification of a one-half-mile strip near the perimeter of the City of London, from the General Post Office building, along Newgate Street westward past the Spier and Ponds hotel and the main generating station at number 57 Holborn Viaduct, across Farringdon Street, by St. Andrew's Church to the juncture of Holborn Circus and Hatton Garden. This was a proving-ground effort, deliberately undertaken by Johnson and Hammer in tandem with the Edison exhibition at the Crystal Palace in Sydenham, to heighten the Edison profile in England. It was an extension of Edison's relationship with London which went back almost a decade to his first trip there in 1873 when he demonstrated his telegraph system to the Post Office. And in early

1878, William Preece had demonstrated the first phonograph ever seen in England at Faraday's venerated lecture theatre in the Royal Institution on Albemarle Street, an instrument given to Preece by Edison himself. Alfred, Lord Tennyson had been in the audience that evening to hear his own verse, "Come into the garden, Maud," phonographically recited.

With substantial support from Drexel, Morgan and Company, the English Edison Electric Company, Ltd., installed a "Jumbo" dynamo, appropriately named after P. T. Barnum's immense elephant, at 57, Holborn Viaduct, shipped over from the Edison Machine Works. A second dynamo, driven by an Armington and Sims steam engine, was soon added in April 1882. Again, Joseph Swan's earlier efforts to illuminate the streets of Newcastle-on-Tyne were slighted, even in the British magazine, *The Electrician,* where it was asserted that "to the energy of Mr. Edison we owe the first installation upon a scale large enough to once for all show what can be expected from such lights," all 938 of them. *The Metropolitan* hailed the setup as "the greatest that has yet been made by Mr. Edison's representatives in this country." For the first four months, in a goodwill marketing ploy, electricity was provided without charge. Subsequently, it was offered at par with gas. Lighting the Post Office building was likewise a strategically apt move, in view of past problems there with the enfranchisement of Edison's telephone. Under the supervision of Mr. Preece, the gas jets in the Post Office press room were extinguished as sixty incandescent lamps were switched on. An even, shadowless light was thrown upon the tables, to much admiration by the workers.

These were expansive times for Thomas Edison, overseer of a Fifth Avenue property and growing workforce, a player on the international scene, a worldwide celebrity whose enemies had even been converted. Imagine the whirlwind life: morning meetings with J. P. Morgan to discuss the plans for the isolated lighting system being installed in his mansion and library on 37th Street and Madison; a quick trip down to the Pearl Street site to pay a visit to the men in the trenches, roll up his sleeves, dig and lay cables with them, then ride back uptown on the Third Avenue El with Insull, dust-laden, sweaty and disheveled, laughing as they read the laudatory dispatches about the Paris show in the New York *Herald.* A short catnap, answer letters at his rolltop desk, then on to dinner followed by a few cigars, share more laughs with "the boys" at Delmonico's on 26th Street between Fifth and Broadway.

Mary and the children—Marion, nine; Tom, Jr., six; and little Will, four—had been living temporarily at the Lenox Hotel across the street from 65 Fifth Avenue. Marion was taken out of Bordentown Female College near Trenton, New Jersey, when the family moved into New York, and began taking English, French, drawing, dancing, and music at the Misses Grahams' School for Young Ladies next door to her papa's office, at number 63. But she still scarcely saw him even then. During their second year in town the family relocated to a magnificent four-story townhouse which they rented on the south side of Gramercy Park, number 25, and Marion and young Tom switched schools again, to Mademoiselle de Janon's English and French Academy for Young Ladies and Gentlemen at 10 Gramercy Park. The Edison family doctor had become increasingly alarmed at the reclusive Mary's emotional troubles: "She seems very nervous and despondent, and thinks that she will never recover," Dr. Leslie Ward wrote to Edison in the winter of 1882. "She seems so changed physically and mentally of late that something ought to be done." How ironic that Dr. George Beard, Edison's colleague from etheric force days, had just published his classic treatises, *On Nervous Exhaustion (Neurasthenia), Its Symptoms, Nature, Sequences and Treatment* and *American Nervousness.* He practically defined Mrs. Edison's condition: "the anxiety habit," caused by the trials and tribulations of modern urban civilization, too much travel, too many new ideas being developed and introduced too rapidly into the culture, excessive commercial activity, the repression of emotion, domestic trouble, the demands of religion, inordinate worry—all, according to Dr. Beard, far greater in America than in any other country in the western world. American nervousness was characterized by "fleeting, transient, metastatic and recurrent" symptoms including excessive drinking, especially by those "who live in-doors," dyspepsia, increased sensitivity to fluctuations in temperature, anxiety among the privileged, and the "wasting of cerebral force" in the confined classes—especially women burdened with "increased brain effort."

Dr. Ward persisted in his appeal to Edison: "I believe that an entire change would be of benefit and if you could take or send her to Europe for a few months she might return improved in health and be better pleased with her surroundings here." To which Edison scribbled cursorily, "I am going to take her off on a trip soon, etc." But he did not, sending Mary and the children to South Carolina and Florida for a month while he stayed behind, the captain of industry who would never leave his ship.

CHAPTER

12

Edison's First District lighting system, a multitenta-
cled octopus, radiated symmetrically outward, from adjoining gutted and
steel-reinforced, refitted warehouses on Pearl Street just south of Fulton and
three blocks west of the fish market. Two subterranean tiers of Babcock and
Wilcox boilers, fueled with eight tons of cheap anthracite slab coal per day,
heated two thousand cubic feet of water, working up pressure of 120 pounds
per square inch to drive at 325 rpm. the shaft linking each thirty-ton
Sturtevant blower-cooled "Jumbo" dynamo and noncondensing Porter-
Allen steam engine, bolted together onto cast-iron base plates. It took three
months of operations to eliminate the bugs before the full complement of all
six engine-dynamo sets were cranking in staggered shifts.

Placing electrical cables underground in the same mode as gas and
water mains was a sagacious move. Lower Manhattan was crisscrossed with
an unregulated forest (Edison called it a "wilderness") of privately owned
utility poles up to 150 feet tall, mounted with festoons of barely insulated,
drooping wires for all manner of telephones, stock tickers, and burglar
alarms, subject to the wear and tear of the elements, showering sparks, prey
to heated arc street lighting below. More than one municipal repair man
had been roasted far above, to the morbid fascination of assembling crowds,
which in one recent instance took up a collection on the spot for his be-
reaved wife and children. Edison's direct-current copper conduits avoided

DYNAMO ROOM AT THE PEARL STREET PLANT

these hazards as they were laid down in a "tree" system, so called because the cross section was of necessity much wider close to the originating source of power via the common main, and tapered down through secondary feeders as wires entered individual buildings where drops in voltage would be effected. The main conducting cables originated at the Pearl Street power plant and then proceeded under cobblestoned streets immediately contiguous with the sidewalks. The cables were wrapped first in Manila hemp, then inserted into twenty-foot lengths of cast-iron pipe, which were in turn insulated with what can only be described as an exotic recipe developed at Menlo Park: refined Trinidad asphaltum boiled in oxidized linseed oil with paraffin and a touch of beeswax. This concoction hardened as it cooled. Edison predicted his conduits would last more than half a century, and indeed they did. In the transition from street to indoor wiring, cotton impregnated with shellac, paraffin, or resin served to protect the copper, the whole encased in cardboard and then run up the surface of the walls.

In the First District system, each lamp had its own circuit breaker, a small "safety wire," designed to burn out if the amount of electricity designated for that lamp was exceeded. To measure the use of electricity per customer, Edison the retailer devised a unique meter of constant resistance, based upon electrolysis; he actually assessed *use* in terms of *weight.* The meter consisted of two zinc plates immersed one-quarter inch apart in a glass jar containing zinc sulphate. As the current passed through the incoming circuit on its way to the consumer, a tiny, sampler amount was diverted to the meter via a German silver wire, causing zinc from the solution to be deposited on the plates. Monthly, the cells were removed and weighed, and thus the electricity expended could be calibrated relative to the amount of zinc coating.

As for the bulbs (or "lamps"), although they retailed at one dollar apiece, each customer in the First District received an initial premium offer of one dozen free, each "globe . . . in the shape of a dropping tear, broad at the bottom, narrow in the neck," to be twisted into its uniformly threaded porcelain socket. By turning a thumbscrew at the base of the socket, the circuit was completed and the lamp, shrouded by a ground-glass shade, was illuminated, "mellow and grateful to the eye."

In the final weeks before the Pearl Street launch, Edison made much of the archetypal struggle between ancient, "evil" gaslight and his new, "good" electric light. Edison was the knight in shining armor who was going to seal the fate of "the barbarous and wasteful gaslight" once and for all. "It is a light for the dark ages," he intoned. "The distribution of gas through the city is done by means of an immense system of sewerage pipe, through which it is forced and kept under pressure—gas reeking with impure material and made by a dozen different processes. This gas is allowed to escape through holes into our apartments where it is burnt, taking oxygen from the air to support combustion, the products of which are carbonic acid, carbonic oxide, sulphuric acid, sulphuretted hydrogen, and a host of other substances, which vitiate the atmosphere . . . The result of the vile poison is almost entirely heat and only incidentally a little light. It is a nasty, yellow light, too, and far removed from the color of the lovely natural light." It was time to sound the death knell for the threatening, volatile, "nauseous, dim flicker of gas, often subdued and debilitated by grim and uncleanly globes," and make way for the consistency and steadfast character of Edison's lamp, with its "soft radiance . . . singularly powerful and even . . . perfectly steady."

By late summer of 1882, sectors of the First District were being selectively lighted. Early in the day on September 4, Edison allowed himself to be seen "in his shirtsleeves in the workshop" of 257 Pearl Street, surrounded by the grime and hubbub of preparation. But after lunch, he donned his frock coat and in midafternoon accompanied by a few trusted lieutenants strode over to J. P. Morgan's offices at 23 Wall Street. Precisely at 3:00 p.m., John Lieb, chief electrician at the central station, threw the master switch, and Edison personally connected the lamps in the Drexel, Morgan & Company suite. "In a twinkling," said the New York *Herald,* in true fairy tale style, "the area bounded by Spruce, Wall, Nassau and Pearl Streets was in a glow." At that first moment, only one-third of the square-mile target area was alight, with about four hundred lamps in operation—including at the New York Times building, where by seven o'clock in the evening the editorial and accounting offices were still bright as day, but "without any unpleasant glare." The journalists marveled at the transformation, as did the workaday crowd heading homeward down Fulton Street on their way back to Brooklyn by ferry. Edison and Johnson were ecstatic, found by a reporter from the New York *Tribune* to be "in a high state of glee at the opening of their system."

It seems hard to believe it was merely fortuitous coincidence that the day after Edison's triumph the first "Labor Day" parade was celebrated in New York City. "Five thousand jewelers, masons, machinists, carpenters, iron puddlers, bricklayers and printers" started off with a rally at City Hall, just a few blocks uptown from the Pearl Street station, keeping in step to the tune of "When I First Put This Uniform On" from Gilbert & Sullivan's new operetta, *Patience.* The guilds gathered to hear exhortatory speeches at Union Square, and then continued up Fifth Avenue, broadcasting their identity against fears that traditional crafts were becoming devalued, lowly workers enjoying a moment of pride in the face of encroaching, systematized capitalism.

Thomas Edison, however, treated his laborers well. The payroll for the Central Station in its first season reveals that John Lieb as superintendent was receiving $2,400 per year in compensation. The general pay scale ranged from $5.00 per day for Charles L. Clarke, the chief engineer, who reported to Lieb, to $2.00 for "Coal Passers" and $1.50 for the "Regulator Men," which still compared very favorably with the typical standard of one dollar a day for most manual labor at that time.

Three decades later, on December 12, 1913, looking back on these first heady moments, Edison pithily clarified his motivations: "Electricity is not power; electricity is a method of transporting power."

And it was precisely *his method,* exploited through the strategic perception of grass-roots and corporate needs for convenient access to clean environmental energy from a central source, superseding raw or partially
processed fuel and the hazards of the open flame. It was his method, not just
for lighting, although at first, as Reyner Banham once again has shown, "the
fascination of the new clean light was such that further thought seemed unnecessary," but for running sewing machines, fans, and other small appliances on an intimate, domestic scale, supported by the continued enthusiasm of the money men and spotlighted in the great urban centers of
America and Europe. The whole network was finally so well deployed, despite Edison's chronic runaway promises, that simultaneous with Pearl
Street going on line, the Edison Company for Isolated Lighting, as subsidiary of the Edison Electric Light Company, had already reached licensing
agreements for the use of Edison's 216 patents "securing the fundamental
principles of incandescent lighting" in two hundred sites across America—
mills, factories, hotels, steamships, stores, and residences—burning an aggregate total of nearly *forty-five thousand lamps every day.*

From the Baldwin Locomotive Works and the Stetson Hat Company
in Philadelphia, to the Oregon Railway and Navigation Company in Port-
land, to the printers Hinds, Ketcham & Co. in New York City, to the Clark
Thread Company in Newark, to the Wamsutta Cotton Mills in Fall River,
Massachusetts, to the bankers Spencer Trask & Company in Albany, to
Rand McNally Maps, Cyrus McCormick Harvesters and Marshall Field's
Dry Goods department store in Chicago, to George Eastman's Dry Plate
Photographic Company in Rochester, to the Whiting Paper Company in
Holyoke; and to the homes and estates (and yachts) of the rich and famous—
James Gordon Bennett, J. P. Morgan (who, "entranced by Edison as by no
other inventor of his time," immediately purchased an additional one thou-
sand shares of Edison Company stock in the fervor following the Pearl Street
District illumination), Henry Draper, and William H. Vanderbilt—Edison
sold the means to make and use the power.

And via the same technology-transfer dynamic in Europe, through the
Compagnie Continentale Edison, the Société Electrique Edison, and more
recently the Société Industrielle et Commerciale Edison, in Ivry, local manufacturing site for Edison inventions, 158 isolated plants were humming in
Berlin, Manchester, Paris, Bordeaux, Amsterdam, Munich, Bologna, Rome,
and other cities.

"We have never had a plant rejected," trumpeted the Edison Company in its high-pitched, international marketing crusade. "No fire or accident of any kind has ever occurred from the use of an Edison plant"; and "Many of our plants have been largely increased [*sic*]." Furthermore, the seemingly indestructible Long-Legged Mary Ann, more officially known as the Z Dynamo, the veritable heart of the isolated lighting system, was a bargain at only $1,200, complete with sixty lamps and sockets; it needed "only a few minutes attention each day for oiling, &c., and does not get out of order."

With growth in financing at home and ever greater demand abroad for the franchise, Edison turned to another member of his team to join Charles Batchelor in representing his European interests: John William Lieb, who at the tender age of twenty-two had already more than earned the right to throw the first ceremonial switch at Pearl Street. Born in Newark, he attended Newark Academy and Stevens Prep, and graduated from Stevens Institute of Technology in Hoboken, New Jersey, with the degree of Mechanical Engineer. Edison hired Lieb away from one of his major competitors, the Brush Electric Arc Lighting Company in Cleveland, and brought him in as a draftsman. Lieb was instrumental in drafting the plans for the "Jumbo" dynamo and then moved on to draw up concepts for the complex switches and regulating equipment installed at Pearl Street. He also put in a year at the Goerck Street Machine Works, helping to perfect Edison's electrolytic metering system, and then graduated to become general supervisor of the Pearl Street effort. At the end of November, on the basis of this comprehensive experience with all aspects of his operation, Edison selected Lieb to go to Milan and direct the installation of the Edison central station there, which was to become for several years after its opening in June 1883, the largest electric light and power supplier in Europe. Lieb became chief engineer of the station with the establishment and licensing of the Società Generale Italiana di Elettricità, Sistema Edison.

As excavation work for the immense, three-story Milan generating station got underway on the site of the Theatre of Saint Radegonde, an ancient section of canal construction put down four hundred years earlier by Leonardo da Vinci was unearthed, kindling what was to become Lieb's lifelong obsession. Teaching himself Italian—as well as five other languages—Lieb published many monographs on Leonardo, lecturing at major universities and art museums on "The Master Mind of the Renaissance." Through the next decade in Italy and continuing on until his death in 1929, Lieb set his sights on collecting Vinciana, eventually accumulating more than five hundred items. This great archive of first editions and facsimile manuscripts, the largest in America, was purchased in its entirety by Samuel Insull, who donated the material to Stevens Institute, where it is now housed in the Samuel C. Williams Library.

Like Charles Batchelor and Samuel Insull, John Lieb stands as an important case study in the rapid growth of Edison's corporate interests during the 1880s. Thomas Edison was the entrepreneurial force, the unstoppable imagination driving this expansion. J. P. Morgan and Eggisto Fabbri, Grosvenor Lowrey, Henry Villard, and others we shall soon meet were the "industrial statesmen" (or "robber barons," depending upon which side of the dialectic is preferred) with whom Edison coexisted uneasily but realistically. Insull's expertise took him onward to supervise the Chicago Edison interests. Returning to America in 1894, Lieb committed his talents to Edison Electric in New York City even beyond the point after the turn of the century when Thomas Edison was no longer directly involved with the company. These loyal Edison men and others like them who flourished under the "Old Man's" wing and became wealthy because of their staying power were not controlling financiers, nor were they visionaries. They represented the beginnings of a new echelon, what Alfred D. Chandler, Jr., in his landmark study *The Visible Hand,* has called "the ascendancy of the manager" in the rise of modern American business enterprise. It is one of the most exquisitely ironic themes in Thomas Edison's story that the very "managerial hierarchy" (to employ Chandler's term) he had to establish and cultivate— as his artisan-oriented shops inevitably became companies and therefore of necessity more integrated in their functions—rose up to alienate and confuse him in the years ahead, eventually and most sadly involving management team members of his own family.

In the mirror of nineteenth-century fiction, Thomas Edison might have encountered Silas Lapham, the protagonist of William Dean Howells's classic Gilded Age cautionary tale. The erstwhile tycoon, manufacturer, and international exporter of "the best paint in God's universe" was possessed with "the pride which comes of self-making." Raised by a self-sacrificing mother and an ill-educated father, young Silas had overcome the boyish trials and struggles of farm life, ultimately reaching a social level where he was "supremely satisfied with himself," smug and cigar-smoking in his Boston brownstone. By turns preoccupied and intoxicated by commercial advancement, Lapham ignores the underlying tensions in his marriage and gradually becomes emotionally alienated from his beloved wife and daughters, entering a lonely world of "isolation brought on by adversity" from which he is rescued only by coming face to face with failure.

The Rise of Silas Lapham (1885) reminded its contemporary readers that attaining the grail of accomplishment left inevitable casualties. In the aftermath of Pearl Street, the equally successful Edison divided his days between business affairs at "65," as the Fifth Avenue headquarters was called, where demonstration rooms blazed with light deep into the night so that potential customers could stop by and discuss the particulars of acquiring isolated systems with Lighting Company salesmen; sessions at the Avenue B laboratory; and reading marathons in the sitting room at Gramercy Park while Mary and Dot went out to the opera. Edison was well into his usual heavy mixture of fiction—Victor Hugo, Dickens, Hawthorne, George Eliot, James Fenimore Cooper—and poetry, sticking faithfully with Longfellow.

Mary's chronic emotional and health problems made it impossible to maintain any semblance of normalcy on the home front. In the fall of 1883, Dr. Ward insisted that Mrs. Edison relinquish housekeeping responsibilities immediately. Unable to break their lease at Gramercy Park, the Edisons continued to pay rent there, but moved residence again, over to the Clarendon Hotel on Fourth Avenue and 18th Street. That winter, Edison finally did take his wife, sans children, on a Florida excursion. She happily picked strawberries and orange blossoms by their hotel while her husband hunted and fished. Dottie, as her mother was fond of calling her, remained under Insull's watchful and responsible eye, while the two little boys stayed, as they did more and more frequently, with their Stilwell grandparents, who had moved in with Mary's brother-in-law and older sister, William and Alice Holzer, at Menlo Park. (William had been hired by Edison as a glass-blower at the now booming Lamp Factory.)

Mary's winter respite was short-lived. No sooner had she returned from the southern trip when her beloved father fell ill. Rushing to New Jersey to be at his side, she became hysterical with grief: "Send trained man nurse who is not afraid of person out of mind," Edison wired Insull from Menlo Park. "Send as soon as possible." Nicholas Stilwell died on April 11, 1884, at the age of sixty-one. "I am so awfully sick," Mary wrote to Insull two weeks later. "Just now my head is nearly splitting and my throat is very sore."

At the height of the July heat, holed up on the top floor of Bergmann's factory a short hop from his Clarendon digs, Edison was putting in more and more night hours at the laboratory, experimenting with varieties of electrochemical decomposition reactions in lead-acid battery cells and noting with relish that the sulphuric acid smelled "as bad as 3 square miles of hell." Test tubes were exploding violently and with regularity, and the resulting fumes were overpowering, burning him and his assistant, who at one point, Edison's notebook states clinically, "spit blood."

These nocturnal endeavors were interrupted by a telegram from Menlo Park, where Mary Edison lay tossing and turning. At two o'clock in the morning on Saturday, August 9, she died, according to the doctor's announcement, "from congestion of the brain." She was twenty-nine years old. Dottie awoke to find her father by his wife's bedside, "shaking with grief, weeping and sobbing." The next day, a huge earthquake shook the entire east coast from Maine to Delaware.

Funeral services were conducted on Tuesday at noon in the Presbyterian Church of Metuchen. A large company of mourners attended, including Edison's brother, William Pitt, who arrived from Port Huron, and Mary's brother, Charles, from Hamilton, Ontario. Sister Marion received Edison's telegram too late to make the journey in time; besides, she pointed out, "Brother Al" had not answered any of her letters in recent years, and she had never even met William Leslie, the youngest Edison.

After the funeral, a special train took the coffin into Newark, and from there to Fairmount Cemetery, where it was not interred, but rather, laid in the receiving vault among the ancient birch trees. Samuel Insull and Charles Batchelor, just returned from France, were among the pallbearers.

In early October, Mary Stilwell Edison's body was removed to Mount Pleasant Cemetery and buried under a huge headstone in a special plot on a low hillside by the Passaic River. Established in 1844 on Broadway in Newark, behind Mount Pleasant's Gothic, brownstone arches from the old Belleville Quarry, the 36-acre cemetery held the graves of members of the city's most important families, the Frelinghuysens, Murphys, Ballantines, Drydens, Farrands, Rippels, and Westons. Evergreens, horse chestnuts, dogwood, sycamore, linden and maple trees bordered Mount Pleasant's winding gravel paths. It was a resting place suited to "the wife of the great inventor," who, the Newark *Daily Advertiser* somberly reported, "had suffered sadly in health . . . [enduring] a long illness, during which her life was almost despaired of. . . . A trip to Florida was taken when the convalescent could bear the fatigue of travel, but in vain. She leaves behind three children, two boys and a fair-haired little girl, Dot, who has inherited her mother's blonde beauty."

In Marion Edison's memoirs (which, we must remember, were written seventy-two years after the fact) she recalls that her mother died of typhoid fever, asserting with pride that "from then on my Father and I were inseparable." For the first year or so following Mary's death, this seems to have been the case. His twelve-year-old daughter became Edison's traveling companion while her younger brothers were looked after by the widowed Mrs. Stilwell. Dot trotted Papa around the countryside during those waning summer days in her pony cart; the adult Edison *never* drove any vehicle, be it horse and buggy or motorcar. Once the school year began, Madame Mears, who had been Dot's instructor, inquired as to whether Miss Edison would be returning to her Academy on Madison Avenue. In due course, Insull replied on Edison's behalf, the child would indeed once again take up residence in New York, where her father had rented a new flat on 18th Street between Madison and Park.

Meanwhile, however, Dot accompanied Papa—who had recently been elected a vice-president of the American Institute of Electrical Engineers—to the International Electrical Exhibition in Philadelphia. Nothing in the Great Hall attracted more attention, of course, than the elaborate display mounted by the Edison Company. Lights twinkled within a circle of calla lilies, roses, japonica, and chrysanthemum. A towering mica cylinder adorned with colorful bunting stood ablaze with incandescent lamps. The Jumbo and Z dynamos were in a section all their own at the northern end of the building. Edison's men also built an actual replica of their copper main underground "three-wire" cable system with wrought iron piping for all to see. But the pièce de résistance of the show was undoubtedly the Edison Electrical Darky, which *Scientific American* described with attempted restraint as "an illuminated colored gentleman who wore a helmet crowned with a suddenly flashing electric lamp and politely distributed cards to astonished visitors."

Descriptions in Dot's ill-formed cursive penmanship can be found in Edison's chemistry notebooks dated October 1884. After "helping" Papa in the lab, she would tag along with him to Delmonico's past midnight. Edison went straight to the smoke-filled "men's restaurant," where he was a familiar figure. "No ladies were allowed there," Marion remembered, "but Father had no trouble getting *me* in." She feasted on squab and vanilla ice cream. She bought five-cent cigars for Papa and cashed checks for him; they sat huddled together in the front row of the minstrel show so that Edison, hard of hearing, could get the raw jokes properly—this was *his* kind of entertainment!

"I think I must have been my father's favorite child," she wrote wistfully, "because I was the oldest and most mature. . . . I loved being wanted and never left his side."

CHAPTER
13

THE WIDOWER WOULD NOT PERMIT HIMSELF TO BECOME IMMO-
bilized by grief. Within weeks of Mary's death, Edison enacted a complex
merger of the Thomas A. Edison Construction Department with the Edison
Company for Isolated Lighting, responding to the increasing demand for
lighting facilities across the land. Following the successful Edison company
presence at the International Electrical Exhibition in Philadelphia, which
raised his profile a few more notches, Edison negotiated a contract with the
American Bell Telephone Company in Boston, which agreed to subsidize
his expenses and salary at $12,000 per year for the next five years in pay-
ment for options on the results of resumed telephone experimentation. The
two communications competitors had long since accepted both home and
abroad the viability of sharing their self-interests.

In addition to feeding his cash flow base, this deal moved Edison back
into collaboration with Ezra Gilliland, old pal from teenage telegraphic
years. The two had met during the Civil War back in Adrian, Michigan,
then roomed together for a stretch in Cincinnati during Edison's *"Richard
III* period," his flirtation with thespianism, when he paraded up and down
the hallway of their cramped flat reciting speeches from Shakespeare, and
practiced transmitting and receiving skills using passages from the Bard.
Gilliland had gone on to work with Edison a decade later in Newark on the
manufacture of the electric pen, and then briefly at Goerck Street, before

moving up to Boston, where he was now head of the Bell Telephone Company Experimental Department.

The reunion with Ezra Gilliland must have been more than an expedient business move. We can only imagine, because Edison has left no signs to the contrary, that after coping with personal loss, he may have felt the need to reestablish reassuring links from simpler, adventurous days. During the winter of 1884–85, another in the seemingly endless chain of technological extravaganzas characteristic of an era intoxicated with electric power was mounted in New Orleans—the grandiose World Industrial and Cotton Centennial. The Boston Bell Company booth featured several of Edison's sending and receiving devices. The combination of commerce and climate was irresistible to the footloose Edison. Faithful Dot in tow, he picked up his valises again and headed for the temperate southern city, agreeing to rendezvous at the Centennial with Gilliland and his wife, Lillian, with the plan to segue onward to Florida for a vacation.

Promenading up and down the aisles at the Exposition one day, the Edisons and the Gillilands "ran into" another millionaire mogul of American industry, Lewis Miller of Akron, Ohio, accompanied by his daughter, Mina. Again, as with many matters having to do with Thomas Edison's inner life, it is not altogether clear just how *unplanned* this meeting really was. Young Mina, at nineteen having recently graduated from Akron High School, was on midwinter break from the exclusive Mrs. Johnson's Ladies' Seminary, a finishing school in Boston, and, because of Lillian Gilliland's prior friendship with the Miller family, she had already spent many welcome evenings in the Gilliland parlor, singing and playing piano at her home away from home. Mina was as petite and dark as Mary Stilwell had once been buxom and fair. Like a gypsy, her thick hair pulled up and fastened with a golden

MINA MILLER AT EIGHTEEN

barrette, she had huge, deep, liquid brown eyes and flawless olive skin, set off against ruffled lace collars and pearls. Like Edison, a seventh child—of Lewis and Mary Valinda Alexander Miller's eleven children—the lovely Mina was at the moment when she first met Thomas Edison engaged to

George Vincent, son of the Reverend John Heyl Vincent, cofounder with Lewis Miller of the Chautauqua Assembly in upstate New York. She had meant to attend George's graduation from Yale that coming spring; however, her watchful father was less than enthusiastic about the economic implications of a union with a preacher's son. To further confuse the tangled skein, even though Edison had rather absentmindedly reneged on Reverend Vincent's insistent invitation to demonstrate the phonograph at Chautauqua seven years earlier—preoccupied with the response of the tasimeter to the eclipse of the sun in Wyoming—Mr. Miller had already met Edison in Menlo Park during its heyday; thus, the Wizard's reputation preceded him in New Orleans.

Lewis Miller's career is one of the great and oddly neglected phenomena of American industry, more so because its convergence with Edison's makes eminent sense and serves to shed more light upon our protagonist.

Lewis Miller's grandfather, Abraham, had emigrated from Zweibrucken in the Rhineland in 1776, joined the colonial army, and served with George Washington at Valley Forge. His son, John, in turn, was a pioneer farmer who made the traditional trek from the Pennsylvania hills into unspoiled land, where Lewis Miller was born in Greentown, Stark County, Ohio, midway on the major north-south stagecoach thoroughfare from Akron to Canton, on July 24, 1829. As a boy, he watched the Conestoga wagons roll by, "Ships of the Desert" drawn by three pairs of horses, high-paneled sides painted blue, rounded canvas hoods rippling in the wind. He heard the "Gee" and "Haw" calls of the drivers and the pistol-shot crack of blacksnake whips in the air. Like Al Edison, Lewis was soon distracted from direct work with the land, preferring instead the mechanical applications of an agricultural milieu. Graduating from the district school, he entered Plainfield Academy, learning the plasterer's trade and then teaching for a while.

At the age of twenty, Lewis came to rest in the burgeoning reaping-machine industry, and never looked back; the first five reapers in Ohio had been made in Greentown, and Lewis joined the firm of Bell, Aultman (his stepbrother, Cornelius) and Company there, mastering the mechanic's trade for a starting wage of fifty cents a day. Moving with the firm to Canton, where the business thrived, Lewis became a partner and founded a Methodist Sunday school, still a new concept in America. Here, he introduced a graded system of classes paralleling the children's levels in normal school, so that they would not become bored with the instruction—another innovation, this time pedagogical. A resourceful and competitive young man, Miller became known for his favorite aphorism, "Let's see how we can beat ourselves!"

Miller followed his own exhortation, and ever the empiricist, took to walking methodically behind the mowers observing them as they cut swaths through the grain. Well before he was thirty he had introduced into the machine its most revolutionary improvement since Cyrus McCormack and Obed Hussey first overwhelmed the market. He noticed that the knife-bar, which extended out perpendicular to the two-wheeled, cast iron cart in which the farmer sat, was always rigid, and therefore of necessity slanted downward *away* from the machine as it moved forward. Thus, the grain farthest away was cut shorter than that closer in. With hay-mowing, this defect caused much of the harvest to be lost. In response Miller devised a floating cutter bar. He built a hinged joint to connect the knives with the side of the machine so that the bar would follow the contours of the ground, "accommodating to low patches or down grain." As an improvisation on this innovative touch, he constructed a lever that enabled the farmer to raise the entire bar vertically to avoid stumps, boulders, "rough and boggy ground," and the like while continuing to move ahead, instead of stopping entirely or having to back up. And finally, in a simple but brilliant move, Miller shifted the bar assembly forward (front cut), away from its dangerous position directly to the side of the driver, so that if by chance one of the wheels went over a bump and the farmer was thrown from his narrow perch— heretofore the cause of hundreds of accidents a year, many fatal—he would fall well behind the moving blades, and avoid injury.

Lewis Miller's Buckeye Mower and Reaper, as it was now called, was a great success. Facilities in Canton and Greentown were no longer adequate to meet the demand. Akron, twenty-five miles to the north, with its strategic Erie Railroad connection, was the perfect place to expand. In 1863, Aultman, Miller & Company, with Lewis Miller taking the preferred title of superintendent to emphasize his affinity with the working men, produced eight thousand Buckeye mowing machines and five hundred threshing machines. By the mid-eighties, the huge Akron plant was manufacturing more than twenty-five thousand harvesting machines annually. "The old and tiresome methods of securing grass and grain crops by scythe and sickle are clearly within the memory of the middle-aged," said Miller & Co.'s opulently produced trade catalogue, "and it seems incredible that in so short a time the Mower and Reaper have become a necessity with the American husbandman. It is not claimed for the Buckeye that it was the first mowing machine brought into use. The history of mowing machines shows that many others were introduced prior to it, and yet we claim for the Buckeye the position of the *pioneer machine* [italics in original]."

To their established reaper, in 1865, Miller added a table rake, and a binder, automatically landing sheaved grain tied with twine and ready to be gathered in back and to the left of the horses so that they would not damage the grain by stepping on it when they came around the field in subsequent cycles. Improving upon its competitor, the reel rake, Miller's apparatus could be stopped and started at will by the driver. Miller then went yet further, making the cutter bar and the table rake fully detachable and interchangeable, so that the farmer could shift modularly from mowing hay to reaping grain at will. All in all, over a span of nearly half a century, Lewis Miller registered ninety-two patents for his Buckeye machines.

Lewis Miller was a rich inventor when he moved to Akron in 1863 with his "strong and stern" Scotch-Irish wife, Mary—native of Plainfield by way of Litchfield, Connecticut—and their five children—six more would follow in the next dozen years. Even though Lewis was a commuter, forced by the pressures of market demand to keep in touch with both Buckeye plants and drive the Akron-Canton route every day, the Millers remained a close, religiously observant family; Lewis still found time to stop off at the old family homestead in Greentown to have a quick drink with his father in the stone house still standing today by the stream where he grew up. In Akron, Miller immediately became a presence in the spiritual and commercial life of the city. Always a proponent of close Bible study, never ceasing to strive to be "as good a Methodist as [he] could be," he expanded the Akron First Methodist Church Sunday school curriculum to embrace love of the open air and the deep woods, "the mystical ways of nature," creating a revolutionary semicircular interior plan for the new school building and adding a comprehensive music program, both prototypes for Chautauqua's amphitheater; he eventually rose to become superintendent of the school. On Sundays, Miller would appear at the school smiling, with armfuls of flowers for the children. With his unquenchable passion for popular education, Miller was also president of the Akron board of education and was active on the board of directors of Mount Union College, the first college in America to give equal educational rights and privileges to women and men, and also the first to establish a four-year course of study.

He bought twenty-five acres of land on a knoll overlooking downtown; Oak Place, as the Miller estate was called, became renowned for its endless hospitality. The patriarch made it a ritual to bring a guest, or two, or three, home to dinner every evening, including Sundays, for full-course meals followed by extended family colloquies on the important issues of the day. And into this wealthy, crowded, purposeful Christian home of high intellectual and social expectations, Mina Miller was born.

Their business in New Orleans concluded, Edison, with Dot and the Gillilands, proceeded by train to St. Augustine, as planned. Northeastern Florida was familiar to Edison; his final holiday with Mary, only a year earlier, had been to nearby Palatka on the St. Johns River. Unfortunately, the weather in St. Augustine was chill and dismal. In desperate search of the sun, the intrepid party forged on to the less traveled southwestern coast, by train to Cedar Keys and then by yacht through the Gulf of Mexico, reaching the telegraph terminus of Punta Rassa in early March. The story goes that the manager of the little station, "while sitting on the veranda smoking cigars" with his fellow "colleague of the Key" (and doubtlessly trading old tales with Edison, who was in his favorite element), told the inventor about a picturesque village seventeen miles up the Caloosahatchee River from the coast called Fort Myers, scene of the surrender of Chief Billy Bowlegs during the Seminole War. Ever curious, Edison took the trip.

OAK PLACE, THE MILLER FAMILY HOME IN AKRON

He found lodging at the twenty-room Keystone Hotel (thus named by its builders, who hailed from Pennsylvania) and around it a rustic, old world cowboy town of unpaved streets with a population of three hundred people, complete with a church, livery stable, dry goods store, drug store, and even a spanking-new newspaper, the Fort Myers *Press*. Diminutive though it was in size, Fort Myers professed no small measure of local frontier pride, the town fathers considering it "The Eden of Florida . . . the Only True Sanitarium of the Occidental Hemisphere, Equalling if not surpassing the Bay of Naples in grandeur of view and health giving properties!" And indeed, with its mild air, infinite variety of tropical vegetation, pineapples, coconuts, pines, palms, sugar cane, and bamboo; grazing cattle, forests abundant with wild turkey and deer ("a veritable Hunters' Elysium"), fertile soil, and promise of solitude, Fort Myers seemed at once ideal. Just at the bend of the river a mile out of town, a thirteen-acre tract of land was up for sale. Edison signed to buy the property for $2,750 within twenty-four hours of his arrival. Part of the land would belong to Gilliland, and the two friends (Edison likened their intimate relationship to that of Damon and Pythias) planned to build a pier extending out into the river, and to situate matched, neighboring prefabricated houses of Maine lumber within view of the water. Across the road that ran through Edison's land a laboratory housing a forty-horsepower dynamo, as well as a workers' residence, would soon be constructed; could the electric light be far behind? Even in vacation times of supposed respite, the year-round inventive urge could never be stilled.

The Caloosahatchie—derived from the Choctaw language for "strong and black," domain of the Calos Indians who had until the eighteenth century dominated the surrounding land—extended all the way from giant Lake Okeechobee to the Gulf. Like the blue waters of the frigid St. Clair in Edison's Port Huron youth, the river exercised a powerful allure, at once pulsing with steam transport, freight, and commerce, and symbol of natural forces, a combination Edison always favored.

If we rely upon Dot's memories, Edison upon his return north to New York was engaged in an overt, systematic campaign to regain domestic stability by finding a house (not an apartment) and a wife to be a mother for his three children. Therefore, he asked Mrs. Gilliland to help him do so by arranging for a series of young women to assemble at Woodside Villa, the Gilliland seashore cottage outside Boston, while he was a guest there, so that he could survey social prospects at his leisure. However, correspondence from Ezra Gilliland indicates that he and Edison were involved in new business ventures together during the spring; he implored Edison to "come to Boston as soon as convenient" and stay with him, combining work with play. By the end of June, Edison was writing from Woodside Villa to Sammy Insull, who was likewise seeking female companionship, "Could you come over here [Boston] at Gill's there is lots of pretty girls here."

For a scant nine days during this visit, from July 12 to 21, 1885, Thomas Edison kept a formal diary, the only such journal by him that has survived. The manuscript is impeccably written in his trademark hand, as if it had been copied over from a rougher version; be that as it may, we can still discern he had renewed acquaintance with Mina Miller at the time of his letter to Insull, before the diary's beginning date. "The Belle of Akron" had returned to Boston for the spring term, and was staying on with the Gillilands for several weeks prior to her traditional midsummer move to the family cottage at Chautauqua. "Awakened at 5:15 a.m.," Edison writes, in the first entry. "Thought of Mina, Daisy, and Mamma G[illiland]." Just before going to bed that night, he resumes the obsession, albeit somewhat ambivalently. "I will shut my eyes and imagine a terraced abyss, each terrace occupied by a beautiful maiden . . . Worked my imagination for a supply of maidens. Only saw Mina, Daisy and Mamma. Scheme busted—sleep."

The major subtext of Edison's floridly written diary is that he is at odds with his adolescent daughter about his romantic intentions. "I picked out the stepmother I wanted right away," Marion wrote in *her* version of their soap opera summer at the seashore, "more because she was a blonde like my Mother than for any other reason." This fair-haired damsel "with hair like Andromache" was Louise Igoe of Indianapolis, who would within two years marry one of Mina's older brothers, Robert. "I had the impression, however," Marion conceded, "that my Father was in love with the Ohio girl, Mina Miller, whom he had previously met." Indeed, Edison's love was more akin to distraction. Spending July 15 in Boston on business with Gilliland, Edison "saw a lady who looked like Mina. Got thinking about Mina and came near to being run over by a street car. If Mina interferes much more will have to take out an accident policy." There was danger lurking in the turbulence of Dot's feelings as well. Edison helplessly "uses Mina as a sort of yardstick for measuring perfection" and watches in dismay as his daughter in her thwarted affections pouts like "Lucretia Borgia." To Marion, Mina was "too young to be a mother to me but too old to be a chum."

No analogy seems too farfetched to the lovesick Edison. One evening, over music and after-dinner conversation, "Dot asked how books went in the mail. Damon [Gilliland] said as second class mail matter. I said Damon and I would go at this rating—suggested that Mina would have to pay full postage. Damon thought she should be registered." In the idealized sense, the object of Edison's admiration (at one point he baroquely refers to Mina as "the Chataquain Paragon of Perfection") becomes revered to the level of fantasy. He "slept so sound that even Mina didn't bother me, as it would stagger the mind of Raphael in a dream to imagine being comparable to the Maid of Chautauqua, so I must have slept very sound." Finally, on August 11, Mina and her ardent pursuer were reunited in the stately pine groves at Fair Point, the quiet promontory by Chautauqua Lake in the rural southwestern corner of New York State.

By the time of Thomas Edison's arrival on the scene, the celebrated Chautauqua Assembly—spiritual haven for America's prosperous middle class—had been going strong for more than a decade. The brainchild of Mina Miller's father and his Methodist partner, Reverend Vincent, the debate, as with every great invention, persisted over who actually came up with the idea first.

Alabama-born John Vincent began his formal career in the church at the City Mission in Newark after having spent a brief year as a teacher and then hard time as a circuit-rider in the rural area around Lewisburg, Pennsylvania. Active pastoralism and the lure of the open road won out over John's ardent wish to complete his college education, a thwarted goal that only made him a more devoted believer in lifelong learning, bitterly regretting the chance he had missed: "the dream of higher education," as he called it, "the gateway to the wide, wide world . . . The earliest Methodists were college men," Vincent wrote, "believers, as John Wesley was, in the broadest and most thorough education." Even as Lewis Miller was assuming the role of superintendent of the Canton Sunday school, John Vincent had begun teaching in Irvington, New Jersey. In the late 1850s, he was transferred to Joliet, Illinois, where he refined the idea of what he called "Palestine classes," introducing the geography of the Holy Land firmly into the foundations of Sunday school lessons. At the end of the Civil War, he was appointed to the ministry of Trinity Church in Chicago. There, Vincent continued his work on the improvement of Sunday schools, and created a central institute for coordinating Methodist curriculum materials and uniform standards worldwide.

In the course of his travels to important Methodist centers of learning, he naturally enough passed through Akron, where in 1868 he met Lewis Miller for the first time and the two began their lengthy debate over the best way to train teachers and enhance the value of the Sunday school environment. Miller, the wealthy, modest layman, was a democratic, cultural generalist who was convinced, for example, that the study of such subjects as music and belles lettres was not out of place in the church. Vincent, the eloquent platform speaker and entrenched minister, considered religious instruction as above all else a natural pathway to membership in the church and a better understanding of the Bible's teachings. Vincent rose to become the chief Methodist pedagogue and publisher, general agent and secretary of the Sunday School Union, and editor of the influential *Sunday School Journal*. Working out of New York City and Plainfield, New Jersey, he attended a phonograph demonstration, visited Menlo Park—and made the initial invitation to Edison to visit Chautauqua.

Meanwhile, by 1872 Lewis Miller, among his myriad of other civic and religious activities, had joined the board of directors of the Chautauqua Lake Camp Meeting Association. He was well aware of the pastoral charms of the place, and its excellent location midway between New York City and Chicago, eight miles south of Lake Erie. Small revivalist groups in spring and summer evangelistic camp meetings had been convening in tents at Fair Point for about a year. "The idea [for an expanded meeting program] came to me as an invention comes to a person," Miller later recalled; in fact, it arose during a conversation with two sisters who had been teachers in his Canton Sunday school. Miller envisioned a way to synthesize his all-embracing view of the mission of the Sunday school with his love of the natural world, "a Convention, or an Assembly, where all persons interested in Sunday School work could meet—say, for three weeks—for Bible study, normal class work and general instruction. There could be musical entertainment, lectures, and pleasant recreation, all to be mingled with appropriate devotional exercises. The next time I see John Vincent," Miller ebulliently promised Kate and Lydia Patterson, "I shall take this thing up with him! . . . There would be no end to the possibilities of the development of such a place!"

Seven-year-old Mina Miller was kept awake listening to the vehement voices of Lewis Miller and John Vincent (by this time an exalted bishop) extending well into the night, as the ideological arguments in the library at Oak Place extended intermittently over the winter months of 1872–73.

Vincent's personal approach came in the form of a novel request to use the Akron Methodist Episcopal Church, one of the largest congregations in the country, and by this time nationally renowned because of Miller's pioneering work, as the springboard model site for a wintertime Bible normal school, a prolonged course of study for the development of the art of Bible teaching, for Sunday school teachers drawn from congregations across the land. Assuming this experiment was successful, these enrichment courses would become a moveable feast, going from city to city, year by year. But Miller wanted to "take it to the woods," transcending the boundaries of the at-times over-heated camp meetings with their chants and testifying, reaching out to make "a whole-some, Christian educational re-sort" that would extend a wel-come to public school teachers as well—and even invite *women* to deliver speeches, an idea from which Vincent recoiled with particular skepticism. Such a

THE GROVES OF CHAUTAUQUA

gathering in a grove, the bishop believed, with emphasis on extraneous sub-jects beyond the Good Book, would only serve to encourage attendance by "the rowdy element . . . and hysterical half-wits." Vincent strenuously and suspiciously held out against what sounded to him like "emotional extrava-gance," too much attention to the vulgar and sinful world of daily com-merce and too far from the austere educational scene he had always favored.

In order for the Chautauqua experiment to succeed, Miller needed Vincent's establishment imprimatur, educational influence, and eloquence as a religious leader; Vincent, in turn, needed Miller's deep pockets (for there would undoubtedly be deficits in the early years of Chautauqua, de-spite the modest entry fee), pragmatic authority, and vision. To sell his ad-versarial friend on the idea, Miller took Vincent to Fair Point by steamer late in the summer after the camp meeting had closed for the season. It was indeed a breathtaking site, several hundred feet above sea level, the air crisp and invigorating, the gentle hills adorned with bright flowers sloping down, dotted with "grand old forest trees," their "leafy temple" festooned with tangled vines, pressing close to the lake sparkling with "glorious sunsets and bright dawns."

The key to the settlement between Lewis Miller and John Vincent, allowing the first two-week Chautauqua Assembly to go forward in August 1874, was to set a structure. Their common mission was to study the Word of God, *and* His manifold "works" in nature, history, and mind. The Sabbath was strictly observed. Liquor, card-playing, and dancing were outlawed. Exhortation and calls to sinners to repent had no place on the program; the disappointed evangelists quit in a huff. Voluntary and spontaneous gatherings were not permitted. Only scheduled speakers were allowed to sit on the platform. In fact, no meetings could take place without Vincent's permission.

Thanks to Vincent's brilliant publicity machine—the Methodist *Sunday School Journal,* which he edited and published, had a circulation of well over one hundred thousand readers—more than four thousand participants a day converged on the eighty-acre site from twenty-five states during the first several seasons. "The Trumpet of the gospel has been so well blown," proclaimed one enthusiastic convert, "that the walls of bigotry have all come down . . . Why not have *all* our churches and denominations take a summer airing? The breath of the pinewoods or a wrestle with the waters would put an end to everything like a morbid religion. One reason why the apostles had such a healthy theology was because they went a-fishing."

Every year, Chautauqua grew from strength to strength: curtained tents lined up by the lake gave way to five hundred wooden cottages; shops sprang up as "the milk-man, ice-man, and newsboy made their morning rounds"; the great, colonnaded Hotel Athenaeum was built, its genteel, broad front porch adorned with potted ferns a favorite spot for visitors to idle in wicker rocking chairs and scan the white sails flitting past; Bishop Vincent built his beloved, topographically correct scale model of the Holy Land by the lakeside (in the mind's eye transformed into the Mediterranean Sea), where children played tag when school was not in session; benches set up along dirt aisles among the trees for sermon audiences were replaced by a magnificent amphitheater and gravel paths; Fourth of July fireworks reflected gaily on the lake to the tune of evening brass concerts; Hebrew classes were introduced to the burgeoning curriculum, soon followed by French, German, Latin, and Greek; a scientific congress was convened during the 1876 Centennial year; classes in photography and calisthenics were added to the mixture; a Scientific and Literary Circle was founded, precursor of the university extension movement, presenting a four-year "life-direction" reading and correspondence course for home study, with a full-fledged curriculum; there was even room for home economics courses on health, ventilation, and cooking.

By Edison's visit, Chautauqua's grounds had grown by fifty more acres. The steamboat landing and ornate pier bordered upon Miller Park, the grove renamed in tribute to one of the two founding fathers, where the original tents had long ago been pitched. There were stables, wood yards, highways, and paved avenues neatly laid out. The two-week season had stretched to seven weeks extending all the way to the end of August. And the national movement had begun: more than three dozen other Chautauquas had sprung up across the land, from Siloam Springs, Arkansas, to Canby, Oregon. Small-town Protestantism asserted itself into direct confluence with social reform; common, everyday domestic mores and devout Christianity found alliance in Chautauqua during a crucible time in post–Civil War mainstream American culture, in a place where respectability could act as a workable forum for real ideas, even as the society at large tried to absorb the repeated hammering of technological and economic change. Chautauqua was a safe haven, a campus where the plain citizenry celebrated their preferred way of life and their value of knowledge as every man's refuge.

"The world-renowned inventive genius," as the Chautauqua *Assembly Herald* anointed him, was duly impressed. While not a religious man in the conventional sense, although he did value the spiritual, even otherworldly aspect of things, Edison at this juncture in his life had social pretensions, even if he lacked the wherewithal to articulate them. "His conceptions of the place," he conceded to the *Herald*, "which were too narrow, have received a very great expansion. . . . As a benefactor of his race, he is welcome to Chautauqua."

Welcome indeed, but he did not see very much of Mina Miller. Despite its woodland aura, Chautauqua was, after all, about the church, and not a very suitable place for romance to flourish, let alone exist properly. The place was crowded, noisy, full of hubbub and constant activity. Bishop Vincent's son, George, sulking in the background, was still purportedly Mina's official betrothed. It is not clear when or if she formally ended the engagement; however, we do know from studying George Vincent's diary that by early September, he had left on the Cunard *Auramia* for an extended tour of England and France with his mother, effectively out of the picture while Edison pressed his advantage and, for his part, managed to convince Mina to take a chaperoned rowboat ride on the lake with him, but little else.

Depending upon whose version of events we wish to accept, at the conclusion of the season, Edison asked the Millers if Mina might drive with him and the Gillilands, and Dot, of course, through the White Mountains of New Hampshire; or, conversely, the Millers invited Edison to accompany them on their traditional Vermont excursion. Marion insists in her memoirs that while Edison was sitting next to Mina in the horse-drawn carriage they shared, he tapped a marriage proposal in Morse Code onto the hand of The Belle of Akron, to which she tapped "yes," in reply.

The more likely scenario is as follows: that Edison was in relentless pursuit of Mina there can be no doubt. From the time of their initial meeting in New Orleans and for the ensuing half-year, they were in sporadic correspondence, although none of the letters, not surprisingly, has survived. After the Vermont trip, Edison returned to New York City and Mina went back to Akron for a couple of weeks prior to the beginning of the new school year. On September 17, 1885, the ever watchful Gilliland sent Edison a quick letter describing Mina's plans for her return to Boston: "I understand she [Mina] is to leave there [Akron] next Tuesday, that would bring her here [Boston] Wednesday, 3 p.m." Next to the days of the week in Gilliland's note, Edison, planning his next move, has written the corresponding dates, "9/22" and "9/23." The plot thickens with the recent discovery in Edison's personal library of a book ßcalled *Character*, by Samuel Smiles, inscribed "To Miss Mina Miller, Boston, *September 24, 1885.*" This slim gift volume, a manual of prescribed Victorian behavior, includes up-standing, caution-ridden chapters on The Influence of Character, Home Power, Companionship and Example, Work, Courage, Self-Control, The Discipline of Experience, and Companionship in Marriage—in short, advice to the uninitiated.

The following week of September 30 finds our hero back at his famil-iar laboratory desk in the Bergmann & Company building on Avenue B, penning an epistolary appeal to Lewis Miller, which reveals that the smitten Edison had properly proposed to Mina and that she told him to write di-rectly to the presiding eminence of the Miller clan: "My dear Sir: Some months ago, as you are aware, I was introduced to your daughter, Miss Mina. The friendship which ensued became admiration as I began to appre-ciate her gentleness and grace of manner, and her beauty and strength of mind.

"That admiration has on my part ripend [sic] into love, and I have asked her to become my wife. She has referred me to you, and our engage-ment needs but for its confirmation your consent.

"I trust you will not accuse me of egotism when I say that my life and history and standing are so well known as to call for no statement concern-ing myself. My reputation is so far made that I recognize I must be judged by it for good or ill."

As we have seen before, even in his earlier, preachy letters to financial supporters, behind Edison's ornate rhetoric invariably resides an in-domitable, implicitly self-confident message. Now that he had become in-volved in the competitive chase for this woman, he would not rest until he had achieved his goal. Edison had convinced himself that as noble and lovely as Mina was, he deserved her, because of what he had come to represent. And Mina Miller, in turn, aside from her unambiguous desirability, stood for a cultured, respectable world Edison heretofore experienced only through its trappings, intermittently and half-consciously. The time had come for her to permit him to gain rightful entry into her world—so that he could share it without question.

CHAPTER

14

To Lewis Miller, the match made perfect sense. His daughter's suitor was a kindred spirit, a man of commerce and imagination; and, unlike George Vincent, Edison had the wherewithal to provide for Mina in the proper fashion. However, Mrs. Gilliland feared that her dear friend Mary Valinda Miller thought that Mina had been unduly influenced by the elaborate matchmaking arrangements made to bring the young woman into Mr. Edison's acquaintance, which, in fact, Mary and her oldest surviving child, Jane Eliza, did believe was the case.

Ever since the death by sudden illness at age sixteen of the Millers' firstborn, Eva Lucy, Mina had turned to Jane for solace and support. As Mary was so far away and by nature more quiet and restrained, Jane (familiarly called "Jennie"), who was unmarried at thirty and living in New York City, was Mina's deepest confidante and surrogate mother. From Mina's early adolescence, Jennie, who lived and traveled abroad for two years after graduating from high school, had been a constant and vigilant guiding influence, urging her sister to "improve each moment" toward becoming an "accomplished young lady," to perfect her penmanship, embroidery, sketching, piano playing, and French; never to go barefoot or associate with "rude girls"; keep her fingernails clean, her hair long and in order, and always wear a clean collar and cuffs. By twelve years old a somewhat stout child, tending at times to gain weight, Mina remained self-conscious about her figure, but

Jennie reassured her that it was the inner, cultivated person that should matter most to any man who eventually came her way.

Jennie found "Mr. Edison" intimidating at first meeting and was certain he did not like her. Jennie frankly told Mina that she thought Edison was looking for "a child" rather than a mature wife, to which he replied, "You are not a child by any means, on the contrary a fully developed woman and with some experience superior to them all." Edison wanted to get married in February, but the protective Jennie, fearful lest her sister jump in too hastily and make a mistake she would regret, did not understand what all the rush was about. Why couldn't Edison permit his love to become more seasoned, and wait a year or six months before marriage? But then, she did not know the man very well, or she would never have posed that question.

A person inherently tough on herself, like many raised to be achievers, and naturally inclined to periods of self-doubt, Mina fell into what she called a "blue" period in the aftermath of her father's quick consent. She was worried about the reaction of the Vincents to her decision. The two families had been so close for so long, and Mina did not want to appear responsible for any rift between her father and the bishop that might arise out of her breakup with George. She worried about the kind of relationship she would be able to establish with her soon to be stepdaughter, Dot, after all, a teenager like herself, who had received several conciliatory letters from Mina but, according to her father, "seem[ed] to be at a loss to know what to say." At times Mina became confused about whether she loved Edison for *himself,* or because *he* loved *her.* He leapt feetfirst into a Miltonic riposte, reassuring Mina that "after February, the world will look changed. Comus will be the presiding deity, and all melancholy things whatsoever shall retire from the foci of those divine eyes of yours."

Toward the middle of December, Mina joined Jennie in New York City and they assembled the last details of her trousseau—sheets, towels, underclothes, night dresses, and fourteen suits trimmed with German lace—struggling to keep the cost below $1,500, but requiring, of course, only the finest things. Mina talked with Edison about "housekeeping," and they surveyed several properties, within the city and beyond. Just before Christmas, Edison was invited to Akron to spend his first visit there and to meet the rest of the Millers.

The Buckeye Mower and Reaper Company's sprawling, L-shaped fifteen-acre complex on the east side of the Akron and Great Western railroad tracks at Center Street and Middlebury Road, with an operating surplus in excess of $1.5 million, and the largest payroll of any company in town, was not by any means the only important corporate presence in Akron in 1885. Ferdinand Schumacher's rolled oats fed the Union army during the Civil War, and Quaker Oats became a giant in the cereal industry. George Barber's Diamond Match Company dominated its field, and would for decades. Benjamin Franklin Goodrich established the only rubber company west of the Appalachian Mountains, specializing at first in fire hoses, and ten years later making a wise move into bicycle tires. The J. F. Seiberling Company emerged as another contender in the farm machinery market, spawning all manner of suppliers—knives for cutting bars, chain drives, binder twine, and castings. Seiberling was also a player in the straw-board (composition board) and rake-manufacturing sectors. The Akron Rolling Mills—eventually purchased and relocated by Lewis Miller—operated twenty-four hours a day, providing iron from ore mined locally in neighboring townships for the fabrication of mowers and reapers. The Akron Boiler Works, expanding to become the McNeil Machine and Engineering Company, eventually supplied machinery for the rubber industry, Akron's greatest final claim to fame. There were more than enough capitalists in Akron for Edison to feel at home.

The city's name was aptly derived from the Greek *akros,* "a high place." Its home county of Summit sat at the loftiest point along the Ohio Canal. And immediately to the west of downtown Akron and overlooking the canal and the city, epitomizing Victorian Akron at its most exalted and ornate, the Miller mansion was situated.

Picked up at the train station in the family sleigh drawn by two high-stepping plumed mares, tucked under thick robes and blankets, crossing the canal bridge to the accompaniment of tiny bells, ascending Ash Street Hill along a curved driveway to its very crest, Thomas Edison at first would not have been able to discern the three-story gray chalet hidden in the distance behind dense copses of trees. Only as he stepped from the sleigh, kept in a stable large enough to be suitable for a Rhine castle, was Edison finally able to survey the scene. Under Miller's direction, the estate had been laid out with an eye to aesthetic harmony reflecting his abiding interest in horticulture. Gravel walks proceeded past greenhouses (his pride and joy, where Miller cultivated exotic crotons and philodendrons), a gazebo, and peacock pens. Across a spacious, snow-shrouded lawn past two granite lions guarding its stained-glass front door, the house was in all senses immaculate and generous. A restorative cup of coffee before the coal fire in the library, the center of the home, adorned with fresh, hothouse flowers, was the first order of business. Across the hall was the parlor with a piano where Mina and her sisters played and studied their music; next to the parlor the dining room, where breakfast was served every morning at 7:00 a.m. sharp and morning prayers were de rigueur; then, the kitchen, connected to basement baking ovens by dumbwaiters of Miller's devising. In earlier years, when more of their brood were still at home, Lewis and Mary Valinda kept their bedroom on the first floor hallway, to say good night to each of the children as they went up the circular staircase to second floor bedrooms. On the third floor was a huge attic recreation room with miniature railroad, printing press, and gymnasium, and the shared bedroom of the two pillow-fighting, inseparable youngest sons.

At dinner with Thomas Edison, these two, the "irrepressible and roguish" twelve-year-old John Vincent (named for the bishop) and ten-year-old (nearly eleven) Theodore Westwood, as was the family custom, sat to their dimunitive mother's immediate left and right. Mary called them her "Petty Boys." Within a few years, they would be off to St. Paul's in Concord, New Hampshire, for prep school, then Yale, and then into battle during the Spanish-American War, always together. Big sister Jane, her mother's backbone, was "the family authority on the proper thing to do . . . and a stern upholder of social conventions." Twenty-nine-year-old Ira Mandeville Miller was next in line. He had taken a job at the Buckeye plant working for his father in the counting-room, and was instrumental in establishing the first street-railway system and commercial electric light works in Akron; Edward Burkett at twenty-six had graduated from Ohio Wesleyan and then Stevens Institute in Hoboken, and likewise worked for his father. Robert

THE EIGHT OLDEST
MILLER CHILDREN, 1870;
MINA THIRD FROM LEFT

Anderson helped manage the Aultman & Company plant in Canton, as assistant to his uncle Jacob; Lewis Alexander at twenty-two worked at the Aultman, Miller & Company twine mill and dabbled in real estate. Next in line came Mina, and then Mary Emily, who though still at school in Akron, would be off east to Wellesley in the coming year; Grace, the youngest sister, was home for the holidays from Miss Porter's School in Farmington, Connecticut. Of all the Miller children, Mina was the first to become engaged, and the first to be married.

John and Theodore were enthralled by Edison. They stuck close to the famous inventor and did not allow him very much time alone with their sister. In turn, he seems to have lavished more attention upon them than he ever had upon his own two sons, just a few years younger. As soon as Edison returned to New York, and just in time for Christmas, via express mail, he shipped the Miller boys a telescope "for viewing the rings of Saturn," as well as a pair of telegraph instruments with battery and wire, and an electric shocking coil, accompanied by a battery cell, a supply of bichromate of potash with which to make the battery fluid, and an exhaustive six-page letter of irreverent instructions for its use: "The [generator] wheel should be turned about 200 times per minute for [shocking] a black cat and 199 ½ for a cat with a sanguine temperament. . . . Be sure in pouring the sulphuric acid that you do not let any of it spatter into your eyes. I think your Brother Robert understands batteries and he will doubtlessly show you how to make the fluid, etc."

Although Edison's direct influence cannot be traced, he may have also motivated the Miller boys in another direction. Within a year of Edison's visit, John and Theodore began to publish their own weekly (later monthly, and finally quarterly) newspaper, *The Jumbo.* Even as Edison had named his isolated lighting dynamo after the celebrated P. T. Barnum elephant, the boys took the name in tribute to the great beast's memory; in 1885, the unfortunate pachyderm had been struck and killed by a freight train as he climbed into his private circus railroad car. Like the Grand Trunk *Herald,* the Millers' newspaper suffered in its early numbers from problems of typography and spelling, no doubt caused by having to hand-set the type upside down and backwards: "Mr. Lewis Miller is intending to start for Euroqe [*sic*] the 10th. of this month. Bishoq Foster was visiting Mr. Edison's last Sonday [*sic*] . . . Mr. Lewis Miller has spent this week in Chicago attending a pattent [*sic*] suit. . . . The French Rayalists [*sic*] deny that they are in league with General Boulanger."

However, once the lads got the rhythm of it, the newspaper became a faithful chronicle at the bargain subscription price of $1.00 per year, primarily detailing the myriad comings and goings of the extended Miller family, as well as presenting Local Notes, national and international news, cartoons, riddles, and jokes: "It is painful to see family relics sold at auction, but the most painful thing under the hammer is generally your thumb-nail . . . 'Ain't that a lovely critter, John,' said Eliza, as they stopped opposite the leopard's cage at the Zoo. 'Well, yes,' said John, 'but he's dreadfully freckled, ain't he?' . . . The height of curiosity: To kill oneself in order to see what is in the other world," and so on. Edison was a faithful subscriber to *The Jumbo* from its inception, and saved all of the issues sent to him in New Jersey, having Brentano's bookshop bind them in red moroccan leather.

A spacious, proper home suitable for raising a family and for entertaining, set upon broad grounds, far enough from metropolitan travail to provide tranquillity, yet close enough for the appropriate degree of social involvement—such were Mina Miller's accustomed circumstances growing up in Akron, and so it is not surprising that she would desire much the same atmosphere for her life as a married woman. And during their wintertime house search in New York and environs, the couple had toured just such a spot—Llewellyn Park, America's first exclusive garden suburb, 420 landscaped acres twelve miles west of the City in the hills of West Orange, New Jersey, little more than an hour by train and ferry from Manhattan. Laid out between 1853 and 1857 by Llewellyn S. Haskell, nature lover, pharmaceutical tycoon, Fourierist, and Swedenborgian mystic, and Alexander Jackson Davis, eclectic architect, poetry fan, Gothic and Classical revivalist par excellence—this "middle landscape" was, like Chautauqua, an absolute haven for the popular, balanced nineteenth-century ideal of cultivated primevalism, brought within the acquisitive purview of the well-to-do. Six hundred feet above sea level, Llewellyn Park commanded an impressive vista eastward, over the Oranges to Newark and beyond, past the dun-colored Jersey marshlands to the Hudson River and New York. From the heights of this enclave, a contemporary observer rhapsodized, "[t]he eye measures a strip of clean and distinct landscape, a hundred miles long and thirty wide . . . to stand on the crest of it, looking down into the great, green depths below, and then looking up into the illimitable blue above, is to be at the midway point between heaven and earth, between life and death." Llewellyn Haskell was a great admirer of the works of Frederick Law Olmsted and Calvert Vaux; within his residential park, as in Central Park, Haskell set aside a purposefully untouched and uncultivated wildness area, a romantic, dark, and deep ravine with jagged, rural contours shaped by a meandering brook and miniature waterfall, shrouded under evergreens, oaks, beeches, and chestnuts. This cool, artfully rustic glen was offset by landscaped villa sites insularly separated by fieldstone gate lodges, thick rhododendron rows in lieu of fences, and ample, sloping lawns for the "country homes of city people."

Conveniently, one of the largest estates in Llewellyn Park was on the real estate market in early 1886. Built with embezzled funds during 1880–82 by Henry C. Pedder, a since-deposed executive at Arnold Constable department store, "Glenmont" was available completely furnished, on thirteen and a half acres of land, for the cut-rate liquidation price of $125,000. Edison put down 60 percent in straight cash and took out a mortgage on the rest. On January 20, 1886, the house was his. A massive multicolored-timbered and scallop-shingled conglomeration of steeply pitched gables, paired windows with striped awnings, and balustered porches curved like ship's prows out toward manicured gardens, Glenmont had been designed by the renowned apostle of the Queen Anne style, architect Henry Hudson Holly. Here was "the natural building style for America . . . an architecture of our own . . . worked out in an artistic and natural form . . . expressing real domestic needs of which it is the outcome," he announced. In wholesale rejection of "slavish conformity" to Gothic, and advocating in its stead utilitarian and decorative purposes at the same instant, Queen Anne was honest and true. Every detail had a functional as well as determinedly eye-catching appeal. With its "free-classic lines . . . smooth, workmanlike roofs, galleries, great chimneys . . . large habitable halls well opened into adjoining living rooms, stairs always very carefully contrived," and other seemingly inexhaustible details (every piece of furniture was likewise calculated into its master plan), Glenmont was a veritable Queen Anne palace, a high Victorian tribute to the raging vogue of vernacular country estates. More than a house for Thomas Edison and his bride to be, it was the perfect statement, securely vast enough to encompass the lifestyle and constant remodeling and redecorating Mina would come to impose over the next half-century.

As the designer himself declared, "The mansion, as compared with the cottage, is like a full-grown man compared with a child." The luxuriously furnished entrance hall, illuminated by a glass and silver chandelier and adorned in oak wainscoting, featured built-in settee and armchairs, Japanese vases (believed by the architect to be more suitable than a mundane rack for depositing umbrellas), and landscapes on the walls. The adjacent library, which Holly insisted should reflect "the master-mind of the household," was made entirely of mahogany bookcases, their doors lined with box-pleated silk, while its walls and ceilings were stenciled decoratively; the mantel, in true Holly aesthetic, had little compartments for ornamental display. "A house without books is like a room without windows," the architect wrote. The drawing room was furnished entirely in rosewood and adorned with popular oil paintings of the period. Lambrequins (valances) of maroon and old gold satin hung on the walls, and the parquet flooring was protected by Turkish rugs. The conservatory, approached via french doors from the reception room, was lush with tropical and subtropical plants, arching palms, and ferns, and furnished simply in willow and wicker. Across the first floor on the west side of the house was the dining room, covered in peacock-blue wallpaper on dull yellow ground with Persian fig-urings; one walked through this twenty-three-foot-long formal scene into the den. Around to the rear of the first floor, of course, were the servants' dining room and the kitchen. "Except in houses of very small dimensions," said Holly, "we consider the back staircase indispensable . . . it keeps the servants retired."

The second floor was more informal domestically, but still opulent. There would be soon another living room here, extending over the porte cochere, which became Thomas Edison's favorite retreat in the evening, his "place of quiet repose" where he sat at his "thought bench"; it was more comfortably used as a family room, for game-playing—especially Parcheesi—reading, needlework, or quiet letter-writing. A few steps away was the master bedroom with its canopied mahogany, four-poster, and Ped-der's favorite little engraving, "Girl Gathering Daisies," by A. H. Bellows and Arthur Small, which the Edisons kept close by in the early years. When Mina's mother and father visited, they would stay in the south bedroom, and when Edison came home very late at night (or in the early morning hours) from the laboratory, he would settle in there as well, rather than dis-turb his wife. Birch and birdseye maple were the woods of choice in the south bedroom. Three additional bedrooms would serve as nurseries and guest accommodations in the decades ahead.

The lofty third floor was set aside predominantly for the children's and governesses' accommodations, and for domestic staff (of which there were no less than ten in help during the Edison heyday): three servants' bedrooms, to start, and a sewing room which also doubled-up as a tutoring spot for the children and as a piano-practice room. There was even a billiard room where Edison sought to improve his game—a mahogany cigar box, two cuspidors, champagne cooler, liqueur cups, and decanter were close by the carom table, cues, and bridges—but aside from its recreational purpose, Henry Hudson Holly again regarded this room, not as a frill by any means, but rather as a socially obligatory and integral dimension in all of his important houses, regardless of the owner's predisposed inclinations, especially "in a country mansion, somewhat isolated, so that the occupants must, to a great extent, depend upon their own resources for amusement . . . Why not accept the billiards pastime as one of the social amusements, and give it the same prominence as music or cards?" As soon as he moved in, typically thorough, Edison ordered and read the popular classic by Albert Garnier, *Scientific Billiards, and Practice Shots, With Hints to Amateurs.*

The four weeks leading up to the wedding day were jittery for Edison. A recurring fear was that the minister would not enunciate clearly, prohibiting the nervous, nearly deaf groom from hearing the command that it was time to slip the ring over his bride's finger. "I'm afraid our 'binder' will not speak loud enough—in fact I'm getting pretty scared—I wonder if I will pull through," he confessed to Mina. "I know you will, women have more nerve than men." Among his papers from this period we find many letters from the hard of hearing, pleading with Edison to invent a hearing aid, and he replies deceptively that he is "working on it, I need it as bad as anyone." This was a common stalling tactic, and he never did perfect one. Over and over again in private conversations Edison said that he preferred being deaf, because it helped him to think more effectively. This may have been the case, but his infirmity—most convincingly argued to be a congenital arthritic condition of the three small bones of the inner ear—always made him skittish at the prospect of appearing before or among large crowds, and the wedding was shaping up to be an extravaganza, to be sure.

Edison practiced writing his beloved's married name over and over again, covering several sheets of notebook paper, "Mina Miller Edison . . . Mina Miller Edison," and up until the final days was trying to lose himself in experiments, working with John Ott and Martin Force on mixing different formulations of lampblack carbon for arc lighting. Buried within the lab notes and listings of hard and soft insulators are several surprising pencil sketches for a new laboratory building, much more grand in scale than anything Edison had ever imagined before. The structure rapidly but clearly defined by Edison was square, three stories with a mansard roof, a tower at one end, and a large archway with double carriage-doors entering upon a central courtyard—details that make it resemble a castle more than a factory. A second sketch of the inner plan from a different perspective reveals that around the open quadrangle would be arranged Edison's "private" laboratory (so marked), chemistry lab, machine shop, library, and power plant. In the expansive spirit of Glenmont, Edison was thinking ahead to a way in which he could likewise consolidate his scattered business enterprises. That inevitable day was not far off.

At last, on Saturday, February 20, Charles Batchelor threw a farewell stag bash for Edison at—where else?—Delmonico's. In attendance were most of the inner circle "boys"—Ed Johnson, Sammy Insull, Alfred Tate, Sigmund Bergmann, and best man Lieutenant Frank Toppan of the Navy—with "general regrets that Kruesi and Upton had somehow been overlooked."

"The cords of love and troth reunited [Thomas Edison] with his native State . . . [in] the most notable nuptials in Akron's history," on Wednesday, February 24, at Oak Place, the bride's home. In the hour before 3:00 p.m., a procession of horse-drawn carriages brought more than eighty guests up the hill to the elegant house, where they followed a red carpet within, to see a floral bell suspended from the ceiling at the foot of the central winding staircase, and fragrant smilax vines entwined in the chandeliers. In the parlor was an altar canopied with palm leaves and calla lilies. As three chimes rang from the clock in the hall, piano strains of the wedding march from *Lohengrin* began, and the bridal party entered—Edison (in Prince Albert coat, black tie, "ungloved") and Toppan; then Mary Valinda and all the Miller brothers and sisters, forming an aisle through which, finally, the radiant Mina glided, on her father's arm. She was dressed in a simple, square-necked

MINA MILLER EDISON, THE BRIDE

white silk gown trimmed with Duchess lace and a plain veil, and wore around her neck a gift from her betrothed, a pearl necklace with a crescent-shaped diamond pendant over her bosom. The Reverend Dr. E. K. Young of the Akron First Methodist Episcopal Church performed the ceremony as the couple knelt before him on a velvet cushion. Mendelssohn's wedding march signaled the conclusion of the rites, and all adjourned to the library where, under a great wishbone of roses with the monogram "E-M" spelled in pink carnations, the newlyweds stood for the reception. On behalf of the Edison family, father Sam, sister Marion, and brother William Pitt made it to the event. A congratulatory telegram from all the men at the Pearl Street Power Station was read aloud to great applause. A seated dinner fully catered by Kinsley's of Chicago and followed by helpings of mammoth cake was served by fifteen waiters.

Edison and Mina left that evening for Cincinnati, and stayed for their wedding night at the Hotel Gibson before moving on by train the following morning to Fort Myers, where the Gillilands and Dot had gone some days before the wedding to prepare for their arrival.

"The years following were the most unhappy of my life," Marion Edison confessed, and how could they not have been? The interloper Mina had usurped Dot's role as closest to papa's heart, and the tension between the two young women was exacerbated when it became clear that the newly-weds would not yet be able to move into their new home by the Caloosahatchie. Construction of the paired houses and laboratory was held up by the delay of building materials shipped from Maine. Instead, everyone, including Gilliland's sister and her family, stayed at the Keystone Hotel for a month, and then the elder Millers arrived from Akron to add to the pot-pourri, at which point the in-laws and everyone else moved into the one house that was finished.

It was not the most romantic of honeymoons. There was an old story that, on the day of his wedding to Mary Stilwell, the preoccupied Edison had to be roused from an experiment to get to the proverbial church on time. True or not, here was a man who, although he did give in to the trop-ical tranquillity and sleep later than his accustomed dawn-breaking hour, needed to be close to his work. The intellectual nature of that labor does seem to have been shaped by the major changes in his emotional life; it had returned, for a while at least, to a more *spiritual* plane, the kind of thinking that had been so seductive to Mme. Blavatsky and her followers. Ever since Mary's death, Edison had spoken on occasion publicly of metaphysical concerns, his belief that within every atom, every subdivision of nature, there could be found "a certain amount of primitive intelligence . . . Look at the thousand ways in which atoms of hydrogen combine with those of other elements, forming the most diverse substances. Do you mean to say that they do this without intelligence? When they get together in certain forms they make animals of the lower orders. Finally they combine in man, who represents the total intelligence of all the atoms."

"But where does this intelligence come from originally?" he was asked.

"From some greater power than ourselves," was the reply.

"Do you then believe in an intelligent Creator, a personal God?"

"Certainly," said Mr. Edison. "The existence of such a God, in my mind, can almost be proved by chemistry."

"Think of it! A man in this skeptical century who dares believe in a discovery beyond all discoveries," Edison's shamelessly editorializing visitor concluded. "Here is a student of nature who is not afraid to have the spirit of a Galileo or a Kepler or an Isaac Newton . . . And so we discover down on Avenue B, in the prosaic city of New York, a philosopher who believes in a personal God."

In the wonderful opening scene of the 1940 MGM classic, *Edison, the Man,* Spencer Tracy, playing the elderly but mellowed curmudgeon to fairy-tale perfection, is surrounded by a group of adoring children, smiling benignly upon them as the innocent questions come thick and fast, until with ultimate humility he smiles and points heavenward, saying, *"That's* the *real* Inventor!"* This was a statement from life, not a line concocted by the scriptwriter.

Mina was invited to join her husband on a few occasions in the new lab (little more than a shack at first) across the road, certainly a breakthrough beyond belief. There they indulged in whimsical and mundane experiments together, "shock[ing] an oyster to see if it won't paralyze his shell muscle and make his shell fly open," and refining her Morse Code skills. This comradeship, however, did not last very long. Edison spent most of his days, once Mina's parents had come, writing, by himself, pursuing abstract, even otherworldly speculations, wondering if "the solar system is a Cosmical Molecule [TAE's capitalizations]. . . . all substances we call elements," he reasoned, "are composed of molecules of different atoms. All atoms are primal hence matter is composed primarily of one substance, *the primal molecule.*"

Edison was exercising his mind upon a theme that, like most, if not *all* of the concerns in his career, he never definitively abandoned.

Magnetism as the governing force of life was a concept that never failed to intrigue him: What if, he wrote to himself, "counter-lines of force outside of our planetary system," including forces generated by the stars, helped contribute to the elliptical orbit of the "molten" earth and the planets around the "incandescent" sun (a heavenly body he explicitly compares to a dynamo) by counteracting the gravitational pull emanating from the sun? Half of the sun's gravitational "lines . . . pass through space from pole to pole forming a vortex." It followed that we were essentially surrounded by competing strains of electrically generated magnetism; thus, by extrapolation, weight (Edison calls it "electrical attraction" or "earth lines") of matter diminishes as we recede from the earth. By further extension—and these ideas proceeded by organic thought, Edison allowing himself to indulge in a kind of automatic writing, but in a scientific mode, over a series of entire days—*"Perhaps our solar system is rotating as a whole."* This supposition sends him tantalizingly close to the fringe of a Big Bang theory, when he imagines—in a practically illegible blur of excitement, ten days after his first, tentative notes—"detached matter . . . thrown out into space" from the sun and "getting a rotary motion" by virtue of this expulsion, this "eccentricity diminish[ing until] it took on its regular orbit." Edison then swings in analogizing manner from macro to micro, contracting from the undefined parameters of a dynamo-like, electrically charged universe to an irreducible atom "rotating on its axis with inconceivable velocity." The refreshing thing about these lines of thought is that they are most decidedly not utilitarian, not about marketing or selling. Rather, they reveal Edison equally at home as the mental explorer.

As respite from navigating these uncharted paths, there was the pressing business at hand of landscaping the entire Fort Myers estate, now that the Edisons were settled in one house and the Gillilands' would soon be ready. Once again, Edison the consummate conceptualizer stepped to the fore. In this effort he may have had Mina's help, since she had grown up in a house full of flowers, and the word "we" crops up here and there among the plans. Unlike Glenmont, which came into Edison's possession ready-made, Seminole Lodge (as the Florida residence would be called) was a blank slate aching to be filled in—and so it would be, with a twenty-foot-square banana bed; one thousand pineapple plants from Key Largo; pecan trees from Jacksonville; peanuts in great profusion, soft shell almonds, Brazil nuts, date palms, English walnuts, and filberts—Edison was a nut-lover; fig trees (to be planted out of sight of the main house, since Edison considered them to be "an ugly tree"); half a dozen peach trees, the early-blooming variety; grapes, red and black raspberries and strawberries, the latter in a patch twenty by one hundred feet; mangoes, alligator pears, sapodillas, and Spanish gooseberries; pomegranates, mulberries, orange trees (including mandarin orange), and guavas; plums, pears, and crabapples; "1,800 varieties" of flowers ("every variety I can procure, to be fertilized with four tons of guano and the best manure, costing not more than $50 per ton . . . the more tender shoots to be protected with one thousand yards of common print cloth soaked in boiled linseed oil and hung out until dry to prevent radiation"); cotton plants; cabbage palmettos; cherry and grapefruit trees; a twenty-foot-square tobacco patch; double rows of sugar cane fifty feet long—and, of course, two beehives for the flowers.

Leaving nothing to the imagination of his groundskeepers, Edison drew an exhaustive sketch of the property, keying in the exact location of every structure, path, lawn, shrub, tree, and plant named in his list, then taking it all a few steps further, specifying the precise height and grade of fence he wanted built around the property, the shape of every flower bed, the composition of the lower terrace leading down to the riverbank (crushed shells), and the appearance of the sea wall, which he directed must be whitewashed.

Mina, meanwhile, was enjoying the balmy weather, but not sleeping well at night, still weary from her acclaimed performance at the wedding, and a touch homesick, especially when Jennie wrote that "the canopy of flowers still stands in the library," and admitted that when she went into her closet and saw Mina's silk foulard dress that her sister had worn so much last summer she "could not keep back the tears." Mina's thoughts were taken up with the coming obligations of Glenmont, where two little stepsons eagerly awaited her return: "We kiss our new Mamma every night before going to bed," said ten-year-old Tom, Jr. Meanwhile, Mina's father warned her sternly that she must feel as loving toward Marion, Tom, Jr., and Will as she would toward any member of her own family. She had an obligation to Mr. Edison to "take them to [her] heart as a mother"; she must guarantee their happiness cheerfully, without giving any hint to the Edison children that they were a "bother, or a burden." Lewis told Mina that Marion had confided in Mary Valinda that "her own mother was not concerned about her doing things just right as you [Mina] were, and that her own mother let her do things which were not right and that got her into bad habits." His daughter therefore had an even greater calling now, to realize "that [her] life and character will make itself felt on the three you have accepted as in your own blood. You must not look for perfection," he said, "only for improvements."

This was easier said than done.

CHAPTER

15

"HAVE YOU HAD A GREAT MANY CALLERS? AND IS LIFE VERY fashionable or is it something like you lived before you were Mrs. Edison?" Mary Miller wrote breathlessly to her big sister. "Tell me all about Dot and the little boys—do they like you?" There was much to get used to at Glenmont for the lonesome young bride, and yes, Mina had been settled in her new home barely two months before it became quickly clear to her just what it meant to be actually married to Thomas Alva Edison and living with him, morally obligated to share her new husband with the society at large because, as Mary pointed out, he was "working so hard out there to benefit the world." To be supportive as best she could, Mary supposed that for Mina, having Mr. Edison to herself in the morning and the evening would very likely make up for "having to spend so much time alone." She pleaded with Mina not to "borrow trouble . . . [because] in four months time [when the new laboratory would hopefully be nearer completion] he will be just as true and loving as he is now if not more so . . . I am sure he loves you deeply and would love you more if you would let him."

And there was the endless task of maintaining the huge premises, as well as Laura, the "upstairs girl," and Mary the cook to watch over. Seeking further solace, Mina, turning to sister Jennie, confessed she was waging a "battle" within herself. However, over time, the once adversarial Jennie had had a change of heart about Mr. Edison, telling Mina that she would not have been happy with a different choice. Her husband was "loving and devoted. I think he loves you perfectly. I think the children love you! You must be happy and not think of the past. I often look at you in wonderment to see how nobly you are filling your position. . . . *Every* housekeeper makes mistakes, even after years of experience," Jennie wrote, soothing Mina. "Return your calls as soon as you can and then do some studying, practice [piano] and do a little fancy work." Brother Ed likewise joined the Miller chorus: "What a good housekeeper you are!" he exclaimed upon returning home to Akron after a stay at Glenmont. "You should be just the happiest young lady in Orange. Why do you let little things worry you so? . . . . It seemed you wouldn't need to touch a thing but once a month in the house and still everything would be prim and pretty."

As for Mr. Edison, the honeymoon was truly over. No sooner had he arrived back at his desk at 65 Fifth Avenue and begun to catch up on business affairs and attack the mountain of papers that had amassed in his absence when he was faced—unbelievable as it seemed—with his first labor strike, an ironic consequence of success. As Major Sherburne B. Eaton, chairman of the Edison Company, put it, "The danger is not that we won't have enough business, but that we won't be able to grow fast enough to handle it." Consumer demand for Edison isolated plants across the country had increased to such an extent that the New York dynamo-generator manufacturing facilities were overloaded to sheer backlog exhaustion; finally, the men at the antiquated Goerck Street plant in lower Manhattan and the Bridge Street "Tube Works" facility in Brooklyn just walked out.

These men were not born yesterday and they could see the escalation in fortune of the Edison empire. Three hundred fifty strong—laborers, pattern makers, and machinists—they presented their roster of demands to Charles Batchelor: Management could continue the nine-hour work day, but at pay for ten hours' work. They wanted double pay for overtime. Stop all piece work subcontracting. Limit one man per machine. And unionize the shops. "We know that we are in the right in this strike, and we mean to hold out until the end," declared the titular head of the work force, one D. J. O'Dare. "The Edison Company have been getting their work out at cheaper rates than any other company by at least twenty-five percent. Do you know that if the Edison plant machinery gets out of order that none can repair it but us?" This was the fear that spread through New York like wildfire; would the strike make the precious lights go out? Nonsense, laughed the clerks at headquarters, where an unidentified spokesman scoffed, "We can have the machines made and repaired where we will. Any machine shop can do the work!"

Batchelor, in the thick of the fray, tried to interject a note of reason, insisting (and rightly so) that no laborers in New York were better paid than Edison's; some mechanics took home upwards of $40 a week. As far as one man being assigned to rotate among several machines, Batchelor felt this was necessary for efficiency's sake; why should a worker sit by a lathe and watch it spin for two hours when he could be off starting another task? And where piece work was concerned, "When a man is making the same thing over and over again, we make a contract with him and he goes on, on his own hook. If he loafs or gets drunk he loses his own time, not ours."

As the costly weeks of idleness went by until the end of May, it became clear to Edison and Batchelor that there was a deeper issue at stake. This strike was a warning signal, but unionization was absolutely out of the question. Edison said bluntly that he had for a long time been contemplating some tactic "to get away from the embarrassment of the strikes and the communists to a place where our men are settled in their own homes." Perhaps it was finally the moment to relocate and consolidate major manufacturing functions in a true *factory,* out of New York City, to move farther afield, with a decent railroad and natural resources nearby, where real estate and labor were less costly, and there would be more breathing room for growth. Scattered, autonomous shops at the root of current company structure, while nostalgically appealing, were too difficult to control, and no longer the best way to operate.

Outside Schenectady, in upstate New York, the Edison men found what they were looking for, ten acres of property with two large but unused buildings already in place, originally erected for the McQueen Locomotive Works. The good citizens and businessmen of Schenectady chipped in $7,500 of the $45,000 asking price as a gesture of their interest in attracting the Edison Company to their town. By the end of 1886, Edison had moved his personal laboratory out of Bergmann's building on Avenue B and over to the Lamp Works in East Newark (Harrison), nearer to Glenmont. The Goerck Street shop and the Tube Works were permanently shut down, and the twenty-seven-year-old wunderkind, Samuel Insull, was promoted out of his secretarial duties at the home office on Fifth Avenue, given a raise to four thousand dollars a year, and transferred to Schenectady, where he and John Kruesi supervised the work of the two hundred men willing to make the move from New York. For the next several years, the loyal Insull spent by his reckoning as many as six nights a week on the train between New

York, New Jersey, and Schenectady, as he still had to report to Edison and the other directors of the newly configured Edison United Manufacturing Company.

Despite all this movement, Edison and Gilliland continued their close collaboration in two important areas. The most immediately exciting was the perfection of "air-telegraphy" between moving trains and fixed stations, making it possible to send wireless messages. For a change (and perhaps as a tribute to their friendship), Edison actually credited Gilliland as having approached him first in 1881 regarding a method of telephonic message transmission back and forth between railroad cars, accomplished by the process of induction. The current was generated through a wet-cell battery in the car. Via a wire connection from the battery and through the Morse instrument, the roofs of several cars were electrically charged, acting as one polarity of a giant condenser, the other pole being the ubiquitous telegraph wires running alongside the track. At the receiving train station end resided a similar battery and a connection extending to the regular telegraph wires. A "musical" transmission imitating dots or dashes, short or long notes, was sent from the train at a much higher frequency than normal telegraphic signals, utilizing the same wires. The telephonic receivers in the train stations picked up the high musical vibrations—six hundred per second.

It was an ingenious concept, piggybacking onto current technology for another purpose, with greater efficiency. In Edison's imagination (a century ahead of in-flight credit card telephone calling) further manifestations were naturally possible: "A traveller may send back to his office in New York, San Francisco or Chicago any information which he happens to pick up, or business decisions which he was not able to make before starting." He saw the possibilities for newspaper correspondents filing their stories from a train in advance of their arrival back at the editorial office, and police authorities suspecting that a culprit might have fled by railroad could send instant descriptions up and down the lines. For that matter, perhaps ships at sea could make use of the same dynamic by using a kite coated with tin-foil "soaring several hundred feet above the deck and controlled by a fine wire. The song of the Sirens would have no charm for any mariner who could hear from home and send word to his sweetheart by this method."

The second joint project also beginning at Ezra Gilliland's instigation would prove to be far more significant and disruptive over the long haul: the revival of the phonograph. Gilliland reasoned that now that his comrade was pulling all the areas of his personal and scientific life together, he should return to his "favorite," his "baby," lain dormant for the past ten years. Edison countered with exasperation that his initial patents had expired, the concept would never grow beyond the novelty stage in the minds of consumers, the Edison Speaking Phonograph Company existed on paper only, and besides, the field was was wide open; the Volta Laboratory Association run by Alexander Graham Bell had in fact built several "graphophones" during Edison's downtime. These machines were designed for courtroom stenographic application. "This [graphophone] is very similar to Mr. Edison's phonograph," A. B. Dick, the mimeograph manufacturer, wrote in curiosity to Edison's secretary. "Perhaps it is the same machine under another name. Will you please inform us if such is the case?"

Edison's apathy aside—the inventor was absorbed with, among other matters, the wholesale reorganization of his corporation—Gilliland took it upon himself to make some notes in the fall of 1886 toward what he called a "standard phonograph." Once more, the intention was for office use. Gilliland exploited the considerable experience gained during his own years with Bell, envisioning an instrument that would feature a telephone-like receiver held to the ear. This time, instead of being manually run, the phonograph would be driven by an electric motor at uniform speed, which meant that the cylinders could be stopped, started, and even "backed up"—in a real sense, rewound. The "cylinders or plates . . . would be made of glass or polished steel," surfaces with the greatest ability in Gilliland's view to retain the integrity of the grooves, a far cry from tinfoil. Forty or fifty threads to the inch would provide a capacity of ten thousand words per cylinder, and the cylinder, stylus, and diaphragm would be detachable and interchangeable—a truly modular machine. According to Batchelor's assiduous *Diary,* on the evening of December 21, 1886, Gilliland and Edison came to his house, "talking up" experiments. Was the phonograph on their agenda, or was it still too early for Edison to begin feeling the old competitive pressures rising up within, moving him to act? We shall never know for certain.

Ten days later, Edison came down with a severe case of pleurisy. The stress of constant change had finally got the better of him—in the midst of masterminding the Schenectady move, he had signed the papers for the new laboratory site, fourteen acres of land next to the main road through the Oranges, at the bottom of the hill leading up to Llewellyn Park—and now he was confined to bed, gravely ill for more than six weeks.

A recuperative trip to the benign atmosphere of Fort Myers was arranged. The family arrived on February 19 on the steamer *Alice Howard,* and "under the influence of Florida's beautiful and balmy weather, [Edison's] complete restoration to health [was] only a question of a little time," the Fort Myers *Press* reassured its readers. The little journal had come to the realization that Edison's comings and goings, no matter what they might entail, were the biggest news ever to hit the humble village. And if there was not much to report for the time being, they could always make up some good gossip. One such story going the rounds was that Edison, even though indisposed, was catching fish by running a wire along the bottom of the river and electrifying the poor creatures, so that they rose to the surface dead. "Fish thus electrified keep indefinitely without ice," said the *Press.*

Much progress had been made in Fort Myers since Edison's last sojourn. The neatly matching Edison-Gilliland homesteads were completed and ablaze with incandescence. Thanks to their illustrious new resident, and the recent arrival of the familiar Long-Legged Mary Anne dynamo, the town proper would soon be electrified. Gilliland busied himself supervising the finishing touches on Edison's laboratory building, strongly reminiscent of the old Menlo Park site, although on a smaller scale—divided into boiler room; laboratory proper, lined with bottles; dynamo room; and Edison's personal sanctum sanctorum. Friend Ezra also had to stave off the ever-present swarm of reporters jockeying for a glimpse of the Great Man, especially with rumors still flying that Edison was "a physical wreck, incapable of movement and scarcely of consistent thought . . . 'He certainly *was* a sick man at one time,' Gilliland admitted, 'as he was being treated for a cold in the muscles of his chest and a slight affection of the bronchial tubes, the trouble somehow reached the heart, and it was found necessary several times to administer hypodermic injections of morphine.' "

Edison was once again up and about, although slowly. He made a surprise entrance, dressed in customary dark suit with white shirt and neckerchief as of old, and sporting a broad-brimmed straw hat; during a leisurely cruise on the Caloosahatchie aboard his new yacht, the *Lillian,* he had found time for a touch of sun, and his face gave off a ruddy glow. "I understand from some reports that you have been compelled to seek rest and forgo your electrical researches for the present?" nervously queried a reporter.

"*Rest!*" Edison shot back. "Why, I have come down here to work harder if anything. I will tell you how I rest. I am working on at least six or seven different ideas. When I get tired of one I switch off only to go to another, and alternate to such an extent that I have a constant succession of new and pleasurable efforts."

"But don't you get confused in that manner?"

"Not in the least. On the contrary, when working on one idea I often drop in something helping another idea, because they mostly tend to one grand result. . . . All electricians work in the dark," Edison lectured his visitor, warming to the theme. "They grope and grope, catching on to the slightest clue, which they work upon until its fallacy or use is demonstrated, and then if useful they follow it persistently until the results are accomplished. There is luck in it, and in truth, we electricians are discoverers, not inventors."

Even though Gilliland was close at hand, Edison seems to have declared a temporary moratorium on further phonograph deliberations while in Fort Myers. Now that the laboratory land had been properly acquired, he was concentrating upon more refined plans for the new building, in constant communication with Batchelor, reminding him to include "a special or secret part to Machine Shop for special things I want *sub rosa.*" True to form, Edison also kept up his rudimentary ore-milling trials, using flour as a substitute for iron filings, and observing its behavior while falling in vacuo. He told Batchelor that he was "feeling a little better although little dizzy in head."

Alas, this vertiginous sensation was the precursor of another health setback. Two successive abscesses were discovered below Edison's left ear, the one that gave him chronic trouble, and emergency operations were required. In early May, Edison and Gilliland and their wives—Mina was pregnant, and had just announced her "expectation"—left Fort Myers, the *Press* heartily looking forward to the great occasion "next fall, when they will come earlier and stay longer."

Although none of the group realized at the time, it would be another fourteen years before Thomas Edison returned to Fort Myers—and Ezra Gilliland never came back again.

What better man to draw up the new plans for the grandiose Edison laboratory than Henry Hudson Holly, the very same architect who had done such a noteworthy job at Glenmont? Batchelor had been down to Holly's office at 11 Bowery in New York City and signed him up at Edison's request. The main building would be three stories, 250 feet long, and 50 feet wide, its west end perpendicular to Valley Road and graced with arching windows for a two-story library. At the other end of this building were the boiler, steam engines, dynamos, and a machine shop large enough for fifty men. Four smaller structures, each 100 by 25 feet, would be situated perpendicular to the main building, in keeping with Edison's earliest, castlelike sketch, in which he wanted a courtyard effect. Each of these numbered "satellite" buildings (as the scholar Andre Millard aptly names them in his important study *Edison and the Business of Innovation*) would have a different function: electrical work, including lamp testing; chemical laboratory; woodworking; and metallurgy. When completed, the West Orange complex would be more than ten times larger than Menlo Park. "I will have the best-equipped and largest facility extant, incomparably superior to any other for rapid and cheap development of an invention, and working it up into commercial shape with models, patterns, and special machinery," Edison declared. "In fact, there is no similar institution in existence. My laboratory will be equipped with every modern appliance for cheap and rapid experimenting," he said, "and I expect to turn out a vast number of useful inventions and appliances in industry. . . . In time, I think it would grow into a great industrial work with thousands of men." Edison was aware of the fact that his proper place was now in the laboratory: "I do not manage shops myself as I am incompetent for that class of work, but I do know how to select the right kind of men to do it for me." Evidently, the esteemed Mr. Holly was not one of them. After barely three months on the project, Edison accused the architect of approving inferior masonry materials, and he was summarily fired.

Back directly under Batchelor's trusted supervision, one boiler was installed at the West Orange site. The main building had a roof. The metallurgy building had walls. The chemical and woodworking foundations were sunk. A gas machine was hoisted in. Even as the din of construction grew louder and louder, Edison, in his mind, was already in a place far, far beyond the still incomplete laboratory. He wanted to build a chain of factories, a veritable industrial park stretching for thirty more acres through Essex County and the Orange Valley.

One of these installations was a plant in Bloomfield where fifteen men stood ready to begin resumed work on the phonograph. It is difficult to pinpoint the exact precipitating factor that finally drew Edison's attention back to his earlier work during that hectic spring and summer, with so much else going on. According to Raymond Wile, who has assembled a meticulous study of the events surrounding Edison's return to the fray, at midyear Edison received a request from a manufacturer in Boston for a license to adapt a smaller version of the phonograph for the purpose of inserting it into a "talking doll." Along with Gilliland's incessant nudging (Batchelor, too, had joined in) and ongoing rumbles from the graphophone entrepreneurs, this letter seems to have tipped the balance.

Edison knew that a decade earlier, he had originated the process of cutting the grooves in a cylinder, which, after all, was the archetypal principle of the machine. In view of sparks of publicity surrounding the graphophone, Uriah Painter and the surviving officers of the Edison Speaking Phonograph Company, meanwhile, attempted to reintroduce the old tinfoil machine onto the market.

Edison therefore instructed Batchelor to accelerate his efforts toward finding the right resin-paraffin mix for the improved cylinder. They would develop their own recipe as well for the spring-driven "talking diaphragm." It was a situation Edison was used to—creating his own distinctive variation on a theme, in this case a theme he had written. He refused to meet with the Bell people, despite their efforts at a rapprochement: "Two [rival machines on the market] would be better, and let the best machine win," was Edison's Darwinian approach. Edison was bitter over Bell's "appropriation of his ideas," and became even more incensed when his agent in England, Colonel Gouraud, wrote to inform Edison that he had taken the liberty of accepting Bell's offer, accompanied, of course, by a sizable amount of capital, to join the graphophone company overseas, even going so far as to authorize the imprinting of Edison's name onto Bell's machines. Gouraud's complex rationalization, trying to play both ends against the middle, was that this allowed him to "protect" Edison's name, to "continue the identification of your company with the apparatus in question" in Britain, since Gouraud had heard only glimmerings and rumors that his boss was about to get back into the phonograph business.

"Under no circumstances will I have anything to do with Graham Bell," Edison scrawled, infuriated. These men were little better than "pirates. . . . I have a much better apparatus and am already building the factory to manufacture it and I not only propose to flood England with them at *factory prices* but I shall come out with a strong letter the moment they attempt to float the Company there."

True to his word, by the end of summer, Edison had formed a new corporate entity, the Edison Phonograph Company, capitalized at $1.2 million, "to manufacture and sell Phonographs and apparatus and devices embodying the same." He formally announced that he would have nothing more to do with the "old" phonograph company of the 1870s, neatly tying up his future interests by stating that his rights would have to be recognized: "I have a strong impression that our [old] patents are not worth a cent," Edison reiterated to Painter, "so I am going on the basis that we are going to hold the field only by making a better and cheaper machine [than Bell]." As for Painter and colleagues, if *they* wanted to sell out to *him*, they were welcome to be absorbed into Edison's brand-new corporation at $1.00 per share. Painter, who once had been an accepted member of Edison's inner circle, was melodramatic in his reply, melancholy that the good old days, simpler times, had changed so radically: "Whether I ever make money out of the phonograph or not is a matter of minor importance with me," he told Edison piously. "I have a private cemetery into which I turn down my disappointment and losses, and I can if necessary tuck the old Phonograph away into a quiet corner and soon forget it, but I have no room for the lodgment of a charge that either the [Edison Speaking Phonograph] Company or myself have treated you outrageously in any matter on earth."

Ezra Gilliland was appointed exclusive agent to market the new machine. As such, he would be given a profit of 15 percent on the sale of each one, and on top of that, a profit margin on acquisition of supplies of 100 percent. Already a wealthy man, Gilliland stood to become even more so if the resuscitated Edison phonograph was successful. The immensity of his vested interest in the new machine being assembled at Bloomfield drove Gilliland to tell Edison, who was in his customary rush to hit the market, addicted to the commensurate "risks and struggles of competition" that energized his daily life, that he did not think the second-generation phonograph was yet up to par, not strong enough to be a real competitor against Bell. Gilliland had carefully staked out a long list of features for the new Edison office phonograph that, once perfected, would be certain to supersede the Bell machine. These were aspects that Bell never claimed to cover, including mechanical provision for repeating sentences; making duplicate copies; erasing unwanted sentences; and the ability to transfer at will from electrical to hand-driven power. In all candor, Gilliland told Edison on December 18, 1887, these abilities had not been successfully built in. For the very first time, a defensive tone invaded Gilliland's communications with his close friend who may in turn have seen him as verging upon opportunism—even, perish the thought, as a rival. "You particularly dwell upon the point that no man could invent and do business at the same time,"

Gilliland wrote, responding to Edison's accusations. "I want to say that I started a new factory, invented and worked out for manufacture magneto bells, switch boards and other telephonic apparatus, travelled and solicited orders, had no assistance but hired help, and made $55,000,000 for my [former] company [Bell] the first year."

"Time is the all important thing now," Gilliland insisted in an emphatic, handwritten postscript, "and if I have said or done anything to displease you it has been due to my own anxiety *to do the most and best I can for you.*"

CHAPTER

16

THE PUSH TO RESUSCITATE THE PHONOGRAPH CAUSED THE
first irreparable fissures in the proud, tightly knit Edison fraternity.
"Damon and Pythias," through friendship's bond, could have forged forward at the new and expanded laboratory, growing old and rich together,
Edison's inexhaustible imagination united with Gilliland's smart marketing. But the hard reality was different. As his working capital stretched
thin, Edison turned once more to Henry Villard, in hopes he would be willing to invest in a phonograph works on the West Orange grounds to outproduce the Bloomfield shop, with the capacity to turn out more than two
hundred phonographs a day once it got rolling. Villard was not interested,
and prototype production slowed nearly to a halt.

Due in large measure to Edison's intransigence and possessiveness
about his "baby," the Graphophone group, the old Edison Speaking Phonograph Company, and Edison's new but stalled enterprise were all at an impasse. Into this breach—sensing a once-in-a-lifetime opportunity to make
some real money from the hunger for new technology—stepped the venture
capitalist Jesse H. Lippincott, a Civil War veteran who rose from humbler
ranks as a grocery store proprietor in Pittsburgh to make his fortune as a
glass tycoon. His Rochester, Pennsylvania, company produced seventy-five
thousand tumblers per week.

Lippincott played the adversaries against one another. He would, he announced, create yet *another* new corporate entity, The North American Phonograph Company, and offered a flat half a million dollars to the Graphophone Company for the rights to assume production of their machine, and then straightaway approached Gilliland (whom Lippincott rightly suspected was sooner or later to become the weak link in the Edison chain) as well as John Tomlinson, Edison's personal attorney, with precisely the same terms.

The three men met on several occasions during the spring of 1888 in Gilliland's office on Dey Street, unbeknownst to Edison, and went so far as to hash out a roster of terms whereby the Edison Phonograph Company as well as Gilliland's separate but allied manufacturing and distribution arm would be completely bought out, including the stipulation that Edison would receive a 5 percent royalty on every machine sold for the next fifteen years, over and above the initial $500,000 purchase price, freeing him to continue to provide modifications and revised patents while no longer worrying about the daily production nuts and bolts. They reasoned Edison would go along with this plan because it represented his enduring dream.

Tomlinson took the temptation to Edison and sat down with him under an old apple tree in the laboratory yard to spell it out, urging his client to sell outright, reminding the Old Man, and appealing to Edison's vanity in the process, that he "needed time to spend in getting up new inventions [rather than] taking the risk of merchandising the phonograph." By deliberate design, Gilliland's name was not mentioned in this conversation.

It was not difficult for Edison to agree in principle to this alluring deal once he overcame his initially instinctive resistance at parting with anything he had created that had potential for exploitation. Edison figured the sale would bring him far more money up front than he had ever received for the rights to any prior invention. He drew up a systematic checklist going back to the old Gold & Stock tickers he had made for Samuel Laws: "$500. Laws sold for $150,000 . . . Automatic Telegraph. Sold by Co. to Jay Gould for $1.5 million, but never got anything—in Litigation . . . Electric Pen. Received about $15,000 in royalties . . . Telephone: Sold to W.U. Telegraph for yearly royalty of $6,000 for 17 years—They sold it to Bell Telephone Co. for ¹/₅ of profits that amounted to about $400,000 annually . . . Electric R.R.: Rec'd my money back only . . . Railway Induction telegraph: Sold the invention for $95,000."

Neither Lippincott nor silent "friend" Gilliland told Edison, however, that *simultaneous* negotiations were in progress with Bell and Painter, because Lippincott, the empire builder, envisioned a combined entity controlled by him that would eliminate any possible threat of future phonograph competitors. The Bell people were not entirely pleased with the initial terms being offered to Edison and came forth with a few modifications of their own that would have to be agreed to before they would do business with middleman Lippincott, including the withdrawal of exclusive manufacturing rights and extended royalties to Edison.

These revised terms were resubmitted to the unwitting Edison by Lippincott as having been recommended by unnamed business associates. Tomlinson again relied upon the argument most salient to any debate with Edison, that the sale was above all else in his best interest because it promised creative emancipation. The bottom-line goal of Lippincott's plan was a simple flip-flop: to buy up Edison's stock and future patent rights, then to turn around and offer the Graphophone Company an option to purchase stock at the same price at any time in the next five years.

Gilliland and Tomlinson's separate, secret allegiance with Lippincott for the sale of the phonograph distribution franchise included a proviso that they could cash in for $250,000 in stock options themselves. Lippincott did not anticipate the cash flow problem that arose almost immediately when these two entered redemption claims far sooner than expected, while the Edison Speaking Phonograph Company stockholders—Uriah Painter et al.—doubled the ante on their own buyout requirements. These liquid demands left Lippincott by September in the rather awkward position of being tardy in his installment payments to Edison, who at that point became suspicious.

Lippincott at first begged off by telling Edison that Tomlinson had told him that Edison would be amenable to time installments rather than a lump sum. His equivocation forced Lippincott by circumstance to reveal to Edison the rest of the true and byzantine tale.

The inventor was beyond fury. It was not so much the money as the excruciating pain of betrayal. And Gilliland for his part was already beyond reach, having escaped by ship with Tomlinson for Europe in early August. The inseparable "twins" had come to the absolute end of their relationship.

But the phonograph fiasco was only the beginning of Edison's agony. He vigorously pursued a case of fraud against Gilliland and Tomlinson through costly litigation for nearly twenty years—how could he not? Four years after Jesse Lippincott's death in 1892, Edison clawed his way back into vindicating control over the North American Phonograph Company by forcing it into friendly receivership. Presenting an ever tougher exterior all the while, Edison nevertheless revealed the different drama of his private suffering in two long unsent, unpublished handwritten diatribes to another seemingly close colleague, Edward H. Johnson, who had accused Edison in front of the board of directors of the Edison Electric Light Company of "stabbing him in the back" and "selling him out" by first going to Henry Villard for additional infusions of capital and then consummating the deal with Lippincott.

Edison in turn revealed that he had long suspected Johnson of harboring feelings of "jealousy" toward Gilliland, and in an acerbic wave, launched into a headlong recapitulation of all he had done for Johnson and Gilliland as well as the other "boys" over the years, especially, as a sign of trust, "taking care of" the inner-circle comrades with a percentage of profits in lieu of salary, Edison reserving for himself what was most important, after all— "the honor of invention."

As plainly hurtful to his pride as this declaration was, Edison defended himself furthermore against the sellout charge by reminding Johnson that he had fought to include a proviso in the Lippincott matter that would cover the interests of "all the boys" who still had a stock interest in the "old" phonograph company, even though Uriah Painter (another betrayer) "tried every device to break me up."

"You, who have no claim whatsoever except that you *was* my friend," Edison rambled on in his complaint to Johnson, losing grammatical control as he lost composure, ". . . in what was previously a happy family . . . as I told you in Tomlinson's office two years ago that it was only a question of time when it [the complexity of competing interests in the company] would break up old associations which it has just about done.

"My plan of dividing up [the interests of the company] is a dam-d failure," Edison concluded bitterly, "and this will explain why young men after reaching 40 or 45 change somewhat in nature. Experience has caused *me* to change. Hereafter I am going to be a hog myself and look out for myself.

"I have written this because I cannot talk fluently."

Over the decades, Thomas Edison constructed a complex network of emotional and financial interdependencies. The Lippincott affair represented the first threat that this same network would ensnare him in a morass of disillusionment. Edison relied upon moneyed interests to subsidize his insatiable inventive thirst and then bring the actual projects to market. In overwrought excitement at the thrill of the chase, he frequently undercut his own opportunities for success in that same marketplace, opening himself up to the inevitability of competition, upon which he thrived even as it undermined his chances to achieve stability. Edison's habit of giving birth to premature inventions, creating demand long before supply was on hand, drove lenders to frustration, their vested interests pulling the resentful Edison toward achieving quicker profits. At the same time, his paternal, domineering personality led him to require fealty from his men in exchange for the potential carrot of wealth suspended at some indefinite distance in front of them. Something had to give in this tenuous equation.

And thus Ezra Gilliland was led astray by the lure of an immediate payoff, and by the affront that his well-meant early warnings were not taken seriously enough by a headstrong friend who cast a long and possessive shadow.

What ultimately carried Edison through this confusing tangle of lawsuits and rivalries was his deep-seated sentimental attachment to the phonograph, rather than any conviction that it had a clear technical superiority or distinction over any of the other machines on the market. Before he had even moved into the West Orange lab, Edison was touting the second-generation machine as the "perfected phonograph," inheriting from the earlier tinfoil manifestation the "germ of something wonderful" that unfortunately had been superseded by the inventor's immersion in the electric light industry. But all the while, the phonograph had been in his mind, gestating there, and still belonging to him by virtue of the fact that he pondered its implications, albeit privately, and "when resting from prolonged work upon the light, [his] brain would revert almost automatically to the old idea."

How appropriate to the epic image Edison strove to create over and over again when he asserted that the entire *Nicholas Nickleby* could be recorded on four phonograph cylinders each two inches in diameter and four inches long! And did not the Assyrians and Babylonians, millennia ago, employ "baked clay cylinders" inscribed with cuneiform characters in order to preserve their history? The phonograph stood in that age-old tradition, except that now, the cylinders were fabricated of wax, and could "speak for themselves, and will not have to wait dumbly for centuries to be deciphered." These wax cylinders could be sent through the mails, providing the ultimate form of communication with a friend or loved one. And how sophisticated that he so often stressed the cultural contributions of the new instrument, its particular ability to replicate orchestral performances—"the strings are perfectly distinct, the violins from the cellos, the wind instruments and the wood are perfectly heard. It is going to work wonders for the benefit of music lovers." It was reported that Edison's old friend W. H. Preece of the British Post Office was moved to tears upon hearing a symphony of Beethoven so "exquisitely rendered" by the phonograph; "but then," the witness conceded, Mr. Preece, "is, as everyone knows, of a very generous and emotional nature." The perfected phonograph would also be a godsend for authors, who would be able to "register their fleeting ideas and brief notes . . . at any hour of the day or night," without having to grope for a pen and paper.

"The *perfected phonograph* is now ready," Edison continued to declare throughout the same troubled period during which he was embroiled with so many adversaries, both within his own camp and without. "It is no longer in a state of infancy. It may be still in its childhood; but it is destined to a vigorous maturity."

Alas, it may indeed have been ready in Edison's imagination, but in fact the phonograph was most decidedly not available to any kind of general consumer public when these sentiments appeared in the *North American Review* during the summer of 1888, and on into the ensuing autumn, when Edison reiterated to Colonel Gouraud in England that the machine was now absolutely "perfect"—in the laboratory, under stringently controlled conditions. On the paranoid suspicion that Painter might now in desperation be spying on him, even sending private investigators to eavesdrop on casual conversations, Edison instructed his secretary, A. O. Tate, to put up a sign outside the Phonograph Works to the east of the main laboratory complex that warned: "Visitors No Longer Admitted." However much he tried to protect himself, the problems only seemed to multiply, as, dead set against Bell and Painter's treadle mechanism (their version, ominously christened by *Scientific American* as a "formidable rival," ran analogously to a sewing machine) Edison kept fiddling with his battery-driven idea. The wax composition in the cylinder was too soft, so it was chemically modified. But when it became more resistant, the stylus wore down, and thick wax shavings descended, making the grooves indecipherable.

Edison could always be depended upon to place an impassive facade before the world, no matter what might have been crumbling behind the edifice, and that was precisely the expected best-foot-forward stance with a laboratory, from which results were heralded when they emerged; within its walls one expected to find a constant buzz of activity, not all of it instantly productive. Just over the hill from the West Orange enclave, past the Llewellyn Park estate, on Eagle Rock Avenue, great-grandfather John Edeson's farm could still be found, the old house on Elm Road standing, a marker for the past. Ancestral history had indeed come full circle, and to Thomas Edison, ever mindful of defining his station in the history books, that continuum would always be important.

Imagine yourself paying a call on Thomas Edison as a curious contemporary visitor in the late 1880s, one who might remember the open ambience and sylvan quiet of Menlo Park of an earlier time. The approach to the laboratory complex on Valley Road is now a distinctly different experience. Walking along the bustling main street past well-maintained lawns, half a mile from the Orange train depot, you will come upon "the huge but not altogether ungainly pile of red brick with its succession of wings, thrown out like the teeth of some Brobdingnagian comb." You ring a bell by the roadside guardhouse. A boy runs up to the wooden gate and leads you through the broad space between the main building and the perpendicular, separate structures. Entering the laboratory about halfway down its length, you are seized by the way it physically epitomizes the schism in Edison's psyche: to the left is the machine shop, "ponderous engines and Leviathan driving wheels" a-spinning, through which grimy, overalled workers pass in constant parade. Benches in front of wide windows are fully occupied by "muckers" in deep labor, "planing, boring, cutting and shaping the hardest metal," constructing the heavier machine parts for prototypes, while, directly upstairs, the more delicate, fine-tooled components are made—levers and cogwheels and needles. Adjoining the machine shop is the stock room, fulfilling Edison's bizarre need to assemble supplies seemingly in anticipation of the millennium, as if the end of the world were at hand and he would be blockaded in the lab, forced to continue work without reliance upon the rest of society. Thousands of pigeonholes and hundreds of drawers from floor to ceiling are filled with skins of animals and feathers of birds, raw and tanned hides, minerals and precious stones, species of dried grasses and herbs, gums and spices, in addition to the more conventional materials, chemicals, sheet metals, hardware of every shape and size, screws and bolts and angle-irons, tools, in short, anything and everything he needs or imagines he needs in practical or hypothetical situations. Access to the precious stocks (Edison fondly calls them "the scrap-heap") is exclusively by signing in and out of a scrupulously maintained ledger book.

To the right is the library, and you are ushered within to sink into a plush chair facing a massive rolltop desk and await Mr. Edison. Ponder the immensity of the space, forty-foot-high walls with cathedral windows reaching toward the pine-paneled ceiling, daylight streaming in on two sides and casting a sheen onto parquet floors, their expanse broken by oriental rugs adorned with lush arrangements of potted palms and flowering shrubs placed by Mina Edison. (In later years, on one of his many health crusades, Edison would abolish the carpets, saying they harbored germs.) Three levels of booklined shelves encircle the room, protected behind railinged galleries connected by a steep staircase. Your eye drifts over row upon row of newspaper-clipping scrapbooks chronicling Edison's every dictum in the domestic and foreign press; the day-to-day history of the incandescent light; the leather-bound patent records. The not unpleasant, lingering odor of dissipated cigar smoke hovers palpably in the air, and in the open fireplace hickory logs glow. It is so still within Edison's hideaway that all you can hear is the tick-tick-tick of the huge clock upon the marble mantel, and through the floor-boards sense the distant throb of machinery down the hall. You are about to nod off into a pleasant doze when the oaken doors open and Mr. Edison enters.

Lately, he has taken to wearing a shin-length, gray gabardine laboratory gown befitting his cultivated role. It is adorned here and there with stains, and his necktie is askew. Edison's bleary, haggard eyes reflect the stress of the unending phonograph struggle, but his handshake is firm. Above all else, he must still be in control.

What activities besides the sprawling and inconclusive development campaign to launch the new phonograph actually occupied the Edison laboratory in its inaugural years toward the end of the 1880s? The place had been constructed with the eventual goal of mass-producing inventions, effecting the final marriage between pure and applied sciences for the purpose of feeding the unlimited consumer appetite of Victorian America and beyond. To gain ad hoc income in the near term, Edison immediately allowed independent inventors and small businessmen to use his facilities for a fee. He also continued to be involved in multifaceted contract work for outside companies related to telegraphy, copper smelting, the electric pen, and telephony, sectors within which he had built up formidable experience and contacts. Edison would take on research projects related to refinement of existing technologies in these areas in exchange for flat fees as well as ongoing royalties against eventual sales by companies he aided. And he had a whole raft of eclectic plans for mini-projects drawing upon existing manpower, which might or might not pan out: a cotton-picker, artificial silk, more efficient submarine cables, electrotyping, "malleabilizing" cast iron, manufacturing finer grades of brass wire, refining copper electrically, manufacturing cheap bolting cloth, manufacturing sheet glass and tubes, developing a magneto railroad signal system.

Moving all of his personal equipment once again, this time from the Harrison Lamp Works, Edison regarded the West Orange facility as a consolidated research and development spawning ground for the multitude of components required for the still growing, still transforming electrical industry around the country and the world. The second and third big growth sectors in Edison's plan would be the magnetic ore-separating and storage battery businesses.

These large initiatives would take hold only within the coming few years as Edison, primarily concerned with holding true on the phonograph, continued to search for its proper foothold in the consumer realm. Despite all his high-flown pronouncements about the proper use of the phonograph as a "serious" machine, Edison gave in to an irresistibly appealing idea endorsed by the ever sagacious Charles Batchelor, father of two girls: the talking doll.

Here again is an important instance where Thomas Edison and the prevailing zeitgeist dovetail perfectly. The idea for an "invention," contrary to folklore, does not arise in some abstract, hermetic realm, and Gilded Age culture was the very time when the earlier rusticity of homemade, cloth-rag dolls was superseded in girls' fantasy-play by ornate, elaborately attired statuettes. Middle-class parents in the years before the Civil War had recommended the use of dolls as the proper, Christian way for girls to learn "useful" mothering skills; the appeal for the girls in Edison's day had become precisely the opposite— the doll had evolved into a recreational toy blatantly reflecting the prevailing values of an increasingly materialistic society. The most elaborate dolls available when Edison entered the fray were predominantly imported from France and Germany; they had pale white porcelain skin; ruby red lips; real hair; sinuous figures; and often were accompanied by complete wardrobes—sewing for one's doll was conveniently no longer required. The doll was ready to attend tea parties and balls, and to pay visits upon other dolls, just like real-life suburban ladies. Priced between $5.00 and $30.00, these toys were pervasive status symbols that could be obtained in the rapidly proliferating department stores beginning to overwhelm little neighborhood shops. And owning one doll was simply not sufficient: A girl had to have at least a dozen to make the hobby worth her while, and to provide the proper companionship upstairs in the nursery.

In view of the interactive play required when girls got together with their dolls to imitate adult social activities, it is not surprising that a doll that talked would exert tremendous allure. Batchelor's idea was actually a regression back to the first phonograph patent, embodying what he called an "automatic return motion." In the hollow innards of the doll was a miniature phonograph cylinder on a screw-threaded shaft connected to a crank protruding from the doll's back. When you reached the end of the message that had been recorded, the cylinder disengaged, and a coiled spring returned the stylus back to the beginning. Leading from the cylinder, a funnel extended upward to a perforated section of the doll's chest, just below the neck, and from here, the "voice" came forth.

Edison committed half the resources of personnel, space, and finance in the Phonograph Works to the manufacture and distribution of the talking doll; he wanted to ship five hundred dolls per day. A visitor to the Works— an entire building was set aside exclusively for the doll business—was bound to leave with a throbbing headache. One entire wall was taken up with partitioned stalls occupied by young women whose sole responsibility was to spend the working day making cylinder recordings of "Mary Had a Little Lamb," "Jack and Jill," "Little Bo-Peep," and other rhymes and stories. Each doll was then packed in a box marked on the outside with the title of "her" song, so that she could be purchased according to the owner's taste and preference. The doll herself stood two feet high and was made of tin, with a pleasant, wide-eyed demeanor, movable arms with dainty fingers, jointed legs, dressed fashionably in a brocaded and beribboned hat, silk dress with fringed shawl, and tiny, black, silver-buckled shoes.

Encouraged by a flood of advance orders for his new creation, Edison had high hopes for the "Toy Doll biz." Following the successful example of the Compagnie Continentale Edison, he planned to set up a 15,000-square-foot factory in Antwerp, Belgium, staffed with "cheap labor," and compete with the Europeans. Parts sent over from West Orange would be assembled there.

Sadly, the domesticated talking doll turned out to be a mute. After the rough wear and tear of shipping, the delicate internal mechanism invariably came undone, and in ninety-nine cases out of one hundred, once she arrived in the nursery, the doll stubbornly refused to speak. By the end of 1890, the entire operation was forced to shut down.

A modest article in the May 24, 1890, issue of *Scientific American* would have been of small consolation to Thomas Edison—bogged down in yet another manufacturing debacle, but of great interest to ethnographers. The eminent scientist Mr. J. Walter Fewkes reported that he carted a phonograph with him into the wilds of the Maine woods outside Calais (which would have been no small feat, since the machine weighed nearly one hundred pounds) where he was able to record on thirty-six cylinders the folklore, popular songs, and "counting-out rhymes" of the Passamaquoddy Indians. The recordings were made so clearly that the Indians themselves recognized the voices of other members of their tribe who had spoken the previous day. The longest story preserved was the tale of "Po-dump, Pook-jin-Squiss, Black Cat and the Toad Woman," heretofore unpublished. *Here* was a noble albeit unremunerative use for Edison's machine!

"Mr. Edison must not work nights any longer," Lewis Miller wrote to daughter Mina, in response to her complaints during the height of the phonograph "wars"—which coincided with the final trimester of pregnancy with her first child—that she did not see the troubled inventor very often these days. "It will most surely break him down. It's very easy to prescribe to others when you speak from experience," Miller said, "but it is much harder to accept and act upon such advice." Sister Mary's counsel, as usual, was much more direct: "I hope you make it interesting for him and try to use your charms to make him stay at home."

Thomas Edison's work-addiction would never be completely resolved, even at the end of his life. During his first marriage Edison lost interest in the insecure and needy Mary, and the two little boys were mere distractions; little did Edison suspect that Tom, Jr., and Will would grow up to become love-starved burdens as a direct consequence of their father's neglect. And even if he did, would the realization have been sufficient to change Edison's behavior? Mina was a woman with a more cultivated view of herself, who wanted to do well by her husband, and therefore needed him nearby so that she could share with him what she deeply cared about: a lovely house, a civil and tasteful domestic scene, the intimations of a proper life. Eventually, and with credit to her special brand of stamina and breeding, Mina accomplished these goals, creating and burnishing the later, mellowed image of

Thomas Edison; as éminence grise, he would be *her* invention, *her* legacy to the myth.

At forty-one, however, Edison was more overwrought than ever, whirling in a vortex of what P. T. Barnum gleefully called "go-aheaditiveness." An observer remarked that "Edison pronounces the words 'work' and 'working' as some do 'prayer' or 'religion.' " Adriano Tilgher, in his 1930 classic *Homo Faber,* insists that work was the central concept of nineteenth-century American culture, "so closely related to modern conceptions of liberty and progress—of practically every activity—that to give anything like a complete account of it would be to write the history of economics, ethics, pedagogy, applied science—in fact of every branch of modern culture." In Edison's personal ethos, industrialism and work were wedded together "in a policy of ceaseless activity" out of a puritan value-system emerging directly from the shop culture he knew and loved best. To be at work was to be a man in the truest sense.

The ingrained nature of Edison's high-tech working life gospel can be instructively measured by contrasting it with the utopian setting of Edward Bellamy's *Looking Backward,* the runaway best-selling novel published by Boston's Ticknor and Company in January 1888. It cannot be determined with certainty that Edison the avowed reader actually read Bellamy's popular work, but he certainly would have had difficulty avoiding its message and vision—an idealized America discovered in the year 2000 by Julian West, a young Bostonian who awakens after more than a century of tranced sleep to discover what his society has become.

To Bellamy, the antidote to the present was a future of "solidarity," a society in which the venerated "unremitting toil" so characteristic of the competititive, unorganized, and antagonistic 1880s would be supplanted by a commitment to equal sharing in the nation's wealth. No longer will "many pull at the rope and few ride." The word "menial" has been expunged from the dictionary. In the ideal future, humanity is no longer defined by work, but, rather, by fairly distributed abundance. Bellamy's millennium has no need of tool sheds filled to bursting; it is a place of abolished inequities and cultural efficiency, not wasteful overproduction and underconsumption, a society where the "small concerns with small capital" once more hold sway, and the "great aggregations" and bankers have become obsolete.

Simultaneous with Bellamy's trip to the future, in August 1888 the Boston-based writer George Parsons Lathrop approached Edison with the brainstorm for a collaborative memoir in which the inventor would share his reflections upon his past accomplishments and Lathrop would polish them. In a reversal, Edison turned this proposal around and set down his idealized projections of life in a fantastic, Jules Verne–like future where technological and human improvements abound. *Progress* was the resounding working title of what would become a novel never published despite more than three years of sporadic work, as Edison became increasingly drawn into his ore-milling operations and Lathrop finally quit in frustration.

Edison's utopia, in contrast to Bellamy's, was a high-tech inventor's heaven. To judge from the thirty-three pages of notes he left behind in chaotic order, it appears that, as the book opens, the earth has been disturbed by a series of cataclysms radically altering its geography, as if to obliterate first of all the conventional expectations that the world of the future in any way bore a physical resemblance to the world of the present. Gigantic volcanoes erupt in the "unknown regions of the antarctic pale, 1100 miles beyond the utmost limits reached by explorers." The lava given off by these volcanoes heats the surrounding oceans, warming the land and making it lush and suitable for cultivation. Meanwhile, in the Africa of Edison's future, the desert floor sinks 2100 feet, creating a "Saharian Sea" and a consequently broader temperate zone there as well. In South America, a "planetoid" strikes the earth south of the Orinoco River causing "gigantic streams to pour down its declivities."

A supremely intelligent race has evolved in some mysterious fashion in response to this metamorphosis. Living in dwellings built with mother-of-pearl walls beneath the surface of the seas and lakes (since it no longer requires oxygen to breathe), this new, war-free, crime-free strain of mankind has perfected sun engines to harness solar energy; underwater photography by use of radiant heat; a uniform, international synthetic paper money system unaffected by water; the plating of all objects with precious metals; vast, shining windows formed from electrically fusible mica instead of glass; and telescopic lenses eleven feet in diameter "and absolutely free from aberration" for viewing "the mechanisms of the depths of space," leading to sophisticated communication with extraterrestrial worlds.

With *Progress,* Edison again broke his early and much publicized vow (or pretense) to employ market viability as the only touchstone for inventive activity. Edison's inventions were not inevitably linked to tangible products. Beyond the one thousand patents that remain a persistent litany in the Edison tale, we continue to discover ideas—and more ideas—as inventions in their own right, with ever greater fascination.

CHAPTER

17

EVEN AS THE EDISON CAMP WAS DIMINISHED BY THE EXILE OF
Ezra Gilliland, another one of the "boys" was also on his way out. Nikola
Tesla, a Croatian-born Serb, had originally joined the Compagnie Continentale Edison in Paris in 1882, thanks to an introduction by family friend and
Batchelor associate Theodore Puskas. Tesla had grown up in the farming
village of Smiljian near the eastern shore of the Adriatic Sea. Storklike, thin,
and pale, with jet-black hair and a triangular face, he was like Edison a nervous, frail child who kept to himself and loved to read, retreating into imag-
inary worlds and cultivating a virtually photographic memory. Nikola's father was a minister with hopes his son would enter the clergy, but the boy's
religious inclinations tended toward hallucinatory rather than spiritual.

While studying at the Austrian Polytechnic School in Graz, Nikola
became intrigued by electrical machinery, particularly the apparent short-
comings of direct-current dynamos. He continued advanced studies in
Prague in electrical engineering, then landed a job as a draftsman in the
Central Telegraph Office of the Hungarian government in Budapest. Its
regular hours and low profile permitted Tesla to develop his nascent idea for
what would eventually become the basic component of the alternating current system.

Direct current of the species exclusively and rigidly espoused by
Thomas Edison—and in his mind forever linked to the proper functioning

of the light bulb—required operation at low voltages in order to be effective and avoid potentially dangerous sparking between dynamo brushes and commutator. In Tesla's prototype, the current *alternated* in direction as the dynamo armature revolved. This method of generating electricity was far more effective for electric power transmission over long distances, measured in hundreds of miles as opposed to Edison's direct current, whose range was limited to the length of city blocks. Tesla's work eventually led him toward greater sophistication in creating what he called "polyphase" current, so that voltages ten times the level of Edison's maximum became possible.

By the time he entered Edison's cadre, Tesla was ideologically in the opposite corner. Charles Batchelor recognized the extreme intelligence of the eccentric twenty-six-year-old, despite Tesla's inability to convince the Edison men that alternating current was the wave of the future. After a two-year stint as engineering troubleshooter, traveling from one Edison installation to another on the continent, Tesla, armed with a glowing testimonial letter from Batchelor at Ivry-sur-Seine to Edison in New York ("I know two great men and you are one of them; the other is this young man") crossed the great divide.

Ever the nicknamer, Edison dubbed the new arrival "our Parisian" and put Tesla to work in the Goerck Street shop repairing dynamos. Impressed as Batchelor had been with Tesla's aptitude, Edison privately belittled his ideas as "magnificent but utterly impractical." For his part, fastidious as Edison was slovenly, Tesla was condescendingly critical of the Old Man's "empirical dragnet" lab methodology, and later observed, "If Edison had not married a woman of exceptional intelligence, who made it the one object of her life to preserve him, he would have died many years ago from consequences of sheer neglect."

The pattern was familiar: Edison, with his uncanny nose for others' talent, moved swiftly to coopt Tesla, as he had Batchelor, Upton, Insull, Kruesi, and Gilliland before him. Eventually, news of the upstart's reputation leaked out from the Edison camp. Approached by two predatory backers, James D. Carmen and Joseph H. Hoadley, the chronically (in his opinion) underpaid Tesla resigned from the Edison Company and set up shop for himself, manufacturing a refined arc light. Tesla's big breakthrough came in the spring of 1888, when he was invited to deliver a keynote speech describing his polyphase generator research before the American Institute of Electrical Engineers. The lecture, "A New System of Alternate Current Motors and Transformers," attracted the attention of George Westinghouse, the Pittsburgh railroad entrepreneur, who was systematically buying up patents in the electrical business, developing a full-scale frontal attack on Edison's direct-current fixation. He hired Tesla as a $2,000-per-month consultant, purchasing the manufacturing rights to his single and polyphase motors and transformers.

The "Napoleon of Invention" sallied forth to war again, vilifying Westinghouse as a "shyster" and routinely referring to his company as "the enemy." Edison's emotional rather than practical attachment to direct current was another manifestation of a desire to have and hold home-grown ideas against encroaching trends. For two years after Tesla deserted him, Edison spent increased time and energy in a morbid public relations campaign to demonstrate that the compelling reason to shun alternating current was because it was dangerous and direct current was safe. In the courtyard of the West Orange laboratory, media-event public executions of animals by alternating current were conducted, escalating from dogs ("criticized because the weight of the animal killed was less than that of a man") to calves to horses. Edison's attorneys Eaton and Lewis suggested in all seriousness that recommended terms for this method (which included "ampermort," "dynamort," and "electromort") were inaccurate in their derivation: "Electricide" seemed more fitting, expressing direct analogies in the English language with "homicide" and "suicide."

But perhaps the best choice of all was to invent a new verb, "to *westinghouse*. . . .* As Westinghouse's dynamo is going to be used for the purpose of executing criminals," Edison's cunning men reasoned, "why not give him the benefit of this fact in the minds of the public, and speak hereafter of a criminal as being 'westinghoused,' " or (to use it as a noun) "as having been *condemned to the westinghouse,*" in the same way that Dr. Guillotine's name was forever immortalized in France?

Edison remained adamant in entrenched opposition to high-tension electric currents as "dangerous to life." He skeptically condemned use of "high-pressure" alternating current as a thinly veiled and crass attempt on Westinghouse's part to save money in the manufacture of conducting wire. In conjunction with the publication in the *North American Review* of his latest polemic, "The Dangers of Electric Lighting," Edison mounted an ultimately fruitless lobbying effort through his affiliated companies nationwide to convince state legislatures and local municipalities out of concern for "protection of the public" to outlaw high voltages: "When the authorities require electrical pressures to be kept within the limits of safety . . . the security which the public demand will be obtained; but until then," he warned, "nothing better can be looked for than a multiplication of the casualties of the past few months."

In the thick of this pitched battle, during the spring of 1889—by which time no less than 1,100 new electric power plants across the country had become equipped by Westinghouse—Edison's perennial backer Henry Villard returned to New York from a three-year stay in his native Germany. Backed by the huge Deutsche Bank as well as by Drexel, Morgan and Company, Villard brought with him the promise of an international syndicate including Siemens & Halske and General Electric of Berlin, adding their financing to underwrite the necessary further consolidation of Edison interests. "The Edison General Electric Company" concocted by Villard was an amalgam of seven existing domestic Edison entities, including the Electric Light Company, the Lamp Company, Bergmann and Company, as well as the Sprague Electric Railway. Villard did not stop there. The restructuring of the Edison interests into regional sales and manufacturing forces segued into a further merger, which took the ensuing two years to effect, with Charles Coffin's competing Thomson-Houston Company—the street-railway giant which had just bought the Brush Electric Company and the Bentley-Knight Electric Railway Company, and which, unlike Edison, was not afraid of becoming a force in the increasingly popular alternating current arena. Edison's personal financial stake in the massive corporate Hydra threatening to engulf him dwindled eventually to 5 percent of the combined stock.

In early August 1889, Edison was off to Paris with Mina on the *Bourgogne* to attend the great Universal Exhibition commemorating the one hundredth anniversary of the opening of the States General, an event that ignited the French Revolution. Edison complained on board ship about the sparse continental breakfasts; rolls and coffee were "very poor for a man to do any work upon. I would like one American meal for a change—plenty of pie." Wrapped in a blanket against the sea spray, he sat for hours in a deck chair watching the waves surge by, thinking of "all that power going to waste. But we'll chain it up one of these days along with Niagara Falls and the wind," he declared. "That will be the electric millennium!"

As in so many past exhibitions in this era of trade shows, the Edison company commanded a huge display space, nearly an acre in the Galerie des Machines between the Avenue Charles-Risler and the Place Joffre. Aside from a forty-foot-high incandescent lamp mounted on a twenty-foot pedestal, the phonograph was as usual the crowd pleaser. A host of machines was set up around the Edison pavilion, with four double-earphone tubes attached to each one, so that dozens of people at a time could be accommodated. Nearly thirty thousand curious "auditors" came by in one month to hear the phonograph, second in popularity only to the new 300-meter-tall Eiffel Tower straddling the Fairgrounds. Two and one half years in the making—constructed over the protests of leading French artists, intellectuals, and composers as a "monstrous thing which *even commercial America* would reject"—it was the

EDISON'S DISPLAY AT THE PARIS INTERNATIONAL EXHIBITION

highest structure in the world and a consummate symbol of technological and touristic success; admission fees in 1889 alone recouped the entire construction cost.

The ex-Parisian Batchelor predicted that Edison would receive "a magnificent reception" as an international celebrity ("I knew what the feeling was over there in regard to you and there is no one who knows better than I how well you deserve it," he wrote to his boss), and he was right. The Edisons were fêted incessantly. At a banquet a few days after the opening of the Exhibition, the French Society of Engineers gave a dinner in Edison's honor. Postprandial coffee was served in Alexandre-Gustave Eiffel's private salon on the top floor of the tower. Charles Gounod played and sang for the assembled engineers.

Edison was magnanimous in his praise: "The Tower is a great idea. The glory of Eiffel is in the magnitude of the conception and the nerve in the execution. I like the French," he went on, warming to the subject. "They have big conceptions. The English ought to take a leaf out of their books. What

Englishman would have had this idea? What Englishman could have conceived the Statue of Liberty?"

At the Académie des Sciences on the Quai Conti, Edison presented the members with a phonograph upon which to record their voices for posterity. At a gala performance at the Opéra, the Edisons were given the use of French President Sadi Carnot's private box, and as they entered before the curtain rose, the orchestra struck up a stirring rendition of "The Star-Spangled Banner."

Before midday banquets and evening events, Edison held court in his first-floor suite at the Hôtel du Rhin beginning at eleven o'clock each morning, doted upon by the accustomed gang of journalists. "Gloire, comme noblesse, oblige," commented the writer for *La Nature.* Compliant, Edison ran through his repertoire of dicta without prompting, always ready to promote the next invention looming on the horizon, alluding vaguely and often to "any number of other projects," and "several new things" as a surefire way of distracting attention from current problems. "This ore-extracting machine is going to be a *great* thing," he promised, leaning with one elbow upon the mantelpiece, a Havana cigar between index and middle fingers. "Then we shall make more money.

"The phonograph—?" he went on. "We have got it into practical form . . . I have also created a pocket phonograph, a small model, the cylinder of which will take three hundred words, the length of an ordinary letter. I have the model here, and you can see it any day you like. [Pregnant pause]. *These are not, however, yet ready for sale.*"

"How are you impressed with Paris, Mr. Edison?"

"Oh, I am dazed. My head's all in a muddle, and I reckon it will take me at least a year to recover my senses," he replied, the puritanical streak slipping in, never the diplomat for long. "What has struck me so far chiefly is the absolute laziness of everybody over here. When do these people work? What do they work at? People here seem to have established an elaborate system of loafing. I don't understand it at all."

When homesickness inevitably struck, there was always "Buffalo Bill's Wild West" (according to the management of the enterprise, it was improper to speak of it as a "Show"; rather, one viewed an "Object Lesson") camped on thirty acres at the Porte des Ternes just outside the city limits for seven months during the spring and summer of 1889. "More than an exacting and realistic entertainment for public amusement," the elaborate program announced, "our object is to PICTURE TO THE EYE, by the aid of historical characters and living animals, a series of animated scenes and episodes which had their existence in fact, in the wonderful pioneer and frontier life of the Wild West of America."

Nearing the conclusion of its first lucrative two-year European tour, "Buffalo Bill's Wild West" was touted as "America's National Entertainment," a narrated, authentic series of spectacles enacting different "Epochs" of American history, beginning with Indian dancers, moving to the Pilgrims at Plymouth Rock, the settlement of the Great Plains, life on a cattle ranch, a buffalo hunt, attacks on a settler's cabin, interspersed with demonstrations of "cowboy fun" and featuring the marksmanship of Annie Oakley and William Frederick Cody himself, "representative man of the frontiersmen of the past . . . to whose sagacity, skill and courage the settlers owe so much for the reclamation of the prairie from the savage Indians and wild animals."

What an epicurean evening it must have been when Chauncey Depew, president of the New York Central Railroad; Whitelaw Reid, the American ambassador to France; Thomas Edison (sans Mina); and Buffalo Bill hunkered down together around a log table for an authentic dinner of rib roast, baked beans, peanuts, apple pie (at last, to Edison's great relief), and coffee, washed down with generous swigs from Bill's silver hip flask!

Mina for her part played the gracious, dutiful companion to her famous husband, but she dearly missed her fifteen-month-old daughter, Madeleine, cared for at Glenmont by doting Grandmother Miller—the baby had begun to walk and talk just before the Edisons left for Europe. Mina pined away for home as soon as she arrived in Paris. By mid-August, she was counting the days, even though they would not return until early

MINA WITH INFANT MADELEINE

October. The hordes of people and social pressure were wearing upon her. "It is terrible to try to do anything here," she told her mother, "everything is so crowded and every body is on the make . . . Somebody is after [Edison] all the time."

The situation was made more delicate by the presence of stepdaughter Marion, flowering into a rebellious and angry sixteen year old. She had not fared well in boarding school that year, and so Edison had taken her out of the academic scene and assigned her to a character-building tour of Europe with Mina's sisters Mary, Grace, and Jennie. No sooner had the young women got together when Dot told Mary and Jennie that she had never liked them from the beginning; she did not enjoy living in *pensions,* preferring first-class hotels; she would not continue with her French lessons; she would not obey her chaperone, but take off on excursions as she wished and where she wished; she would buy and return dresses and hats at will, regardless of the price. Dot had even indulged in some flirtations; a "flattering" young man had spent an inordinate amount of time in her room recently, to Jennie's intense chagrin, and Dot was threatening to go to London by herself to visit him, "and as she doesn't care a fudge for really *seeing* things," Mary complained, "I judge she goes to London for very different reasons."

In summary, "Marion does not seem to want us around," Mary reported to Mina. "She says she is in Europe and I think that is all she cares about now . . . the sooner we tell her that we will not feel hurt if she does not want to travel with us the more she will be thankful that we release her . . . Europe is not the place yet for a young girl of Marion's disposition."

While Dot sulked and pouted at the Hôtel du Rhin, unlike the good old days taking no part in her father's peregrinations, Tom, Jr., and William Leslie languished at Glenmont, dreading their imminent enrollment as "knew [*sic*] boys" at St. Paul's School in Concord, New Hampshire, and threatening, like their contentious big sister, to run away if they could not adjust.

———

Toward the end of his hectic Paris summer, and before quick trips to Germany and England prior to heading home, Edison had shared receiving lines and tilted champagne flutes with more than his fill of "Crowned Heads of Europe," captains of industry, and celebrities. At a dinner commemorating the fiftieth anniversary of the invention of the daguerrotype, he made the pivotal acquaintance of a fellow-scientist whose quiet but revolutionary research would crystallize Edison's as yet uncodified ideas about visual documentation.

His name was Étienne-Jules Marey (1830–1904), the child of a bookkeeper in Beaune on the Côte d'Or, raised in a devout, religious atmosphere, who by the age of eight had built a "Mr. Punch robot," and dreamed of becoming an engineer. Marey eventually settled upon medicine, and entered the interdisciplinary, idiosyncratic fraternity of science where Edison always felt most comfortable. Marey never used conventional professional names to characterize his metier; he was a self-styled "physiologist working at home," a "curio-hunter" surrounded by a "science bazaar" in his cluttered apartment on the boulevard Delessert near the Parc des Princes where he had set up a sprawling laboratory, which Thomas Edison visited the day after they met.

Beginning with his medical doctorate thesis on *Recherches sur la circulation du sang* Marey's consuming passion was to find new mechanical, electrical, and chemical devices to register and then transcribe motion. He began with the internal organs, and in 1860 the beat of the pulse (via a wrist strapon stylus called a sphygmograph), moved to animal locomotion—"the bird's flight, the horse's gallop, the insect's quiver"—and then ultimately currents of air and ripples in water, pushing to illustrate that just as Edison in his phonograph research had theorized discrete sounds to be isolate and sequential rather than flowing, movement in the natural world was also inherently atomized and particular. A cylinder phonograph recording thus was analogous to a sound daguerrotype, a "photograph of speech."

Biology to Marey was an "exigetic science" like archeology, composed of mechanical signs to be read and deciphered. It is no surprise that Marey viewed James Watt, who succeeded in graphically measuring the impact of steam pulses in the engine cylinder, as his most direct influence. By the time Edison came to call, Marey had moved beyond portrayal of wing-beats in pigeon flight via point by point needle-impressions on pneumatic drums to trapping the bird's motion over a proscribed period of time in his chronophotographic gun. In a tiny footnote toward the beginning of his classic *Mechanization Takes Command: A Contribution to Anonymous History,*

Siegfried Giedion makes the understated but telling point distinguishing Marey's ingenuous and disinterested motivations from those of his curious American visitor on that significant mid-September day: "Like most of the great nineteenth-century scientists, Marey was not interested in the market value of his ideas. The practical solutions [to motion pictures] came from Edison in the beginning of the nineties and from Lumière in 1895."

The photographic gun—which was in fact the first portable camera—had been in development for more than seven years. By the time Edison saw a live demonstration, he would have read an exhaustive description appearing in *Scientific American* as far back as June 1882. The gun "barrel" housed a camera shutter. Behind the stationary lens was a chamber with a clock movement linked to a disc on a revolving cylinder moving a strip of film forward by trigger action at the exact rhythm of twelve photos (sequential stills) per second, each film frame held behind the lens by means of an electromagnetic clamp. The cylinder could be reloaded quickly through a dark "hiding box" resembling the housing of a Gatling gun where light-sensitive film was stored.

Edison also saw the experimental building in which *le vol des oiseaux*—by this time having progressed to a seagull, a pelican, and even a duck—was measured by three cameras simultaneously. Marey had constructed a darkened, photographic barn, enclosed on three sides, open to broad sunlight only at the exposed threshold, where the bird was loosely harnessed to the pivoting arm of a merry-go-round, so as to permit free flight, but over a set trajectory. The walls of the barn were coated with lamp-black, the floor lined with bitumen, and strips of black velvet sealed the corners of the structure, to provide an uncontaminated backdrop and capture movements of flight in bright relief. Isolate snapshots were selected from all three strips of film and reassembled afterwards to achieve the most detailed composite representation. Within a year, Marey was applying the same technique to the nuances of a man walking, placing a reflecting mirror at the base of the subject's spine and tracking him as he moved away from the camera, leaving a "luminous trail" in his wake.

To decipher the many strands of stimuli influencing Thomas Edison's involvement with motion pictures, and to measure the nature of his innovations both real and fabricated, we now flash back from the meeting with Marey to an earlier nonconformist character in this act of the drama—the British photographer Eadweard Muybridge, born Edward Muggeridge at Kingston-on-Thames (with, coincidentally, the same dates as Marey, 1830–1904).

Muybridge began his eclectic career in America as an expatriate bookseller in San Francisco before the Civil War. In a stagecoach accident when he was thirty, he received a violent blow to the head which precipitated a condition of sensory deprivation. The recommended treatment for recovery included a physician's advice that Muybridge find a less hermetic occupation and spend more time outdoors in the fresh air. Thus, he turned to landscape photography, and took a job with the war and treasury departments, documenting in dramatic panorama and stereocard the new, far-flung American territories and national parks, and the construction of the transcontinental railroads.

News of Muybridge's accomplishments reached Governor Leland Stanford of California, a breeder of racehorses embroiled in a lively dispute on the possibility of "unsupported transit"—whether or not all four hooves of a horse left the ground at full gallop. Inspired by reading Marey's *La Machine Animal* in 1873, Stanford hired Muybridge to prove the phenomenon. They set up an elaborate system at Stanford's farm in Palo Alto: twenty-four still cameras lined up along the side of a racetrack, linked to trip-wires that snapped shutters sequentially as the horse sped past. The track was covered with India-rubber, so that the images would not be contaminated by dust and cinders thrown up from the horse's hooves. Muybridge next borrowed from the entertainment medium, making image chains from his racehorse footage, transforming the zoetrope, a popular viewing toy of the period, into the first rudimentary projector, by turning the zoetrope on its side and spinning it in front of a light at the rate of ten to fifteen frames per second (a speed identical to Marey's), creating the illusion of continuous movement through persistence of vision. The zoogyroscope (later the zoopraxiscope) became all the rage throughout the 1880s. But despite international wonderment and acclaim, including the praise of Marey, whom he met in Paris in 1881, Muybridge never patented his machine, which a French journalist in a moment of rapture lauded as "a magic lantern run mad."

Under the financial sponsorship of the University of Pennsylvania, Muybridge moved beyond his prototypical horse-motion studies into documentation against backdrop grids of the complete range of human and animal serial physical behavior, including that of the diseased and handicapped. His classic work, *Animal Locomotion,* was published in 1887, with nearly eight hundred photographic plates in eleven volumes.

It was here that Thomas Edison threw his hat into the ring, although for a decade he had been pondering the implications of combining sight and sound, and reading of speculations, in *Punch's Almanac* for 1879, for example, about what might happen if the inventive Mr. Edison were to "create magical moving pictures." Deep within the creased and cracking pages of Charles Batchelor's scrapbook, neatly labeled March 3, 1878, is a clipping neatly extracted from the New York *Sunday Star,* based upon an article by Gaston Tissandier that had appeared in *La Nature* in Paris. "The Kinesegraph" describes an idea for a newfangled machine that proposes to "produce a talking picture which will move and gesticulate as a man does when he is earnestly engaged in speaking." This would be done by "taking instantaneous photographs of the speaker, at intervals of a quarter or half a second, and these, after fixing, are to be placed one below another on a strip of ribbon or paper wound from one cylinder to another. As each picture passes before the eye, it is to be lighted up by an electric spark, and the man will thus be presented to us at successive moments, while the recording phonograph speaks the words which the man uttered."

During the winter of 1888, Eadweard Muybridge, himself a celebrity by now, was on a cross-country lecture trip demonstrating the fantastic zoopraxiscope. At the end of February, after a standing-room-only show in Orange, New Jersey, he stopped by Edison's laboratory for a visit, at which time—according to Muybridge's account in his preface to the third (1907) edition of *Animals in Motion;* as well as detailed contemporary articles in the Orange *Chronicle* and the New York *World*—the two men discussed combining the phonograph with the zoopraxiscope to "reproduce simultaneously in the presence of an audience, visible actions and audible words." Three months after their meeting, Edison had requested directly from Muybridge a selection of plates from *Animal Locomotion* for closer inspection.

At this time, another of Edison's specialist lieutenants was formally assigned to the research and development of a motion picture mechanism. William Kennedy Laurie Dickson, of French-Scottish birth, had originally come over from England while still in his teens to work for Edison at the old Goerck Street plant. After the move to West Orange, Dickson turned his engineering and chemistry expertise to planning for the imminent, gargantuan ore-milling operation. But his real talent was as a photographer. Dickson's images of Edison and his family at Glenmont and in the lab were widely disseminated for publicity and publication purposes over the decades of the 1880s and 1890s.

In October 1888, with Dickson's (anonymous) help, Edison filed his first motion picture caveat, in which he made the unilateral declaration of his intention to enter a field that in fact had already been well-traversed by others: "I am experimenting upon an instrument which does for the Eye what the phonograph does for the Ear, which is the recording and reproduction of things in motion, and in such a form as to be both Cheap practical and convenient. This apparatus I call a Kinetoscope 'Moving View.' " Modifying this statement six years subsequently, Edison employed more possessive rhetoric, moving the inception of his concept for a "kineto-phonograph" even further back in time: *"In the year 1887, the idea occurred to me* that it was possible to devise an instrument which should do for the eye what the phonograph does for the ear, and that by a combination of the two, all motion and sound could be recorded and reproduced simultaneously."

Edison's first realization of the motion picture machine as described in the October, 1888 caveat, manufactured by Dickson and Charles A. Brown, his chief assistant, with help from Fred Ott, was a silver emulsion-coated phonograph cylinder with 42,000 "pin-point" photographic images each 1/32 of an inch wide mounted spirally upon it, to be viewed through a binocular eyepiece salvaged from a microscope; the visual cylinder spun to the simultaneous accompaniment of a contiguous phonograph sharing the same shaft and playing the "sound track." A model of this machine may have been built in spring, 1889, but was never publicly displayed.

Edison sailed for Paris the following summer, and during his two months' absence, Dickson's team worked away in Laboratory Room 5, where they found that rumblings from the machine shop downstairs made their delicate work impossible, and so a separate building was constructed to be dedicated exclusively to motion pictures. As Dickson tells the tale, when Edison came back from Europe to the laboratory in a state of high anticipation, the movie gang was ready with a *projected* image welcoming the Old Man back and "asking him how he liked his projecting kinetoscope." Thanks to the most recent scholarship on Edison and Dickson's rudimentary cylinder research at that early date, we know that it is unlikely that the demonstration, appealing as it sounds, could have occurred as early as claimed. Marey's presentation for Edison just a few weeks previously marked the first time he had ever seen film moved continuously and successfully as a strip behind a fixed lens; a new caveat issued soon after Edison's return assimilated this principle.

Walter Benjamin, in his prophetic essay, *The Work of Art in the Age of Mechanical Reproduction,* reminds us of the novelty images in quaint little "photo booklets which flitted by the onlooker upon pressure of the thumb." And he points out elegantly that as the apotheosis of the accelerated drive to make pictorial representation "keep pace" with speech, the cinema "created its public."

CHAPTER

18

THE SEEDS FOR THE MOST CONSUMING OBSESSION IN THOMAS Edison's career were sown long before the flowering of the inventions for which he is popularly known. In the mid-1870s, working with Charles Batchelor prior to the move to Menlo Park, Edison had noted the scarcity of iron ore with the right composition for smelting in the eastern states. In the post–Civil War era the factories of Newark had come to depend upon steel derived from expensive, unprocessed magnetite shipped by train or river from the midwest, as far away as Lake Superior, and even from Cuba and Spain. This arrived in huge quantities at $3.00 per ton.

Edison's interest in the potential economic rewards of local ore concentrating grew slowly in tandem with the evolving pattern of his material needs. The dynamo era of the early 1880s required the highest quality iron; the lamp filament went through a seemingly infinite number of metallic variations; and the dissemination of electrical systems placed an even higher premium on the acquisition of fine metals. He began as always in his own backyard, testing ore samples for iron levels from William Carman's farmland at Menlo Park even as the laboratory was still under construction.

The technologies of the blast furnace, and the less sophisticated, slower means of screen-jigging and hand-separation, were well known to Edison. But his inclination toward magnetic separation of iron made the most sense. In the spring of 1880, alighting upon the iron-rich properties of common beach sand, first in Long Island and then Rhode Island, and backed by the financing of Sherburne B. Eaton, as we have already seen, he entered the first patent on what would eventually become his prototypical method: passing a stream of sand with 20 percent iron content before the face of a powerful magnet. The magnetite was drawn from the flow into a separate hopper.

Promising, in typically inflationary terms, an output of several hundred tons per month, and racking up demand before the process was adequately in place, the Edison Ore-Milling Company enjoyed a brief and unprofitable first incarnation, ending its corporate existence three years later with a $2,400 deficit. In the end, only one client, the Poughkeepsie Iron and Steel Company, had been adequately supplied.

Edison plowed forward. By the mid-eighties, his separating machinery was off the beaches and on site at various mines owned by other companies across New York State—Port Henry, Lyon Mountain, Brewster—and Pennsylvania—Lebanon, Mont Alto, and Berks County—where Edison conducted field tests with varying grades of crushed and screened ore. Once the West Orange complex was built, an entire ore separation room began operation, and Dickson led the team. Model concentration units were assembled in the laboratory and their technical feasibility evaluated at a small mine Edison purchased in Bechtelsville, Pennyslvania. From low-grade, virgin ore, more than one thousand tons of concentrate were sold to local blast furnaces. At this juncture, Edison reached a paradoxical conclusion: the scale of the processing operation had to be nothing short of mammoth to ensure profitability.

True to form, Edison drew an acknowledged expert into the West Orange cadre to reinforce his chosen direction before the "scientific men." John Birkinbine was a well-known and highly respected technician, and president of the American Institute of Mining Engineers. Some scholars believe that it may have been his idea to encourage Edison to delve into the potential of unexploited eastern mines as a source of iron supply through the concentration method. Edison hired Birkinbine as a consultant for two years, and six months before Edison's trip to Paris in 1889, the men jointly presented a paper, a virtual manifesto, codifying their conclusions for the Mining Institute membership.

"Roasting [smelting] . . . washing and screening [wet concentration]" were summarily dismissed as steam-age and cumbersome refining methods.

These two innovators in contrast were bottom feeders in the mining business. They espoused "re-working dump-piles," going back to supposedly depleted sources, waste, rock, and rejected slag, there to draw forth merchantable ore.

The key to this alchemy was the fineness of the source material, because the more effectively the ore was crushed, the more easily ore could control the desired iron byproduct that would emerge when "Edison's unipolar non-contact electric separator" was finally applied, and—all modesty aside, even in matters of scientific inquiry—"the simplicity of its construction, which is the result of patient and thorough investigation of many different designs and methods, will commend itself."

Currently on exhibit at Edison's laboratory—Birkinbine wrote in his February 1889 essay—was the magnet at the heart of Edison's system, "a mass of soft iron six feet long by thirty inches wide and ten inches thick, weighing 3400 pounds and wound with 450 pounds of copper wire, the coil being connected with a dynamo consuming 2 1/2 horsepower and requiring a current of electricity of 16 amperes and an electromotive force of 116.5 volts." The distance of the surface of the magnet from the falling iron ore material was carefully calibrated so that nonmagnetic particles continued to descend in a straight line, while the desired iron was drawn off to one side. The results, presented in a series of comparative charts, were impressive: a concentrate carrying from 62 to 68 percent iron.

Armed with this data, Birkinbine took the logical approach of recommending to Edison that he begin to bring competing mine owners in the tristate area over to the Bechtelsville site and show them the efficient machine in operation, so as to spark sales of the separator. Edison would have none of this. "I have firmly resolved to waste no time on *proving* the benefits of the process to [them]," he wrote. "I am buying mines *myself* and have lately bought 1/2 dozen and propose putting up mills myself. Some day they [the individual mine owners] will come & side with me without my parleying." Within a year, Birkinbine was fired.

Rejecting conventional wisdom in the engineering establishment, Thomas Edison was about to leap into a dusty, frustrating, dangerous, bottomless, and expensive pit where he would literally dwell for the next decade. Flush with capital from Villard's refinancing of the Edison General Electric Company, and ever the consummate cash recycler, he was intoxicated with the emancipating idea of building an entire industry from the ground up. There would be no outside investors; he did not need them, and, besides, the early nibblers from Wall Street had been more interested in the safer method of mining established rich veins of ore, rather than deconstructing mountains. "I have no intention of taking in partners," Edison wrote wildly in a private memorandum after the receipt of Birkinbine's cautionary advice and just before his voyage to Paris, setting forth boldly this new "industrial undertaking" (the word "scheme" neatly crossed out), "as my principal men and myself have made & have a large amount of money to invest."

Edison's euphoric mood and scope no longer surprise—his intention "to buy up all the Magnetic iron ore mines in the state of Pennsylvania"; his overweening confidence that even though the grade of ore in these mines had long since been judged as "too low to be used in furnaces," his new method would derive an iron product "above that of any commercial ores in the world"; the numbers seemingly pulled out of a hat resolving that at a capacity of five thousand tons per day, "we shall make at least 150 percent on the [initial] investment"; the militaristic assurance that he will conquer the "low grade, *rebellious* ore"; his insistence that the full range of machinery, from crushers to conveyor belts to sifters to ovens, had been *"perfected* so as to be able to cope with every kind of magnetic ore"; and his culminating resolve that "with this *perfect process* I will have a monopoly of one of the most valuable sources of national wealth in the U.S."

Applying Napoleonic zeal to the survey business, Edison assayers fanned out over the entire Appalachian range, from the St. Lawrence River to the Potomac, in search of ore veins suitable for the magnetic refining method. Mining rights were obtained to more than sixteen thousand acres of scattered deposits. The largest single property Edison acquired, wholly independently from the "kings of finance" he wished to avoid (for the sum of $80,000 cash, according to Charles Batchelor, who was present at the signing in Philadelphia), was a 2,500-acre site in Sparta, a town in lake-studded, rural northwestern New Jersey east of the village of Ogdensburg. The spread had been operated as a surface mine since 1772, yet Edison estimated that two hundred million tons of low-grade ore, "a whole mountain," could still be extracted from this seam two miles long and two hundred yards wide on a well-worked ridge in the Musconetcong mountains.

"This venture has all the elements of permanent success," he swore in a marketing notice for his latest corporation, the New Jersey and Pennsylvania Concentrating Works. "All the factors are known."

More than a year and a half of steady labor was required to build the essential structure of the Edison plant at Ogden to the point where all the components were up and running, and the first order could be filled: 117 tons of powdered magnetite for the Lackawanna Iron and Coal Company.

It all began with dynamite, and right away, Edison took a different and intentionally less costly approach. Rock along the Horseshoe Cut edge of the ore deposit was shattered into five-ton masses by strategically placed explosive charges in holes steam-drilled twenty feet deep and eight feet apart. (These separately operated drills were replaced several years later by an Ingersoll compressor capable of pounding fifteen holes simultaneously.)

The boulders were removed by a ninety-three-ton steam shovel nicknamed the Yellow Kid; built by the Vulcan Iron Works of Toledo, Ohio, and previously used to excavate the Chicago drainage canal, it was the largest and most powerful in America. Loaded onto open railway cars aligned on a continuous "V" track, pulled up the hill empty by engine and down again loaded via gravity, the laden "skips" proceeded to the "rock house" where electric overhead traveling cranes with ten-ton capacity picked up the railroad cars and tipped them between the pairs of "giant rolls" for the first step in the crushing chain.

These elephantine crushing rolls—six feet in diameter, slightly more than seven feet apart from center to center, and weighing seventy tons— were studded with protruding steel "knobs." Edison refined the structure of the rolls in a series of subsequent innovations; he built them as concentric cylinders, and between the outer wearing surface of hardened steel that came in contact with the boulders, and the inner driving element of cast iron, he sandwiched a zinc lining as a buffer zone designed to yield to variations in pressure caused by the contours of the rock. This became essential as the gearlike rolls reached a dizzying speed of 3500 feet per minute, at which instant the rocks were dropped between the revolving cylinders, the clutch governing the seven hundred horsepower engine driving the cylinders was released, and the hammering, pile-driving action was accomplished by sheer momentum.

Beneath the first set of rolls was a second set, called "intermediate rolls," four feet in diameter, spinning on exactly the same principle as the first. A bucket conveyor took the further-reduced chunks, now down to roughly fourteen inches in diameter, to two successive sets of thirty-six-inch

rolls that cracked, but did not grind, the rock. Finally, twenty-four-inch
rolls pounded the pieces of ore to half-inch size and less, and they were ready
for the "dryer," wherein surface moisture was removed. This fifty-foot-high
tower was heated at its base by an open-furnace fire. Within the tower, iron
plates were arranged and angled downward in staggered "salmon ladder"
fashion, so-called because as the rocks were dropped from the top, they descended alternately from one side to the other against the upward flow of
hot air.

Edison's highly automated system depended to a great degree upon
conveyor belts. When the plant first got under way, the crushed material
was moved from one building to another on canvas belts coated with water-
proofing and resting on wooden rollers; due to wear and tear, the material
lasted only two months. Toward the middle of 1891, an industrial product
salesman named Thomas Robins stopped at the mine and offered vulcanized
rubber as superior protection for the canvas. The sides of the belts were in-
geniously angled slightly inward to keep the ore from falling off.

The dried ore was placed on this continuous conveyor system and sent
to a stockhouse three hundred feet long with a capacity for sixteen thousand
tons, serving a holding function in the event that the plant-wide production
chain needed to be slowed down or speeded up. From the stockhouse, the
ore moved to the last phase of its refining trip, the concentrating building
itself. Here, the successive action of four "three-roll" mills, the central soft-
iron roll of each trio fixed, the other two flexibly mounted—predicated on
Edison's ingenious idea to combine overt mechanical power with the force
of momentum—took the ore down to fragments of slightly smaller than one
sixteenth of an inch in cross section.

These bits were "tumbled" at controlled velocity over ten sets of baffle-
mounted, inclined 14-mesh sieve-like screens, slotted in the direction of
particle flow, with .0557 inch openings, so that only the iron-rich "fines"
were delivered to the actual separators: four hundred eighty magnets all
told, in three groups. The ore was dropped in a thin stream past banks of
electromagnets with lower attracting power at the top of the series than at
the bottom, so that particles of higher iron content were removed at the outset. Intermingled ore and rock that made it through the magnetic sorters on
first pass was recirculated back up to the "three-high" mills for repeated
crushing and then rescreened and exposed to the magnet ranks a second
time. Phosphorus, an ingredient detrimental to the Bessemer process, was
"blown out" of the concentrate with a delicate air current, in a pulverized
state inherently much lighter than the surrounding iron.

When all was said and done, the final concentrate came in at a very competitive 68 percent pure iron. (The highest proportion achievable was 72.36 percent.) And yet, despite this marvelous efficiency, by June 1891, Samuel Insull estimated that Edison was running the Ogden plant at a loss of more than $6,000 a month.

While Edison tramped through the mud and supervised construction at Ogden, Dickson worked feverishly thirty miles away at West Orange on the kinetograph and the kinetoscope. Edison spent Tuesday through Saturday at the mines, commuted to the lab on Saturday evening, tried to spend most of Sunday at Glenmont with Mina, toddler Madeleine, and infant Charles, the newest Edison, then checked back in with Dickson on Monday before racing back north to the din and clamor of pulverizing.

Dickson and his new assistant, William Heise, were meanwhile making good strides, thanks in no small measure to the addition of George Eastman's "celluloid rolls," 3/4-inch-wide, flexible "transparent or translucent tape-like film" perforated along the bottom and fed horizontally through the camera, thus reminiscent in its technological ancestry of the familiar telegraph ticker tape. As Edison had predicted in his analogies with the phonograph, and as explained in the patent application signed and witnessed at the end of July 1891, the goal was explicitly "to produce pictures representing objects in motion through an extended period of time which may be utilized to exhibit the scene including such moving objects in a perfect and natural manner."

"All that I have done," Edison said with carefully crafted magnanimity in an interview at the time, "is to perfect what has been attempted before, but did not succeed. It's just that one step that I have taken." Nevertheless, in the very earliest official descriptions of their intentions, Edison (essentially Dickson, the noted photographer, working under the Old Man's imprimatur), ever mindful of present and future competitors, spoke in the most all-inclusive terms possible, inextricably linking *taking* and *showing* pictures. "Persistence of vision" was the central illusion of the kinetographic camera, the idea that isolated, equidistant "fixed and single" images taken at a high rate of speed—forty-six pictures per second, fifty miles per hour— appeared to flow together "when brought successively into view . . . [and] intermittently advanced," and the observer would then be able to witness the original scene reenacted.

By Dickson's calculation, once the camera got rolling, it could take

165,600 pictures in one hour, "an amount," he pointed out modestly, "amply sufficient for an evening's entertainment." In what may have been the world's first film loop, Dickson connected the two ends of the celluloid strip, creating a continuous band of images. When a positive was made from the original negative, and set in motion in exact synchronization with the phonograph, *voilà!*—"the establishment of harmonious relations" was accomplished, and in the rapturous rhetoric of Edison's erstwhile collaborator, George Parsons Lathrop, one would be privileged to see and hear "men walking, trees waving in the wind, birds flying, machinery in active operation."

Edison was still a locked-horns adversary of Jesse Lippincott over the North American Phonograph Company affair, and tenaciously held a vested interest in this oft-touted linkage of sound and sight, the "kineto-phonograph." "All I want is security," the inventor wrote plaintively to Sherburne Eaton that spring, "and even then its character is poor. I have put up more money [i.e., to continue developing the potential of the phonograph] than Lip[pincott] personally as I believe he had partners while he may if he pulls through make a pile of money I wont even if he does & if he dont I shall lose a great deal—This raising several hundred thousand dollars in a panic has been tough on me."

Over the coming two decades, Edison's broad-brush claims for proprietary rights in the motion picture would be challenged in an endless chain of litigation. He was constantly challenged with nineteenth-century precedents for his many assertions, among them (and others that have already been noted) the French inventor Augustin Le Prince's filing in January 1888 for an ornate camera with no less than sixteen lenses that took alternating photographs on two sets of film strips. It is not within the scope of this study to delve into the myriad wonders of Le Prince's contrivance, nor to assess the relative merits of William G. Horner's zootrope, Emile Reynaud's praxinoscope, Henri Joly's photozootrope, Raoul Grimoin-Sanson's phototachygraph, or Coleman Sellers's kinematoscope. It was ultimately Dickson who refined the critical mechanism upon which the final, viewed cinematic image was predicated: moving the film forward in equidistant and exact increments.

The Chicago World's Fair Columbian Exhibition, epitome of fin de siècle American commercial culture, was a mere two years away; Edison had every intention of making a big splash there, as he had at almost every exposition of his time. He began to talk up his new "happy combination of photography and electricity. . . . [A] man can sit down in his parlor and see depicted upon a curtain the forms of the players in opera upon a distant stage, and hear the voice of the singers." How fitting that the first members of the public to experience the results of this prototypical vision were Mina Edison and 147 of her fellow members in the national convention of Women's Clubs, convening in West Orange. After luncheon on May 20, 1891, they were taken to the photographic room to witness the perfected kinetoscope.

The spellbinding days of the magical screen in the darkened theater were still years off. To invoke the prophetic Walter Benjamin once more, what Mrs. Edison and her lady friends saw was far from the fanfare of a collective experience, but rather, hushed and intensely private, more akin to "that of the ancient priest beholding the statue of a divinity in the cella." One bent slightly, even reverently, over a simple pine viewing cabinet equipped with a peephole drilled in the lid and fitted with a magnifying glass. A miniature incandescent lamp within and below illuminated the celluloid strip as it flew by from one coil to another—stretched slightly by two wheels on either side of the viewing aperture, much resembling tapeheads—carrying "a row of little portraits" of a man (none other than W. K. L. Dickson, who can indeed be seen in sixteen filmstrip replica frames in positive reproduction at the top of page 446 of the June 1891 issue of *Harper's Weekly,* and in negative reproduction at the bottom of page 393 of the June 20, 1891, issue of *Scientific American*) wearing a straw hat, dark vest, and white shirt, and looking as if he is performing upper-body calisthenics. The homunculus Dickson "bowed and smiled and waved its hands and took off its hat with the most perfect naturalness and grace." Alas, his was a momentary celebrity. On October 31, 1891, acting through Edison's lawyers, Dyer and Seeley, the dutiful Dickson assigned copyright "of his invention—an improvement in the art of photography covered by [patent] application #408340" over to his absentee boss.

To the unfortunate offspring of his ill-fated first marriage, papa appeared as if in the distance, covered with a thin film of rock dust, stubbled and scowling in calf-high boots, overcoat, and slouch hat, receding into the tunnel of hard labor as they wended their unguided ways through adolescence.

After her stepmother and father left Paris for the states, Marion moved on to Rome with Mary Miller and their chaperone, Elizabeth Earl. Dot stubbornly did not write home except to ask her father for money, relying rather upon Mary's periodic reports to convey news of her increasingly alienated mood. While at the beginning of their travels, Mary felt that Dot did not appreciate her surroundings, she had begun of late to sing a different tune; "perhaps," her cousin conceded, "it would be best to leave Marion here," even though the poor girl did not know "the best thing to do for herself nor does she always do the right thing." In the new year of 1890, Marion fell gravely ill of smallpox in Dresden; even then refusing to consider coming back to West Orange, she remained in Europe for another two years.

Tom, Jr. was having a miserable experience in prep school. Half the letters he wrote home were to one of Edison's assistants, John Randolph ("My dear Jhonney[*sic*]"), asking for spare electric pen, phonograph, and kinetoscope parts so that in his own rudimentary fashion he could take stabs at emulating his famous father. The theme of the others was sad and ominous, a despondency that would plague Tom the rest of his days. His dread of inadequacy in his father's eyes, was at this point in his young life by turns real and imagined, because the senior Edison's affections were never clear. "I try to keep up but I have failed in every attempt," the exiled tenth-grader confessed to his "Dear Papa . . . I worry so I cannot study at all which knocks me all out of sorts . . . If I could only study at home I can look forward to unlimited time, I mean I *will* study as hard as I possibly could, because it is for my own benifit [*sic*] that I should do so, because *then* I could get through a great deal quicker, *which I know is exactly what you want me to do, is it not?* . . . Write as soon as you possibly can papa will you please. *I feel badly.*"

To his "Dear Mamma," Tom catalogued an endless series of health problems: "Having the catarrh worse than I have ever had it before, my nose is all swollen up and my eyes I cannot read at all. I have been in the infirmary ever since Saturday night . . . All the other boys are so strong and well as any body can be . . . What shall I do mamma? Will I have to have catarrh all my life?"

Two years younger, William Leslie, as he was ever wont to do, echoed his brother's misfortunes and the underlying anxiety that he, too, had been sent away to be kept out of the way of the two new "babys" that had lately inaugurated the second incarnation of the Edison family: "I realy [*sic*] cannot stand it up here. I do not like the masters and they dont like me. I cannot study because I am so unhappy. . . . May I come home and go to a public school or any school near home. If you want to put me out of Misery you will say yes if you don't you will say no." There seemed to be little expression of support from the home front, as if the boys had been left to fend for themselves in a heartless setting, where "[I am] set up, and snow balls are thrown at [me] . . . ," William the misfit reported. "Some of the boys are very cruel. They made me run the gauntlet and they used straps and sticks."

Edison's own elder brother, the shiftless William Pitt, never did make it beyond the confines of Port Huron, and lay dying of consumption and unable to pay his bills in a boarding house on the corner of Military and Pine Streets throughout the winter of 1890, finally passing away in January at age 59. Edison made a rare trip home to attend the ceremony at Lakeside Cemetery, where Pitt was buried at the family plot marked with a huge granite slab, facing mother Nancy, sister Tannie, dead now almost thirty years, and her husband, Samuel Bailey.

Meanwhile, the unstoppable Samuel Ogden Edison, tall, erect of bearing, white bearded, blue-eyed, and ruddy cheeked, the family patriarch at eighty-seven was on the road, indulging in the eternally characteristic Edison wanderlust. Accompanied by his lifelong friend James Symington, and subsidized by unhesitating, timely checks from son Thomas, the two old men had been wintering in Fort Myers, traveling in leisurely fashion there via St. Louis, Washington, Richmond, and New Orleans for the past several years. Unlike his grandchildren, Edison père discovered fame and attention in the family legacy; newspaper reporters along their route, Symington said happily, "in their comments universally admitted that [Samuel Ogden] was a great producer." In Washington, Edison and Symington were treated to a private lunch at the Capitol, escorted through the Treasury Building to witness "the making of the greenbacks," then toured the White House and actually shook hands with President Benjamin Harrison, "in his own room." In Richmond, Virginia, the editor of the *Times* took the gentlemen to the statehouse, where they met the governor, "who made us both promise that when we came back we was to come to his house and make a longer stay."

Days in Fort Myers were spent rehabilitating the Edison homestead, since its owner had not been there himself in six years; during the inventor's absence, due to the irregularities of the unsupervised caretaker's schedule, the once lovingly landscaped property had become overgrown, a wilderness completely covered in grass and weeds. Samuel Ogden the octogenarian took it upon himself to spend the ensuing two months laboring in the hot sun, "sweating with the hoe," to bring the land back to its former glory, getting rid of the rheumatism in his knees in the process, and much to the proprietary Mr. Symington's approval, completely abandoning his debilitating inclination to indulge in "rot gut liquor."

A thousand miles and more up north, the elder Mr. Lewis Miller was likewise industriously engaged, albeit in less hands-on fashion than his counterpart. His main passions remained the Buckeye Reaper and the continuing health of Chautauqua. In aid of the expanded marketing of the former, he was on the road almost continually during the warmer months, visiting field sales offices across the land. "I came to Texas to cut wheat," he wrote daughter Grace in September, ebullient in his work. "You will say, dear me, the grass has not started to grow yet [in Akron], how can wheat be ripe? We live in so many different places that wheat is cut somewhere every month in the year—how strange and wonderful! Distance is nothing. You can travel forty-five miles by train while you eat an ordinary dinner, and one hundred eight miles while a fashionable dinner is disposed of."

At Chautauqua, the Literary and Scientific Circle was expanding its list of subscribers, an educational prototype for correspondence–home learning in America. The main amphitheater was under construction, as were a sewer system and wide brick pathways to replace the packed mud of the main streets, all capital improvements and sure signs of progress.

But Lewis and Mary Valinda Miller continued to worry about Mr. Edison's lifestyle. Always taking unswerving pride in their son-in-law's many accomplishments, they knew from Mina's regular letters that the grind of the mines exacted its toll, and was wearing upon him. Their new grandchildren, upon whom the Millers doted, were perceived more as welcome "company" for their daughter while "Papa is away from you" so frequently among the "rough" men at Ogden, "out at the works all night." "It is very hard times," Mary Valinda consoled Mina, as America's nineties' economy began to reveal its weaknesses. "Money money money is the cry all over the land all fear of is how we can quiet down and be contented that is what we will have to do and how shall we do it is the question." About all her mother could come up with as a remedy for Mina's blue moods was that she deck her home with flowers.

In 1891 Glenmont did indeed formally become Mina's when Edison sold it outright to her. She viewed this transfer of title as the signal for the estate to be completely her domain. She went on a decorating binge, revising almost every trace of Pedder's remaining influence. The drawing room walls she covered with embossed paper embellished with shaded, gray-green figures on a dull gold-lined background. She hung an immense beveled pier mirror with a Santo Domingo mahogany frame over the fireplace. She filled the conservatory with a huge variety of palms, dracaena, latanias, crotons, and asparagus ferns. As had been done at Oak Place, the dining room was adorned daily with fresh flowers. For the den, Mina commissioned a painting by Arturo Tojetti, a noted San Franciso artist, to be mounted on canvas in the twelve-by-sixteen-foot ceiling recess. Her husband, as might be expected, proposed the theme, "Mercury with Electric Light," which Mina overruled in favor of "Music and Science." In keeping with her preferred activities, Mina also installed an electric Aeolian organ in the den. Against the wall leading into the dining room, she placed a unique gift to her husband from Friedrich A. Krupp, the German munitions manufacturer, a desk set made out of models of guns and projectiles—according to Krupp one of two of a kind in the world, the other in the possession of the emperor of Germany.

Upstairs, in the second floor living room over the porte cochere, "fronting on the lawn, looking eastward over grand slopes and billowy verdure to the great metropolis," Mrs. Edison had the walls redone in "rich, warm tones of terra cotta, paling from deep shades to those of red cream." Because grandmother Mary Valinda spent more and more time visiting, the south bedroom was refurbished and officially renamed Mrs. Miller's Room. With two new children in the brood (Dickson affectionately referred to Madeleine and Charles as "the younger olive branches"), the "room over library" on the third floor was transformed into a sewing room with dress form, sewing table, and gas stove.

When Mina turned her aesthetic attention to Glenmont's vast surrounding acreage, her horticultural talents were displayed to finest advantage. Aside from an ambitious vegetable garden bordered with raspberries and blackberries, the obligatory croquet lawn, and a greenhouse across the road stocked with winter-flowering roses, orchids, and gardenia, she employed the traditional Victorian bedding-out patterns of teardrop, quadrilateral, and arabesque-shaped flower beds on the southern and western sides of the mansion, ordering most of her seeds from the important firm of Peter Henderson & Company on Cortlandt Street in Manhattan. Tender annuals, begonia, various succulents, sweet alyssum, and mignonette filled the beds, all slightly mounded toward their centers and defined with light-colored gravel. Near the porte-cochere entrance, agave plants were installed in huge terra cotta containers. The foundations of the house, however, were devoid of plantings, in keeping with the prevailing belief of the day that trees and shrubs pressing too closely against outer walls were unhealthy and "the sure cause of consumption."

Behind the building, Mina commissioned an intimate lily pond, stocked with pure-white Hardy Nymphs. Clematis imported from Joseph Veitch's exclusive nursery on Long Island were another enduring favorite at Glenmont. With a nod to fashion, Mina cultivated a host of "staring trees" parading down at the bottom of the great hill toward Glen Avenue, meant to be admired across the broad expanse of lawn.

Japanese maple, dogwood, magnolia, copper beech, and chestnut contemplated in autumn from her favorite easy chair by the second-story window displayed a bright palette against the endless green grass, and in late afternoon, while the children napped, Mina tested her bird-naming knowledge in solitude.

CHAPTER 19

AT MENLO PARK, WEEDS GREW WAIST-HIGH. DOORS HUNG OPEN by a bare hinge and banged shut, pummeled by the wind. Windows were punched out or broken in all the sheds and buildings. Wooden boardwalks buckled. Rusted machine parts were strewn about, their discarded carcasses in brown grass, remnants of a prouder time. Despite the entreaties of his attorney—arguing that old dynamos and electric railway motors were invaluable as archival evidence in ongoing patent contests—Edison stubbornly neglected to maintain the old place while in the same breath he strenuously refused to sell the property, for reasons nostalgic and only half-known to himself.

And when Henry Villard stepped forward in the winter of 1891–92 to propose the ultimate merger, Charles Coffin's Thomson-Houston Company with Edison General Electric, Thomas Edison did not overtly protest, capital-starved by his ore business, admitting that here was a perfect opportunity for him to find a new market for his securities, and indeed, he alluded to getting out of the electrical business altogether in order to concentrate upon the quest for perfect iron ore. With J. P. Morgan's financing, how could the deal go wrong? Yet privately, egocentrically, Edison resented the submerging of his invested *identity* beneath this unstoppable wave in the earliest tides of corporate America. Sure enough, as soon as the matter was completed, Morgan, the loudest voice on the board of directors, pushed Villard out and accelerated the election of Coffin as president of what was now christened, simply, *General Electric.*

In one sweep, this new company assumed control of the patents of Edison, Thomson, Brush, Sprague, Van Depoele, Rice, Knight and other industrial trailblazers. The marriage of its three specialized factories was seamless: fundamental incandescent light, electrical distribution, and railroad traction on the one hand; and longstanding arc lighting, and alternating current motor manufacturing on the other. The Edison Schenectady plant sprawled over twenty-nine acres and employed nearly four thousand men, and a new face would soon appear on the scene in upstate New York, the "modern Jove," Charles Proteus Steinmetz (1865–1923), emigré from Breslau, Germany, by way of Zurich, and an alternating current expert. Established with his own laboratory at the Schenectady plant, the dwarfed hunchback with the salt-and-pepper beard, tweed suits, and domed forehead became General Electric's eccentric luminary, patent-winning engineer, textbook author, and mathematics professor in residence.

At the twenty-three-acre Thomson-Houston facility in Lynn, Massachusetts, were more than three thousand workers and the venerable Harrison, New Jersey, Lamp Works, newly reorganized by Edison himself, carried on with production under the GE aegis. Almost immediately, the Thomson-Houston engineers set to work developing a new, squirted cellulose filament for application within the Edison lamp. The production of electrical locomotives for mines and railroads also took an upswing within the company, featuring Edison dynamos housed in Thomson-Houston chassis.

Within three years, General Electric had set up a mining division and an international division with subsidiaries in Britain, France, Germany, and Canada, and had entered into agreements with more than ten thousand clients. As Alfred Chandler points out, the structure established at General Electric one hundred years ago—full-time, salaried managers in a centralized headquarters reporting to a powerhouse board of directors with ultimate veto authority—"remains today a standard way of organizing a modern integrated industrial enterprise."

Embittered that in the interests of placating corporate rivals his name was summarily dropped from the logo, Edison, although he remained a director of the new company for a decade, never attended any but the very first board meeting. What was the point of ceremonially pretending to possess management authority when he did not *really control* the day-to-day operations of the parent company?

Besides, the tensions of the merger caused the Old Man to break with one of his closest "boys." From the outset of the negotiations with Thomson-Houston, Sammy Insull had advised his boss that joining forces was the only way to go; the combination of companies was going to happen regardless of Edison's attitude. This opinion Edison viewed as unconscionable and disloyal, his suspicions further reinforced when Coffin offered Insull the second vice presidency of the new entity, and soon after, Villard approached Insull with an offer to become vice president of his holding company.

Insull at thirty-two was sensitive to his debt to mentor Edison and guilty about his attractiveness to competitors, yet wanted to avoid becoming stalled on the corporate ladder. From a personal standpoint, there was now obviously no longer the option of taking on a promotion within the ranks of General Electric without appearing expedient—therefore, if Insull could not move up, he would have to move out.

The Chicago Edison Company, capitalized at about one million dollars, was one of hundreds of urban central stations that had sprung up across the land during the heyday of Edison's franchising a decade past, and their board was seeking a president and chief operating officer. Insull tactfully seized the opportunity as a graceful way to extricate himself from the Old Man's shadow; the move to high-visibility Chicago, on the cusp of preparing for the World's Fair the coming year, would mean entering into the politically clean power-distribution side of the business, rather than the manufacturing arena crowded with newer players. Insull took the job at less than half the salary he had been making with Edison.

Sub rosa, Insull had Charles Batchelor's intuitive blessing: *"Chicago Ed. Co.,* Insull has accepted the Pres. of this Co. & gone to Chicago about 2 weeks ago," Batch wrote in his diary at the end of March 1892, underscoring with his customary pragmatism that at this problematic juncture in the life of the Edison industries, *"I think [it is] a very wise move for him."* The muckers gathered for Insull's farewell banquet at Delmonico's three months later. John Kruesi, Francis Upton, even the disgruntled Edward H. Johnson, and fifty others showed up in evening dress, setting down their swords for plowshares in one night of blurry reminiscing. Insull must have tippled a bit too much champagne, because when he rose amidst clouds of cigar smoke for his postprandial valediction, Sammy predicted that the fledgling

Chicago Edison Company would eventually exceed the new behemoth General Electric in size. This observation "caused a great deal of amusement" among the assembled multitude.

In another vicissitude of American cultural life, during the intellectual soul-searching characteristic of the last decade of the nineteenth century, William James deepened his researches into the theory of "split personality," the so-called "hidden" or "secondary self," which "could know things that the conscious self might never realize." We must similarly here attend to the two conflated voices telling Edison's story vis-à-vis General Electric. According to his adoring biographer and former personal secretary, Alfred O. Tate (*Edison's Open Door: The Life Story of Thomas A. Edison, a Great Individualist* [1938]), Edison severed his business relationship with the new corporation when it abandoned his name. The symbolism is, of course, too much to resist, as the homespun Wizard is left behind in the lurch, victimized once again by the avaricious robber barons who discard Edison as an empty, outmoded shell, like the rotting clapboard outbuildings sagging on the Menlo Park hillside.

Edison *was* disturbed by the changes taking place; he *was* uncomfortable having to fit the square peg of his rugged (or ragged) individualism into the round hole of corporate bureaucracy; and he *did* want fervently to simplify and channel his energies toward the "Ogden Baby." However, the miner was still a seasoned businessman under the surface. Thomas Edison's ledger books tell a different story than Tate's, clearly illustrating that the West Orange lab continued to work hand in hand with General Electric on a contractual basis and, in fact, Edison was the huge company's largest single outside vendor, developing, of all things, an alternating current motor during the early 1890s, what H. L. Mencken later nicknamed The Electric Decade.

The broader, more harrowing confusion in the surrounding economy played a role in Edison's leaps between different manufacturing sectors as he, like every other American entrepreneur, struggled to remain solvent during the great Panic, a severe cyclical contraction, the worst since 1837, stretching from January 1893 through June 1894. The American financial community had once more become far too confident, intoxicated by the expansive rhetoric of Social Darwinism, the seeming invulnerability of the railroads "binding the nation together with bonds of steel," and the allure of an epidemic of laissez-faire trusts; and further deluded by Andrew Carnegie's eloquent paeans to the immutable natural law of free competition. Spreading the "Gospel of Wealth," Carnegie broadcast his undying faith in Democracy, a social state in which he guaranteed that the ultimate victor in all of this native ferment, "thundering past the old nations of the earth with the rush of an express train," would be the consumer. But the tenuous economic structure built upon the banks' mania of accumulated, attenuated credit finally collapsed, and the fissures beneath it were revealed, the self-destructive tendencies of capitalism leaving all industries staggered—even, according to Carnegie, "the most spiritual, most ethereal of all departments in which man has produced great triumphs, viz., *electricity.*"

The first intimations of what eventually became paralysis at the Ogden mine can be detected in the late fall of 1892, when Edison abruptly announced a policy banning visitors from the mill, just as he had done at the phonograph works during his temporarily paranoid phase there. Precisely paralleling the shock waves in the marketplace at large, Edison hinted further that all was not well in February 1893, when he spoke tentatively of being able to bring the Ogden plant "up to full blast in the coming summer."

But by mid April, with orders slackening, he wrote, "The mill is not now running . . . I am at the Laboratory in Orange every day." Ogden descended into eerie silence until November 1894, thus falling into lockstep with the Panic timeline. During the summer of 1893, in an unprecedented break forced upon him by necessary idleness, Edison spent several weeks at Chautauqua with Mina and the children, where the Millers and their brood were hosts in the president's Swiss cottage by the lake, and the two industrialists exchanged commiserating tales of economic woe.

For the first time in his career, Edison had no choice but to approach Aultman, Miller & Company for a cash-flow loan—a loan they had trouble scraping together themselves, as Akron and Canton creditors amassed by the factory gates. And for his part, passing through the most trying eighteen months of his life, Lewis Miller, approaching sixty-five, never quite regained his former spark of bonhomie and vitality in the wake of the Panic years. He complained more and more often to his concerned children of not being able to "throw off the load" of anxiety, caught frequent colds and upper respiratory ailments, and referred to the "physical and mental weakness" he felt from the stress and strain of economic hard times, both for himself and his son-in-law. It seemed as if the national gallivanting Lewis Miller

THE EDISONS AND THE MILLERS AT OAK PLACE, 1892

had so much enjoyed until quite recently had come to an end. He took some small solace in the development of a new mower, binder, corn harvester, and thresher, hedging against a time, hopefully not too distant, when the clouds of depression would lift and the farmers upon whose trust and good will he depended so much for his livelihood might start buying again.

The dreaded Panic was not limited to the abstractions of commerce. It dwelled in the humanly emotional realm with equally pernicious force. To Edison's chagrin and mounting frustration, W. K. L. Dickson spent the first few months of the Panic winter on indefinite rehabilitative leave in Florida, alone, without even the company of his American wife, Lucie, "taken with an alarming sickness" of indeterminate nervous origin. "I am now sitting at an open window with coat off enjoying a tropical climate. Palms and the like all round," he reported. "I feel very sure that I shall very soon begin to improve and return better able to *cope* with my duties."

In his absence, however, construction of the first motion picture production studio in America, a Kinetographic Theatre, was completed at West Orange to Dickson's odd specifications. It became known informally as the Black Maria, after the police paddy wagons it purportedly resembled. Lined with black tar paper, the building was also reminiscent of Marey's barnlike studio at the Parc des Princes which Edison had visited in the Paris suburbs. "It obeys no architectural rules," Dickson said with a proud, icon-

THE KINETOGRAPHIC THEATRE ("BLACK MARIA")

oclastic fillip unusual for him, "embraces no conventional materials, and follows no accepted schemes of color." The Black Maria was an irregular oblong some fifty feet long, with a hinged top, so as to allow natural light in,
the entire structure mounted upon a circular track to follow the course of
the sun. "With its flapping, sail-like roof, and ebon [*sic*] hue, it has a weird
and semi-nautical appearance," Dickson poetically wrote. "This distribution
of light and shade are productive of the happiest effects in the films."

Unfortunately, there could *be* no films without the cameraman. Although Dickson left behind at the lab a nearly finished prototype model for
an improved kinetoscope, complete with operating and replication instructions for Edison to bring to fruition—only lacking "the right kind of lamp
and motor . . . both tests are at [Edison's] command"—the boss was gripped
by the unavoidable realization that lacking Dickson's cinematic expertise,
"nothing can be done. It is impossible for me to give it attention just now,"
he conceded helplessly to A. B. Dick in Chicago. Every passing day without
Dickson at hand meant the diminishing possibility that the kinetoscope
could be readied in time for mass exhibition at the World's Fair, set to commence with great fanfare on May Day. Without this vaunted machine, a
once in a generation Edison sales opportunity would surely be lost.

"The Chicago Exposition will contain a building of great size, devoted exclusively to the progress of electricity, and filled with machines, nearly all of
them the work of one man," the reporter for the New York *World* said to

Thomas Edison, chatting with him in the West Orange library. "If you were to try, regardless of space, how big an exposition of your own work do you think you could get up? How many machines have you worked on in your life?"

"Well, it would be hard to say," Edison replied, taking but a moment to reach full acceleration, "I have worked on as many as forty machines at one time. An exhibition of all the machines I have worked at and experimented on, if I had kept them, would cover about twenty-five acres."

As it happened, the largest building at the Chicago World's Columbian Exposition—and "the largest structure in the world under a single roof"—was the Manufacturing and Liberal Arts Building, weighing in at thirty acres, whereas Electricity Hall, adorned with forty-two-foot-high Corinthian columns, was a mere seventy feet per side. What it lacked in expanse it more than compensated for with spectacle. Shooting up toward the heavens at the center of the hall was the Edison Tower of Light, crowned by an eight-foot-high replica of a light bulb fabricated from a mosaic of thirty thousand tiny prisms, surrounded by what seemed to be pillars of fire, illuminated up and down with five thousand incandescent bulbs blinking on and off, and set into a base festooned with mirrors.

Electricity was the common denominator of the world's fair, as "the mightiest agent known to man," an endless necklace of garish displays. Timed squarely into the midst of a crumbling economy inhabited by three million unemployed, the fair was the apotheosis of American advancement with the ironic motto, "I Will": an unreal city. Streams of multicolored lights played over tumbling jets of water in electric fountains producing "illusions of great sheaves of wheat, fences of gold, showers of rubies, pearls and amethysts," foaming into lagoons and grottoes through which subaquatic stars shone, while artificial electric flowers leaned into their reflections, and electrified sea serpents wriggled by, "winking their fiery eyes and blowing sparks from their nostrils." Nocturnal submarine divers finned through the depths, carrying searchlights and setting off electric torpedoes, while above the swimmers, "neat and trim little electric launches," gondolas, and skiffs carried passengers to and fro from building to building, or to alight upon the Wooded Island sanctuary at the center of the artificial lake. Gleaming crystal domes, bridges, porticoes and flag-draped roofs metamorphosed into constellations, contours etched against the night sky by the "great white suns" of electric searchlights probing the gloom "in silvern paths as straight as Jacob's ladder," amidst the sizzle of colored arc lights.

Four hundred years after Columbus landed in *his* New World, more than twenty-seven million visitors filed among a promenade of five hundred neo-Renaissance white buildings, conceptually designed by Daniel Burnham and Richard Morris Hunt and laid out on seven hundred acres by Frederick Law Olmsted, experiencing the most famous fair ever conducted on American soil. An entire park on Chicago's South Side had been dredged, drained, leveled, and rebuilt into a magic kingdom, The White City, where gigantic temples named in highbrow, ostentatious tribute to America's advertised strengths—Agriculture, Fisheries, Transportation, Arts, Horticulture, Machinery, Mines, as well as one building for each state in the union—marched in awesome symmetry around a Grand Basin 2,500 feet long, to form "a fairy land, an enchanted place."

In contrast to the high-culture profile of the White City was the other side of the tracks, the alter ego of the fair, the Midway Plaisance. This was a bustling aisle one mile long and six hundred feet wide behind the Women's Pavilion, where the rest of the world was represented—a kind of hoi polloi Smithsonian Institution with multicultural sideshows, dominated by the 1,200-ton Ferris Wheel, its axle the largest steel forging in the world. You strolled along the lowbrow, decidedly motley, "lighter and more fantastic" Plaisance sipping the latest culinary marvel, carbonated soft drinks, and eating your boxed lunch while viewing theatrical entertainments from the Orient, "the Samoan and Dahomeyan in his stucco hut, the Bedouin and the Lap in their camps and black tents, the delicate Javanese in his bamboo cottage," German beer gardens, reproductions of Amazon villages and ancient cities, wild animal performances, magicians, and exotic dancers such as "Little Egypt, the darling coochie-coochie girl of the Nile"; from a distance, over the din, you might be lucky to hear a young pianist named Scott Joplin hammer out his newfangled, catchy ragtime music.

How fitting that the entire glorification of antiquity's values in the modern world, the antiseptic illusion of the White City, was predicated upon a cheap, artificial substance called staff, an amalgam of jute, plaster of paris, cement, glycerin, and dextrin, a faux-marble that made the buildings and statues glow for the six-month life of the Fair, only to be immediately demolished without a remaining trace.

Henry Adams, "drifting in the dead-water of the fin-de-siècle," with his customary knack for stripping the mythic masks from American culture, saw through the surface of the staff and seized upon the fair's heart. Even though so seduced by its majesty that he visited twice—as a spectacle, as stage decoration, the Chicago fair was admittedly head and shoulders above the 1889 Paris show—to Adams, the Columbian Exposition lacked coherence. Like the melting-pot stew from which it arose, the fair was a Babel of incongruously juxtaposed "half-thoughts and experimental outcries." He enjoyed meandering through the grounds, but, as an American, Adams felt the fair could not truly define him—nor, by extension, the confused psyche-at-large of his country at this contradictory moment in its history. In its "plastic" nature, the Chicago fair falsified history and made unattainable promises, holding up the unproveable, unconvincing vision of a better future. "Helpless before a mechanical sequence," Adams remained finally still skeptical, not sold on technology's ultimate benefits for the common man. Precisely the contrary, he sensed around him an unavoidable, difficult "question—whether the American people knew where they were driving." He could not shake from his mind the realization that economic troubles, "the convulsion of 1893," were just outside the gates, pressing forward on all sides regardless of what was going on within. The fair was to Adams a far too hasty attempt to define the nineteenth century before it was completed, for even the great god Science could not carry forward without money.

At the other end of the spectrum of belief stood William Dean Howells who, like Hamlin Garland, Charles Eliot Norton, and other more mainstream social observers of the time, wanted to invest wholeheartedly in the fair's message, which he read as Altruism, pure and simple, the best example of Christian brotherhood, the perfect antidote for society's ills. Returning from a week spent in Chicago in September, and writing as part of a long series in *Cosmopolitan* of December 1893, Howells ascended to a sublime frenzy, insisting that the fair "gives men, as nowhere else on earth, a foretaste of heaven." The millionaires who spent their resources so freely on the vast pavilions did so in order to clarify America's successful image to the rest of the human family, to hold up a standard of pure, capitalist excellence. In this vast metropolis, the wealthy and self-sacrificing guiding fathers united all the arts, with none seeming to be envious of any other. How egalitarian, and thus, how quintessential a place, where the "economically inferior are . . . the peer of any and every other American." The references to Greek architecture were not, as Adams had written, made with the motive of concealment, but rather, an homage to the standards of Greek society, antiquity's faith in "the intellectual over the industrial, art over business," a direct tribute to our native culture, and a reminder to Americans of the beauty on their doorsteps, a vision of the good life and abundance all Americans must continue to cherish.

Thomas Edison had made his indelible contribution to America's material culture long before the launching of the Chicago world's fair. The inside of the Hall of Electricity looked like a veritable Edisonian attic, with serried ranks of dynamos (of special intrigue to Henry Adams, who lingered before them long and pensively), flat irons, sewing machine motors, dining room fans, elevators, windlasses, speaking dolls, and of course, cylinder phonographs in great measure. Even though the kinetoscope did not make its guaranteed debut, Edison's miniature electric railroads, locomotives adorned with gay canopies, carted visitors about the grounds. And even though the Fairgrounds were illuminated by the flow of George Westinghouse's alternating current, the sublime glow of the myriad bulbs was indelibly etched into the American consciousness.

The greatest tribute to Edison's stature in America's collective value-system at this paradoxical moment—regardless of his parlous economic status or declining inventive activity—came in a series of nationally disseminated newspaper features published as a prelude to the opening of the Fair by the American Press Association, a ready-print syndicate based in New York City. The Association commissioned seventy-four commentaries on the American scene, asking grass-roots public figures and recognized pundits of the time to predict aspects of their society in 1993—an exact century away. A uniform roster of thirty-three questions was sent to all potential respondents, including: *"What American (now living) will be the most honored in the 1990s?"*

The poet Ella Wheeler Wilcox saw Edison in that position, because he had "solve[d] the great question of the true relation of capital and labor." Former governor of Ohio and secretary of the treasury Charles Foster singled out Edison as "the greatest genius of this century . . . If we may estimate his future accomplishments by what he has already done, he will succeed." Van Buren Denslow, distinguished legal scholar and author, likewise marked Edison as "most honored by the scientific class." And so did Cincinnatus Hiner "Joaquin" Miller, grizzled frontier-American artist and author— "Who will be best remembered?" he asked rhetorically. "Why, Edison, of course."

Despite Henry George's well-grounded fears to the contrary, as long as "progress" was still America's most important real, or even imagined, product, the iconic image of Thomas Edison as its human epitome would hold sway. And progress could only come about through "trial, *the ultimate test of scientific theory thus formed,*" declared Lester Ward, in his popular 1893 classic, *Sociocracy,* "and [it] may, in social as in physical science, either establish or overthrow hypothesis."

Edison could have been trying to dramatize Ward's dictum when he encountered Charles Steinmetz for the first time in the Hall of Electricity at the Columbian Exposition. Introduced by a fellow engineer, Rudolf Eickenmeyer, Edison, jokingly pointing at Steinmetz, said, "Pure theory." He then turned to Eickenmeyer and said, "Theory and practice." Finally, Edison pointed to himself with a wry smile and said, "Pure practice."

"This is the attitude Edison has always taken," Steinmetz thought, "declaring himself a mere practical man, and the newspaper men have expanded on this and so created the popular belief that Edison does not know anything about theory and science, but merely experiments and tries anything he or anybody else can think of. There is nothing more untrue than this," the new star in the General Electric firmament testified. "It is true Edison never went to any college—but he knows more about the subjects taught in colleges than most college men."

At the time of the gala opening of the world's fair, there was only one barely operational nickel-in-the-slot kinetoscope prototype in existence. On the evening of May 9, 1893, this valuable item was carted over to the Brooklyn Institute of Arts and Sciences for the first public demonstration of a motion picture (albeit only thirty seconds long), *Blacksmith Scene,* before a professional audience of four hundred hosted by the physics department. Filmed in the Rembrandtesque (as Dickson called it in his customary romantic style), chiaroscuro setting of the Black Maria, the brief film was a throwback to the olden days, when men in the shop culture could give equal time to fraternization, sledgehammer and tongs, and, in this case, intermittent imbibing from a laughingly shared beer bottle.

The president of the department, Professor George M. Hopkins—who had certainly done his homework—gave a scholarly introduction in which he traced the genealogy of the machine from its roots in the phenakistoscope and zootrope, through Anschutz and of course Muybridge, invoking the pervasive theme of "persistence of vision . . . in which the image dwells upon the retina," a key concept that would prove within just a few years to be the linchpin of Edison's interminable patent defense arguments. Earlier presentations of human and animal motion, Professor Hopkins pointed out through the means of actual magic lantern projection, were unrealistic, abrupt, not "continuous," "realistic," or "pure," in comparison with Edison's. Selected still images from *Blacksmith Scene* were revealed upon the curtain-screen to illustrate the clarity of Dickson's photography.

At the conclusion of the lecture, each and every member of the audience was invited up on stage to peer through the kinetoscope peephole and view the entire "perfectly smooth and natural" show. Despite the success of the prototype, Dickson needed to tinker for six months to resolve one more recalcitrant function before it could move into mass-production: the film strip had a tendency to overheat because it passed by too closely in proximity to the light source. This defect was not a major concern as long as the films stayed within their parameters of twenty seconds to one-half minute; however, 150-foot-long features were on the drawing boards. Therefore, Dickson hit upon an interposed slotted shutter just large enough to admit a pinpoint of light focused by a reflector positioned underneath the lamp. The shutter protected the celluloid from excessive heat, which meant that a cumbersome alum tank, serving the same role as a radiator in a car engine, could also be eliminated from the already congested kinetoscope housing.

The Black Maria finally went into full production as a movie studio in early 1894. "The *dramatic personae* [*sic*] of this stage," wrote Dickson, "are recruited from every characteristic section of social, artistic, and industrial life, and from many a phase of animal existence. . . . Of human subjects we have a superfluity, although the utmost discrimination is essential in the selection of themes." Dickson was less than candid, if we take a look at his earliest short-film subjects. The first kinetographic topic of the year was none other than mechanic Fred Ott sneezing—a likely choice, because Ott was known around the Edison plant as an exhibitionistic ham, always ready and willing to cut up before the camera. On another occasion, Ott allowed himself to be swathed mummy-like entirely in white bunting, harnessed to a leash, then filmed as he jumped up and down in enthusiastic imitation of a wild monkey. A phonographic "sound-track" of Ott's "explosive expiration" for the sneeze film was also produced, but unfortunately lost.

Gearing up to unleash the kinetoscope as a wholly commercial venture, which was certainly Edison's vaudevillian intent, Dickson kept his films within a fanciful, humorous, music-hall realm, as if he were the ringmaster of a circus, rather than making the slightest pretense of being an au-

FRED OTT'S SNEEZE

teur or practitioner of the cinematic art. Recalcitrant Hungarian dancing
bears ("one furry monster waddled up a telegraph pole, to the soliloquy of
his own indignant growls"); amateur gymnasts and professional contortion-
ists; boxing matches between cats as well as men, including the screen
debut of "Gentleman Jim" Corbett; Spanish, highland, and can-can dances;
marching bands; Mexican knife-battles; trapeze, tightrope and cane exercises, and fencing were typically entertaining spectacles, wherein the revela-

tion of dramatic movement itself was equal in star quality to that of the actual performers. The famous strongman Eugene Sandow was a special favorite, scantily clad, contorting his rippling muscles front and back. Buffalo Bill's Wild West featuring the spunky sharpshooter Annie Oakley even came to town in one film, marching in full regalia down Main Street. These displays were counterbalanced by vignettes of pursuits especially chosen for the male consumer: cockfights, horse-shoeing, wrestling, speechmaking, and barroom brawls.

In the midst of this fervor, Edison received a note from his old colleague Eadweard Muybridge, inquiring as to how the motion picture work was progressing. His perfunctory and circumspect reply hints that Edison had once again donned his master-marketer hat: "I have constructed a little instrument which I call a kinetograph [i.e., kineto*scope;* Edison often interchanged these usages in conversation and correspondence, perhaps because their operations were so interdependent] with a nickel and slot attachment and some 25 have been made but I am very doubtful if there is any commercial feature in it and fear that they will not earn their cost. These Zeotropic [*sic*] devices are of too sentimental a character to get the public to invest in."

On the contrary, while thus throwing hounds like Muybridge off the scent, Edison was in fact on the brink of establishing a new corporation, The Edison Manufacturing Company, to serve as an administrative umbrella to handle all of his motion picture business under one imprint. Alfred O. Tate, along with Norman Raff and Frank Gammon, became principals of The Kinetoscope Company, the first of several established sublicensees of the Edison Company. On April 6, 1894, the first twenty-five kinetoscopes manufactured at West Orange—the same machines Edison referred to in his note to Muybridge—were shipped, five to Atlantic City, ten to Chicago, and ten to a small former shoe shop in Manhattan, at 1155 Broadway between 26th and 27th Streets near Herald Square. The brothers Andrew and Edwin Holland, who had worked with Edison on the phonograph, labored feverishly to convert this establishment, leased for them by Tate, into the first commercial "kinetoscope parlor."

It was an oblong room with a highly polished floor, adorned with potted palms on pedestals along both side walls and a plaster bust of Edison in the window painted to look like bronze. The kinetoscopes were lined up along the center of the room from front to back in two rows of five each, protected by a brass railing at waist level upon which you leaned so as to obtain the best angle of vision through the peephole. To view one row of machines, you would purchase a twenty-five-cent ticket at the booth, "presided over," according to Tate's gleeful plan, by "an attractive young woman." Two rows cost fifty cents; there were no bargains in the kinetoscope parlor.

All was in readiness on Saturday afternoon, April 14, for the coming Monday's official opening. The first ten strips, on translucent film manufactured by the Blair Camera Company, arrived that day, among which we know from contemporary accounts that four of the most broadly appealing were to be shown: Strongman Sandow, his muscles "swelling and relaxing"; *Highland Dance; Organ Grinder;* and the infamous *Trained Bears.* Alfred Tate; his brother, Bertram, serving as parlor manager; and their colleague Thomas Lombard were lounging in the rear office, smoking cigars and chatting, when they noticed an ever growing curious crowd outside on the street, faces pressed against the glass for a better glimpse.

"Look here," Tate said, pointing toward the window, "why shouldn't we make that crowd out there pay for our dinner tonight?" His cronies got the message. "What's your scheme?" asked Lombard with a grin, imagining the delicacies awaiting them at Delmonico's, just one block south.

"Bert," the cocky Tate said to his brother, "you take charge of the machines. I'll sell tickets, and Lombard, you stand near the door and act as a reception committee. We can run till six o'clock and by that time we ought to have dinner money."

They "all thought it was a good joke all right . . . but the joke was on us," Tate remembered. The throng did not abate until past one o'clock in the morning, at which point, pocketing their extraordinary $120 in cash, the kinetoscopic tycoons repaired "to an all-night restaurant to regale [themselves] on broiled lobsters."

Within a year, The Edison Manufacturing Company had established franchise relationships with two additional syndicates beyond The Kinetoscope Company—one to sell kinetoscopes overseas, and the other to film boxing matches, which would be then shown in saloons. At $200 list price per machine, cash flow surged. According to company records, revenues from kinetoscope and film sales came to $177,847.23 by February 1895.

In the meantime, daughter Marion, home from Europe for a brief hiatus, found herself not surprisingly to be unwelcome at Glenmont, and boarded instead with her Aunt and Uncle Stilwell in Newark, spending much of her time trying unsuccessfully to gain her father's attention in order to salvage what was left of Menlo Park, still in her long-departed mother's name.

Yearning for her European friends, Marion made plans to go back to Germany. One Friday evening in early March 1894, she came by the Edison family estate to say good-bye. Her father insisted she was showing her face only to claim her monthly allowance, and refused to come downstairs. Although Marion was due to sail the next morning at 8:00, Edison prohibited Mina from seeing her stepdaughter off.

Firmly back in the black, and for the time being blithely unperturbed by an increasingly unregulated marketplace and the lengthening specter of hungry imitators, Edison returned to his beloved "Ogden Baby" after too

THE OGDEN-EDISON MINE

long an absence. In a burst of restored ego-centrism, he rechristened his industrial "baby," changing its name from Ogden to Edison. The mining site was by now taking on the characteristics of a small town. More than fifty workmen's shanties had been built, in four hilltop neighborhoods set back from the three major cuts: Mudwall Row, Summerville, Cuckoo Flats, and New City. Seventy-five more dwellings would follow as the plant burgeoned before the end of the century. There was even a shop selling cheap beer to the men, in order to keep them from leaving the grounds during the evening. A schoolhouse came next, to entice some of the managers' wives and children to join them.

In its earliest years, Edison had no permanent dwelling place at Ogden, eating his meals at nearby farmhouses or boarding houses, and sleeping in the oven room at the separating plant. By the time of his return to the rocky landscape, Edison had built himself a two-story house devoid of porch, yard, or landscaping, set into the middle of the ubiquitous rubble behind a meager stand of trees overlooking the railroad tracks on a rise northwest of the main mill complex.

The challenge facing Edison with the reopening of the mine in November 1894, was the necessity to change the form of his refined, crushed concentrate from powder to briquettes, for two primary reasons: The railroads shipped the material for a much lower rate if it was in open cars, impervious to the elements, and sturdy enough to withstand the long trip; and the furnacemen in the iron works clamored for a porous, "agglomerated product" that was easier to handle.

The recipe for Edison's briquettes began in the mixing house, where the concentrates arrived on conveyor belt from the stock house and the powder was dumped into horizontal troughs resembling cylinders with their tops sheared off. Here, the binding element was added incrementally, one-third at a time, to the fine ore. As we would expect from Edison, it went through several permutations, including lime and bricking clay, or rosin and clay, or even molasses, tried because of the ease with which it decomposed during the ensuing baking phase. Edison further moistened the dry ore with petroleum to make it "sludgier."

The rich, black paste was carried along on the conveyors beneath rotating die blocks like giant cookie cutters with cylindrical molds one and one-half inches high and three inches in diameter. Compressed oil was jetted into the dies to coat the molding surfaces and prevent the concoction from sticking inside. Each die was filled at 800 pounds per square inch of pressure, relatively loosely, only to receive a plunger at a pressure of 14,000 pounds per square inch. A partial revolution of the diewheel brought it face to face with a second plunger, where a final pressure of 60,000 pounds per square inch was applied. The resulting briquette, weighing nineteen ounces, was dropped onto another conveyor, meshed so that excess material descending from the molds passed through, leaving the briquette clean and smooth.

On into bucket conveyors, each briquette was baked in one of fifteen vertical ovens twenty-two feet high, making five slow circuits in an hour at a temperature between 400 and 500 degrees.

Edison was immensely proud of the fact that it only took a little more than two hours to process an entire railroad car full of raw ore into finished briquettes. His mania for efficiency knew no bounds; he was determined that cheap, mass production would win the day. Conveyor belts, centralized oil supply, special safety bolts acting as a fail-safe system to shut down strained machinery, a network of telegraphy and messenger boys linking the outlying buildings, time-accounting for each machine to head off potential gridlock, gravity-fed dryers, continuous shifts, greater and greater electrification—every manifestation of the plant reflected Edison's fanatic drive toward automation and the ultimate goal, that the ore would never be directly touched by human hands.

Sixteen long years before his breakthrough classic *Principles of Scientific Management* was published, Frederick Winslow Taylor, an engineer at the Midvale Steel Company in Philadelphia, was developing his theory of a new "task system," even as Thomas Edison worked to streamline his ore milling plant. For both men, piecework was fast becoming a thing of the past—and besides, Edison had never been a real convert to that process in the first place. Taylor declared, "Let each workman's interests be the same as that of his employer." Let the management of the company align reward with overall *rate* of production. Do not dwell upon the value of the whole job of work; divide tasks up into their component parts so that the thing produced has the proper egalitarian, collective flavor. "Pay men, and not positions," wrote Taylor; instruct middle-management to spend more time in observation of personal acuity, punctuality, integrity, and accuracy. "The modern manufacturer," which Edison was absolutely pushing to become, "realizes the importance of system and method" and shifts the emphasis from the appearance of men working hard—"soldiering and slacking"—to the application of men's cooperative energies toward making machines work hardest of all.

At the grimy, noisy, manly Edison Town in the North Jersey moonscaped hills, it was the dawn of the assembly line, of an era of faith in production as an end in itself. The same was true at the quieter, higher-tech world of West Orange, in the outlying factory annexes, where "the long, airy and sunny buildings are filled with mysterious machinery, and the girls at the benches sitting before open windows through which the perfume of cut grass comes pouring in and fresh breezes sweep, meet visitors with happy smiles and courteous manners."

"We like women in all work that is not too heavy for them," said superintendent William Young. "Their fingers are so nimble and their work always neat. Mr. Edison favors women whenever he has work they can do.... Women can do many things better than men. Mr. Edison favors their employ purely from a business standpoint."

The dynamo department was filled with young women fabricating the essential brushes, unwinding wire from spools, straightening it, then cutting it into one-half inch lengths; and soldering wires to governors and commutators with a gas-jet. At the Phonograph Works, "cheerfulness, calm, gay eyes and content seemed the greatest virtues." Drive belts were fashioned from dressed calfskin, cut into strips, stretched and shaved, placed overlapping end to end, then glued together, adorned with blue and red silk threads to keep the finished belts from stretching.

In the Kinetograph Room, a comely teenager named Ella earned $18.00 a week meticulously covering brass film pickup spools with fine velvet. " 'Mr Edison *never* gives the men what we can do,' " she was quoted as saying with pride, "all laughter and saucy remarks," looking upward momentarily from her labors to reveal a fresh, rosy face, "blue eyes, large and merry, long lashes and the curliest, golden-brown hair that ever covered a head."

"We consider ourselves very fortunate to be included among the workers here," said Ella. "It does not seem difficult to wind these spools, but every part must be just so, or the whole thing is useless. One act of carelessness might undo a whole day's work."

CHAPTER
20

"My mother was very capable, and she was very busy.
But it wasn't so much fun for her when [Father] was up at the mine," Madeleine Edison recalled for an interviewer, looking back over eight decades past. "You see, for ten years he was commuting up there.

"We went up there once," Madeleine continued, remembering the bright November day in 1894 when the Ogden mine reopened, rechristened as Edison with great fanfare. "We drove up. It took us two days to get there. That was the first time I think I ever realized I had a father. He used to be away all week except Saturday when he'd come back. He was just a 'presence' to me.

"One time I remember especially that we got all dressed up, and I had on a high collar which I hated, and my mother was dressed in a perfectly lovely summer dress and a beautiful hat—and oh, she looked wonderful. She was very beautiful. We got in the open carriage and we all went down to the station to meet Father. I hadn't realized I had a father at all before that, and when the train came in, of course it was soft coal, it was a hot day, and the most bedraggled man I had ever seen got off that train, just covered with soot. This figure came and dashed into the carriage and hugged Mother, with all her beautiful clothes, and I thought, 'My goodness, is *that* my father?' "

To the wide-eyed child of six in the carriage rumble seat, Father ap-

peared little more than a
smudged and grimy
shadow. And for his young
wife, not yet even thirty,
Edison was conspicuous by
his absence. The conflict
that had torn his first marriage apart remained unre-
solved and now achieved
full flower. Mina's side of a
remarkable five-year correspondence with her hus-
band is lost, but Edison's
voice is all too familiar.

Over and over again,
assuming an immutable
stance, Edison tries to coax
and cajole Mina into com-
ing to visit him: "I do not
see why you can't stay for a
week at a time. You can
talk to the children [by
telephone] at noon and
night and it will be just

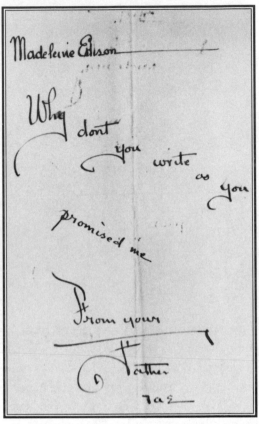

PAPA'S NOTE FROM THE MINE TO MADELEINE

like being at home, if they get a cold or anything wrong you can be home in
3 or 4 hours. *I can work so much better when you are here for then I am satisfied and
my mind is at rest.*" Even Edison's ardent promises to install running water in
the misleadingly named rustic cabin where he lived, The White House, are
inadequate for Mina, accustomed to creature comforts and finding no delight in the thought of roughing it.

Edison had long since come to rest on the assumption that he was the
irreplaceable cog in his business design, and this conviction made it even
more difficult for him to comprehend why Mina could not concede to his
wishes and be the flexible player in their game. Couldn't she understand
how much sheer work there was to be done at this godforsaken place, by
him alone, and no other? "Are having trouble with oiling our elevators," he
wrote. "I was up night before last all night, and last night till 3 a.m. getting
out of the trouble. I think I have it now." Oftentimes with thin excuses he

PAPA AT HOME WITH HIS "TWO BABIES," CHARLES
AND MADELEINE

gave the lie to Madeleine's memory that she saw her Papa even as regularly as every Saturday: "We sent 844 tons through the crushing plant and 1018 tons through Mill Number 1. On Monday I start the briquetter. Tomorrow is Saturday and I feel lost in not going home to see my darling dustless Billy"— Edison's pet name for Mina—"What am I to do without a bath, some smartweed seeds have commenced to sprout out of the seams of my coat, [plant superintendent Walter S.] Mallory is going to send for a package of flower seeds and plant my clothing. Think of it Billy darling your lover turned into a flower garden!" Other Saturdays would often be maniacally consumed by corporate matters, hours plotted out with seamless precision, leaving virtually no space for Mina and the children. "The General Electric Company dinner is Saturday night at Delmonico's at 7:00. So, Batch and I will leave here on the 11 a.m. train Saturday morning. I will arrive I think 2 p.m. at Orange. I can then kiss my darling Billy and her babies, clean up, and go into New York on the 6:10 train, returning home on the 12 midnight."

Thomas Edison drew sustenance from his regressive descent into the workers' world he once legitimately inhabited and still desperately craved. He was proud to rediscover the hands-on way as the only way, plunging into the thick of the maelstrom to solve whatever "trouble" lurked there, "lying down in a closed box with dust pouring so thick that I had to get my eyes within three inches of the object. I became so choked that I was taken with a faint," he told Mina with relish and not a shred of complaint. "It seems the only way to find out just what is wanted . . . How little you realize what I have to do here," he chastised his wife. "I have no-one to help—owing to the immense number of things (details) which must be attended to instantly night and day—every one requiring thought. I could not leave for a moment. It makes me feel blue to think you so little realize the greatness of this undertaking, the whole carried out by one man."

Edison's torrent of language defines a solitary protagonist, surrounded by machines that heed only *his* bidding. We must step back and remember that, on the contrary, during the span 1894–99 he employed a cadre of more than four hundred workers. "I scarcely get any sleep as everything has to be attended to by me. All depends on my self. To come [to Glenmont] now it would be necessary to shut down at a most critical time technically and financially. Had I an intelligent assistant I could have come but while I am trying to obtain such a man I have not yet succeeded . . . I have to watch every move made here so that there shall be no mistake made and the mill will be a permanent success."

What was the character of the marital love within this mind crazed by stolid resolution, dominated by grinding gears? Having taken the position that he could never be detached from his workplace in the mountains, unable to sway his civilized wife to abandon her refuge of domestic order for life among the dirty, ruined hills of North Jersey, Edison tried to breathe romance into the impasse. With astonishing constancy tempered by heavy dollops of manipulation, he tried to assuage Mina's loneliness made larger by their extended separations: "If you knew how much I love you darling Billy you would never fear or worry for an instant about such things [flirtations] or ever be jealous. Don't you know that the fixed law of the organic world requires one man for one woman and that all normal well-balanced men never have the least desire to contravene that law, it's only the ill-balanced degenerate and conceited egotist that does such things." It was a pedagogic tone Edison had employed with Mina since the first time they met, the mature, worldly gentleman recognizing his responsibility to educate the much younger woman about the true ways of society. "Darling Edison," he waxed poetic. "You are mean to doubt me as you did in your last letter. Your image is always in my mind, and no woman has a truer lover, a long distance lover, an unchangeable lover. You never think about your side of the Compact, you are not a lover, only on occasions do you impress me as loving me *in any event it is not a strong deep love like mine.*" Only by insistently building the edifice of his "love," and quantifying his side of the equation, could Edison rationalize the fact that the couple were not seeing each other as often as Mina wished.

Her concerns remained undiminished by his suavities. The devoted Mina felt insecure, unable to measure up to the hardship her husband imposed upon her; at the same time Edison by turns demanded loyalty, or simply dismissed her anxiety: "There isn't a woman in 20,000 that is really so smart as yourself. Their apparent smartness is entirely superficial on account of their gift of gab—their judgement isn't worth a cent, your lack of self-confidence is the trouble. Read the newspapers Darling Billy and stop novels. . . . Blues are from a disordered liver or some long sinuous intestine that blushes from inflammation."

With implications not always subtle Edison could not help countering that he was actually the truer mate, and Mina was not always responsive when he finally did appear in the flesh, and needed *her*: "When I come down Saturday I am going to love you to pieces, i.e., I am going to try, so please don't start in again on forbidden subjects. . . . I am commencing to miss you Darling Billy and will be very strong by Saturday. . . . I will be good and come home early every night. . . . Some day Billy darling you may love me and then you will always be *Grace*, and not *cold Mina*. . . . I wonder if you will be cold arctic like should I come out in three or four weeks."

William James should have been thinking of Thomas Edison in his 1897 essay, "The Will To Believe," when he wrote, "If we look at the history of opinions, we see that the empiricist tendency has largely prevailed in science. . . . I am, myself, a complete empiricist as far as my theory of human knowledge is concerned." And Brooks Adams should have been meditating upon a portrait of Thomas Edison's muddy, oil-streaked face, soiled collar, and dirt-encrusted fingernails when he inaugurated his two new "types," the Imaginative Man and the Economic Man. For the decade leading to the end of the century was Edison's most driven. He was inextricably tied to the mill, and his own wife was only one of many outsiders, pushed to the periphery by a myriad of new Edisonian preoccupations, some rational, some not. There was cobalt, for example. Deposits of the metal ran in streaks in the mine. Edison was fascinated by the divergence of scientific opinion on its value. One engineering journal insisted cobalt made iron "rotten." Another said it "gives great strength. Can it be possible," Edison speculated, filling page after page in pencil-stub writing in his incessant journals, "that there is another iron metal, so close to iron that it gives all the reactions of iron and yet is not iron—like the parafine [*sic*] family, one parafine being so close to the next in the series that they act like a single substance?"

There was the constant challenge of efficiency. Five thousand tons of ore in twenty hours could be mined, crushed and concentrated with only one-fourth of the available manpower at work at any given shift. Yet Edison drove the system harder. As late as 1897, he nourished the dream of building a concentrating plant four times the size of Ogden, to completely dominate the market east of the Allegheny Mountains and north of the Potomac River. Distracted by productivity as an end in itself, pushing cash flow back into maintenance of machinery, Edison did not take into account the concomitant renaissance of activity in midwestern steel mills, especially in Upper Michigan, the Lake Superior area, and the hematite lodes of the great Mesabi Range. These were controlled by larger, conglomerate corporations capable of "vertical integration"—i.e., the acquisition of all links in the commerce chain, including cheaper transportation—leading inevitably to a less costly product. Edison lost perspective as competition encroached because he had long since cut off his dependency upon outside funders and therefore had no one to advise him but himself, preferring instead (employing one of his favorite expressions) to "paddle my own canoe."

Edison's middle-aged tunnel vision—in all senses of the term—became focused upon every area of his life, at work and even at play. Returning home to Glenmont one weekend at the height of summer with superintendent Mallory as his house guest (Mina and the children had departed for the Jersey shore), Edison asked his friend, after dinner, brandy and cigars, and a perusal of the evening newspapers, if he would like to retire upstairs for a game of billiards. Mallory was astonished, having no idea that Edison had the slightest concept of leisure time, aside from his well-known addiction to Parcheesi, the only board game Edison could ever tolerate. They banked for the first shot; Mallory won and started the game. After sinking two or three balls, Mallory finally missed, and a long carom shot was left for Edison, the cue ball and the object ball within a foot of each other on the green baize, the other ball distant nearly the length of the table.

Edison attempted the shot, missed, and said, "Put the balls back," so Mallory replaced them. Edison missed the shot a second time. Mallory continued, at Edison's request, to put the balls back in the same position for the next fifteen minutes—until the Wizard was able to make the shot every time; and then he said, "I don't want to play anymore."

The sanctity of Saturday with the family was further violated for several months of weekends at the beginning of 1896, when Edison went straight from the mine to the West Orange train station to the laboratory, skipping Glenmont entirely, in search of the elusive X ray. The newspapers had been filled with descriptions of the recent sensational, accidental discovery by Wilhelm Conrad Roentgen, a professor of physics at Würzburg University; he noticed that a barium-platinocyanide screen lying nearby on a table in a darkened room began to glow mysteriously (that is, fluoresce) when exposed to an electrically activated Crookes tube—a glass bulb with a partial vacuum within, used for the study of high-tension voltage discharges called cathode rays emanating from it—shielded within a black cardboard box. Roentgen placed other supposedly impermeable obstructions between the Crookes tube and the screen, but the "rays" (in fact, they were high-speed electrons with a shorter wavelength than light, but this concept had not yet been invented) made it through, and the screen still glowed. Roentgen concluded that the fluorescence was caused by invisible radiation more penetrating than ultraviolet rays.

Roentgen's "X-Strahlen" aroused Edison's competitive curiosity, and he set to work to generate them stateside "before others get their second wind," excited even more so when William Randolph Hearst challenged him early in February to make a "cathodograph" of the human brain. "The thing is too new to give definite directions," Edison wrote, "it will require two or three more days [sic] experimenting before proper directions can be given. I use a ten inch spark coil primarily worked by alternating Dynamo—Have trouble with tubes, am making them myself having broken those of Crooke's [sic] make." The next day, much to Edison's amazement, his old colleague Sigmund Bergmann sent Edison an X-ray photograph of a right hand taken in Germany on January 7 by Roentgen himself, bones clearly visible beneath the skin, a thick ring on the subject's middle finger standing out black and defined against the soft contours of the joints. (Since absorption of X radiation depends upon the atomic weight of the exposed material, human bones show up as darker shadows than flesh on a photographic plate.)

Exploiting his expertise in electricity, Edison was on the right path believing that the photographic image could be enhanced by improving the basic Crookes tube, making the exterior walls thinner and cranking up the power of the electrodes, thereby increasing the velocity of the electrons. "I have 3 of the new lamps on [vacuum] pump and cannot leave," he wrote in a note sent up to the ever attendant Mina at the house, "Please Billy darling send supper down here to the lab for Freddy [Ott] and myself." But an image of the brain eluded Edison because the human skull proved too thick for electron penetration. As had happened in years past, when rumors spread that Edison might at last be working on a new hearing aid, letters poured in from all across the land imploring him to find a way to pass a cathodal tube into the stomach, to seek out and destroy the "bacterial germ of Cancer," to once more show the world that the Wizard had not lost his powers. Edison did indeed establish the fluorescing properties of calcium tungstate, and re-layed this information to Lord Kelvin in England; went on to draw up plans for a rudimentary variant of fluorescent lighting; and wrote several articles about his X-ray research for *Electrical Engineer* magazine, published in the first medical textbook on radiology, but he never patented any of his discoveries in this brief flirtation with the phenomenon.

Barely had his X-ray divertissement got under way when Edison received word on February 26, 1896, from niece Nellie Poyer (brother William Pitt's daughter) that his beloved father had passed away at the age of ninety-two at Marion Edison Page's farm in Norwalk, Ohio. Edison felt that the funeral should properly be held in Port Huron, but Marion insisted her husband Homer was too old and frail to make the trip northward. Younger brother "Al" for once deferred to his sister's wishes. After services at the Page homestead near Milan, the body was transported to Port Huron for another sermon at Sam's home for all his friends and neighbors, and then the old man was buried in the Edison family plot at Lakeside Cemetery beside his wife, Nancy, who had predeceased him a quarter of a century earlier.

The years had continued to pass peripatetically for Samuel Ogden Edison. Subsidized generously by Thomas Alva, frugal Sam was booked for a voyage to Europe in a first-class cabin, but to save money he cashed in the ticket and went steerage; he also walked from Paris to Versailles, a distance of fifteen miles, to avoid paying what he thought was an exorbitant train fare. Winters brought Fort Myers sojourns with sixty-five-year-old companion James Symington, who delighted in calling Sam "the young fellow."

Father Sam regaled the Minneapolis *Tribune* with the story that as a lad, Al had fashioned his first practice telegraph instrument out of a whalebone wedged into a crack in a tabletop, but because of the infernal clicking sound, "His father had frequently to arise at 2 or 3 o'clock in the morning to compel the lad to go to bed. . . . He was indeed a peculiar boy, and never played with the other boys for more than a few minutes at any time, seemingly preferring to be alone so that he could give his whole attention to studying the solution of some problem. When about midway between twelve and thirteen, he surprised his parents one evening by bursting in upon them and gleefully shouting, 'Ma, I'm a bushel of wheat. I weigh just sixty pounds!' . . . The father is naturally proud of his son," wrote the *Tribune* reporter, "and readily tells what he can regarding him. When some funny incident in his son's life would flash upon the old man's mind, he would slap his knee and exclaim, 'Oh! He was the darndest kid I ever saw,' and then the old gentleman would chuckle quietly to himself." Sam also spoke at considerable length with the omnipresent George Parsons Lathrop, writing for the magazine *Once a Week* and describing the patriarch as a "sedate, half-rustic gentleman, brimming over with good-humour and a great guyer."

In his last years, Sam had become possessed with the nostalgic desire to return to Digby, Nova Scotia, land of his childhood, to see the rock he used to gambol upon when he was a child: "I've been scattering seeds along the pathway of life ever since I began to live, and now I'm picking the flowers of memory, and they are very fragrant and comforting," he told Lathrop. Sam's had been a full and joyous life: "I am a master of smoking, drinking, and gambling," he brazenly boasted, in marked contrast to his moralistic inventor son. "I have smoked and drank whisky moderately when I needed it, and have known when to let it alone."

Madeleine Edison was right; her mother did indeed "keep busy," because she was expected to, of course—and because running the household made the days move more quickly. "I have something for every day out here," Mina told her "darling Mother and sisters" in Akron, for she wrote faithfully to the family back home every single day of her marriage. Mondays, she attended to "duty calls," receiving guests with other Llewellyn Park ladies at teatime in their homes and hers. Tuesday afternoon was a dancing class for Madeleine and her girlfriends at Glenmont. At the Orange Women's Club, Wednesday was art lecture day. Thursday there were chamber music concerts at the Country Club. And Friday at four o'clock the women of the Odd Numbers Club would assemble in Mina's grand dining room for strawberries and cream. "It is not the undertaking of a luncheon," she conceded, shorthanded in staff because faithful cook Lena was about to go off and get married, "but I thought it would be entertaining for them."

Three months pregnant with her third child in January 1898 ("our little stranger," she called it, siblings Charles and Madeleine praying for a blond baby sister), Mina was fearful to overexert herself yet mindful of the social requisites of her position in the community. Like her husband, Mina had a job to do, a facade to preserve. "Although I am very delicate," Mina wrote, perennially keeping her chin up, "I think I shall be able, with great care, to endure the great ordeal. Some days I have felt rather used up and unable to do anything."

Mina's responsibilities were not limited to her own offspring. Even as she nurtured a new generation, the three Edison children from her husband's earlier life traversed the tumultuous years of teenage and early adulthood, and Mina was *their* mother, too—a situation replete with its own difficult strains, as the Belle of Akron, a mere eight years older than her oldest stepchild, found herself acting more and more often as a buffer zone for an aloof and irascible father.

Marion "Dot" Edison at twenty-one had returned to Germany under an ambiguous cloud. It was gradually revealed over the next two years that the break with her homeland was partially instigated by a failed relationship in New Jersey with a young man named Charlie Levison. Once back in familiar territory, Marion resumed her preexisting "true" romance with Karl Oscar Oeser, an army lieutenant stationed in the town of Chemnitz, forty miles southwest of Dresden.

When Oeser wrote to Edison in the summer of 1894 requesting Marion's hand in marriage, he stated he had just "made the acquaintance" of the inventor's daughter, a slight blurring of the facts—but his intentions were honorable, presented with numerous character references from upstanding citizens in military and civilian life. The permission of the king of Saxony was required for an officer in the kaiser's forces to marry a foreigner; Oeser soon discovered that this bureaucratic formality would be easier to obtain than the blessing of the father of the bride. Marion had finally found utter happiness. She pleaded with Edison for forgiveness, especially regretting the "unjust" way she had left him and the rest of the family: "Dearest Father, please send me a little love when you write, for it hurts me to see you send me none." Instead, Edison sent his uncle Simeon, father Sam's half-brother, over to Germany to delve into Oeser's credentials, and he found them impeccable. Eight months later, the couple were married, but despite Marion's fervent wish, she was not walked down the aisle by her father, but rather by younger brother Tom, Jr., who arrived two days before the wedding and left immediately thereafter.

Then the cards, letters, and telegrams requesting money began to arrive at Glenmont with unabated regularity, not only from Marion ("I conclude that you are unwilling to do anything for me beyond paying my trousseau bills") but, unbeknownst to the new bride, from her mother-in-law, Eugénie. Surely, "My dear Mr. Edison, Esque.," there must be a more generous monthly allowance possible from such a distinguished father in aid of his daughter, "the american lady who is to much accustom [*sic*] to what we call a high life." Mrs. Oeser was perhaps not aware of the $66.66 Marion received monthly via Edison's J. P. Morgan account; she felt that "at least 50 dollars pin money" would be of immeasurable help.

Before long, Marion descended into an agoraphobic haze, unable even to pick herself up and go out of the house to buy her brother Tom a Christmas card. She dwelled morosely upon memories of her childhood smallpox attack, when her face had been so covered with scars that she hid from society. "I think there was something wrong when I was made," Marion told Mina in the midst of one especially blue period. After his perfunctory consent to the marriage, Edison simply did not write to Marion again. "Having no letter from you to answer," Marion wrote across the distance, "I suppose the one-sided correspondence will have to go on without it." Hungering for some species of family contact, she turned in desperation to Mina. The two women had never embraced—had not Marion blamed Mina the interloper for essentially ruining her childhood and taking away her beloved Papa after her mother died?—but now there was no choice but to try and heal the old wounds, "although a second mother can never take the place of our own," Dot said, "and my poor Mother had left us before she was as old as I am now." It was as if Mina's critical raison d'être as far as Marion was concerned was to act as a conduit; there were complex matters still to be dealt with regarding the sale of the Menlo Park estate: "Please show this letter to Papa as it is meant for him too . . . Now dear Mina please try to help me. If Papa will not hear me directly, then [tell me] what does he advise?"

And what of Tom, Jr.? While Marion was in Germany having gained a husband at the expense of distancing a father, her younger brother was in the inevitable process of dropping out, first from St. Paul's School and then the Hawkins School on Staten Island, while yearning to take up another path of endeavor (not correct spelling, however): "I have stoped studying as it is too hard for me . . . I am but a youth with youthful desires and ambitions to become a lanscape [*sic*] artist, but nature smiles and says, 'Oh! No! My boy you can't paint me upon your dirty canvas, I am to proud and I don't want to be insulted.' . . . Ah, Mother," Tom wrote from the tranquil lake shore at Chautauqua, "I must see and try to win my way, see if my true and willing ambition cannot pacify [Nature's] heart . . . which is the one and only one desire and aim I have in this cruel and unappreciative world. I am still painting, mother, but I have such a hard time conquering my brush."

If only life could have remained this idyllic and Wertherian for young Tom. In the fall of 1894, when the Ogden mine reopened, he traveled there with the rest of the family and stayed behind with his father, who put Tom to work, ostensibly to learn a trade, since it was clear to all that he was going to follow in the Edison footsteps and abandon all hope of formal education. Tom was installed, however, as a common laborer, assigned to the storeroom, stacking grease-laden castings, with no special privileges—he was required by his father to report directly to Mr. Mallory instead—and more often than not he strayed off into the surrounding albeit skimpy woodland, where he wandered lonely as a cloud and furtively killed red squirrels, making paintbrushes from their tails.

By the following spring, there was a discernible shift in Tom's attitude to his father, from distanced reverence to outright resentment. The painting ceased, the complaints flowed instead, and, again, Mina was the recipient. Tom seemed to have established in his mind a foundation of filial respect for her, perhaps in order to be able to rage to her against his father for "thinking I am striving to be a self-made man. How father can think that I fail to see. . . . He is the only person known to me that ever accused me of striving to be what he has made himself." Can it be that there would be room for only *one* Thomas A. Edison in the world, and that eighteen-year-old Tom, Jr., in the naturally confused process of self-definition, had finally come up against an immutable force?

To strengthen his position in the pecking order, Tom began a systematic, Hamlet-like character-assassination campaign against Marion, sympathizing with Mina's complaints about the hardships Dot had caused by placing her in the middle of the conflict with Edison *père* (again, the spelling is Tom's): "If I was you, drop her forever. . . . Every time I here a word against you, my ears are closed, for it sends a shiver through my heart. Although I am young yet and ignorant, please except this youthful advice, and the moment you finish reading this, ask yourself quitely, 'Is this all a dream?' and answer silently, 'yes,' and think of it no more—*nor her.*" Although guilt-ridden with the knowledge that Mina had her "own dear little ones" to take care of and cherish, Tom could not prevent himself from burdening her.

Tom's efforts toward "self-making" were Whitmanesque rather than Carnegie-esque. He craved nothing more than to camp under the stars in the great outdoors, now and again jotting notes in his tent and composing tunes for the phonograph, to "study nature in its most rudimental form . . . where not a sound is heard, except perhaps a hooting of an owl on yonder treetop or the roar of a restless panther up in the crevices. Could nature afford, with all its ficilities a more glorious scene then I have taken the liberty to picture you in? . . . I am sorry for those unfortunate souls whose sensitiveness equals mine—who studies each day with the closest scrutiny, and the most minute reserch to find the hidden tyrant—whoes cause is their unhappiness . . . I plead with you to kindly think of me—one who sees this life in a different view and one who loves you all so dearly—but love in return is cold." Tom drew upon peculiar, pastiche-Wordsworthian rhetoric not only to impress his stepmother but also to bolster his courage against Father's silences with words dishearteningly echoing sister Marion's. From Canada's north woods, Tom's plaintive voice cried out from the forest: "I wrote to him—but no answer came."

Sustained by summers in the wild, and the refreshing pastoral and spiritual hiatus of Chautauqua, Tom managed to hang on at the mine during the toughest winters, determined to find a way within himself to work there while at the same time giving solace to his bohemian tendencies. But he became more and more distracted at Ogden, drained by competing tensions. Try as he might, Tom did not have the nerve to face his father directly, instead haltingly channeling the expression of his fears toward Mina's sympathetic but equally powerless ears: "Mother, I wan't a change—some thing that is knew to me that is full of adventure. . . . Why should I write to you about this—for it is a father's claim—but I write to you—with a feeling that—being as father is and situated in this world—it would not be advisable to tell this to him—for he has no time for consideration . . . America's fore-most inventor—I often wonder—and think—am I his son . . . what have I done? to be deserving of such—a great and honorable Father . . . and it is his genius—I have to blame for these doubtful and restless thoughts of mine. . . . I probably never will be able to please him—I have no genius—no talent and no accomplishment—I am afraid—it is not in me—but I shall never give up trying—if I could only talk to him the way I want to—I have many ideas of my own—which sometimes . . . I would like to ask him—or tell him about—but they never leave my mouth."

Even while opening this emotional terrain to his stepmother, Tom remained convinced that the power to confer meaning upon his life resided in his father's authority. Stymied by Edison's distractedness, passing him on the muddy road without a word or a glance in his direction, Tom finally fled the mine "heartbroken," plagued by fleeting, acute suicidal thoughts in the late spring of 1897: "I shall *never* ask Father again for something to do." He found a flat in the village of Orange, and then in Asbury Park, sulking, shiftless, unable to respond to his mother's worried and repeated entreaties that he settle at Glenmont until he was able to find a better work situation. Tom made vague noises about seeking medical help "to know if my mind is affected in any way. . . . I am afraid there is something the matter with me"—and even alluded to returning to undefined "studies," pledging to think again about entering college. ("My plans of course are subject to change and are no means final.")

In the new year, 1898, on the advice of his New York physician, Tom—suffering from mental and physical exhaustion as well as delusions of grandeur about his own potential inventive prowess, hatching an all-too-familiar-sounding "great scheme for utilizing the forces of the ocean for power purposes"—moved to Fort Myers for two months of rest. Separated more dramatically from home and family, Tom finally let his anger break through to the surface, and he declared his bitter alienation in a letter to younger brother Will, who, in turn, seeking his own salvation, passed the word along to Mina. With the Ogden mine dipping slowly into decline, Tom found difficulty preserving vestiges of reverence for the icon of his father, and as Edison senior's public image as a businessman and leader came into question, satirized in the daily press, Tom suffered renewed bouts of confusion: "If my name was Smith—I would be a rich man today—He hasn't a good man with him nor never could keep a good man with him—look at Insull, Johnson, Batchelor [who had in fact finally retired from Edison's employ after twenty-seven years of loyalty]—he couldn't keep these men with him a minute financially. You may think you are alright in your decision to stick to your father," Tom told Will, "but you will pay dearly for it later on—he will throw you down like a dog—like he did me."

Mina was shocked by Tom's torrent, having heard furthermore through tattletale Will that his brother "was having a fine time" basking in the luxury of Florida's balmy weather, hunting, fishing, and taking full advantage of the plush family estate by the Caloosahatchie. "Tom is either crazy," Mina permitted herself a judicious observation in her daily epistle home to mother, "or a very bad character."

William Leslie tagged along two years behind Tom, faring even worse at Mr. Hawkins's Academy. His friendly letters home to Mother, "with love and kisses to all," actually received a response from Father, but not in the reciprocal manner Will would have wished. Edison took his son's cordial notes, marked them up for errors as if they were exams, and brusquely returned them, "Wm.—*See marks*—Your spelling makes me faint—TAE," underscoring such apparently barbaric infelicities as "wouldent," "allowence," and "half" (for *have*).

The headmaster had his hands full with Will, who seems to have been a rather undisciplined lad. Mr. Hawkins appealed to Mr. Edison first for just "a few minutes" personal conference; receiving no reply, the tutor was advised by Will that all matters pertaining to school behavior were to be addressed to Mrs. Edison. "It is impossible for a teacher to effect any change for the better in a boy's moral character unless he has the full sympathy and earnest support of the parents," Mr. Hawkins insisted, but had to settle in this case for only the maternal side.

Suspiciously resembling a chip off the old block, Will was careless about his person, his clothes slovenly and disheveled, and he was lackadaisical in preparation of the daily recitations. "He is boisterous and rude," Mr. Hawkins opined frankly, "and this sometimes approaches disrespect." The boy's temper at times was uncontrollable; he found fault with the lessons, openly questioning Mr. Hawkins's intelligence, who in turn cautioned Mina, "These are the faults of a child, and he must *repress* them in order to cultivate dignity of character." In Mr. Hawkins's view, college (the prestigious Cornell, of all places!) looked to be a very long shot unless Will could tame his baser instincts.

But this reform was not to be. By his senior year, Will had added profligacy to his roster of vices. He disbursed one month's allowance for school sundries within one week, and then was caught borrowing heavily from his school chums. "William has a talent for spending money," Mr. Hawkins reported. "Whatever amount he may have, he may be trusted to get rid of it instantly . . . and his conduct in the school-room has been frivolous."

This habit continued into the after-school years, as Will began to drop promissory notes in the Edison name up and down the northeast coast, and attorneys for the creditors, seeking payment, solicited T. A. E. himself, who responded angrily in turn that his son had no permission from him for such actions.

Tom and Will had nowhere to hide. Their name was a banner, unfurled regardless of any professed desire to establish their distinct identities. Rather than follow through on impotent threats to disown their inherited notoriety, the boys launched new businesses. The Thomas A. Edison, Jr., Improved Incandescent Lamp Company set up headquarters in the Schermerhorn Building at 96 Broadway in New York City, and William L. Edison, Agent for Phonographs, Kinetoscopes, Roentgen Ray Apparatus, Edison-Lalande Batteries, etc., was founded, against Father's wishes, in Rochester, New York. Both storefronts were short-lived.

As fissured generations of Edisons jostled and disturbed each other, a trio of vagabond children scrambling to secure a foothold in the wider world, the Millers struggled with tribulations ironically brought about through closely shared love.

Lewis and Mary Valinda missed Mina terribly, and she them, especially around holiday time, when Mary would become, as she expressed it, "ancious and nerveis" without her favorite daughter at Oak Place. It had been ever so long since they had extended the dining room table to its full length, and the clan had gathered together. Would that day ever come again?

His devoted sons Ira and Robert having entered the business of subsidiary supply sales—binder twine, iron frameworks, gears, and the like—that grew up around the sprawling Buckeye Mower and Reaper Company in both Akron and Canton, Lewis Miller drew heavily upon them for emotional and corporate support during the Panic years. They rallied indignantly to their father's defense when the Treasury department early in 1897 proposed that portraits of Eli Whitney and Cyrus McCormick be placed upon new issues of paper currency; the Millers pointed out that there were other noted inventors equally entitled to credit in the development of reaping and harvesting machines. "It is a great comfort to me to realize with what fortitude the family are bearing this awful downfall," Lewis told Mina.

The rights to another invention equally close to Lewis Miller's heart were likewise challenged. By the mid-nineties, he had risen to become presiding officer of the International Sunday School Convention; his beloved Mount Union College in Alliance, Ohio, was flourishing; and most gratifying of all, the Chautauqua season had reached its peak, extending for a remarkable fifty-nine days, almost the entire summer. On the brink of Chautauqua's third decade of institutional life, three new denominational headquarters were in place (Episcopal, Baptist, and Disciples of Christ), in addition to a complete College of Liberal Arts and a month-long Teachers' Retreat—not to mention the first subterranean community sewer system in America. Lewis Miller's hard-earned pride in his dream of a pastoral setting for lifelong learning and spiritual enrichment was abruptly punctured by the publication of president of the University of Chicago William Rainey Harper's article in *The Outlook* for September 26, 1896, "John H. Vincent: The Founder of the Chautauqua Movement." The reverend was given the complete credit, and Miller's informing concept was ignored, a slant that had developed over the course of some months. It seemed that through his son, George, now also an important player on the Chautauqua administration, John Vincent had been pressuring his disciple, Harper—for indeed, the bishop had hired Harper as principal of Chautauqua's college of liberal arts and head of its summer school—"not to emphasize the liberal side of my views over-strongly . . . to insert somewhere the recognition of Chautauqua as a product of the Methodist Church and as reflecting credit on the church."

"No man ever heard the Bishop speak without respecting him. No man ever came into close touch with him without loving him," Harper concluded the *Outlook* piece, a laudatory paean to his mentor and "honored Chief." The Millers were shocked. Daughter Mary demanded of the bishop if he had approved the article. He replied that he had had nothing to do with it, which of course did not still the family's anger. "The Millers will, I am sure, be on the warpath," the bishop cautioned son George. "Now see Dr. Harper and tell him for the present to be perfectly quiet." A month later, the situation had only worsened: "If Mr. Miller can legitimately regain his influence as *a business man* on the Board, he has a perfect right to do so. To interfere with him now in any way would, as you rightly suggest, be unfortunate. Give him a rope," Bishop Vincent went on. "If he hangs himself we are not responsible. If he pulls himself up into security we certainly have no reason to complain."

The muddied waters swirling around the inspiration for Chautauqua had been intrinsic to its identity since the inception of the place. Lewis Miller and John Vincent had always remained cordial allies on the surface, letting the creative tensions of commerce and religion successfully drive them forward for many years. When their latent disagreements about the identity of Chautauqua, precipitated by Harper's article, rose to a public forum, Mina leapt into the fray, inspired to draft her own *apologia* for Chautauqua. Her handwritten effort, found buried among her papers in the Chautauqua library, and never published, set the tone for her undying and increasingly strident advocacy on behalf of her father's legacy over the coming half-century: "Not America alone has felt the influence of Chautauqua but it has reached out to all nations. The work begun for the advanced Sunday School student now belongs to all classes and conditions of men . . . This happy combination of pleasure resort and university with all its departments has not been the work of a moment but a gradual and natural outgrowth of the original idea through hard labor and earnest thought."

Lewis Miller had final and chastising words for Dr. Harper, reminding him that they had consulted periodically during the formulation of the *Outlook* article to discuss facts pertaining to the origin of the Chautauqua movement, but nevertheless, despite these meetings, "You in your article have undertaken to give the entire credit for the origin of Chautauqua to Bishop Vincent." Although a correction ran in the succeeding number of the magazine, Miller was far from pacified, since in this conciliatory followup essay the bishop still persisted in typecasting businessman Lewis Miller as providing only advisory and financial support, "leaving the impression that I had but little if anything to do with originating the project."

The complexity of Miller's ideas eluded Harper's reductive approach, and intervening years of growth and popularity had obscured the impetus for the place, especially Lewis Miller's sophisticated vision of Chautauqua's summertime venue and year-round mission. He reminded William Rainey Harper:

> IT IS THE PURPOSE THAT THE CHURCHMAN AND PHILANTHROPIST, THE SCIENTIST AND STATESMAN, THE ARTISAN AND TRADESMAN SHOULD BRING THEIR LATEST AND BEST TO THIS ALTAR OF CONSECRATION AND THAT TOURISTS AND PLEASURE SEEKERS SHOULD HERE STOP AND FIND THEIR BEST PLACE FOR REVERIES, AND WHEN THUS STRENGTHENED RETURN TO THEIR RESPECTIVE FIELDS OF LABOR AND THERE THROUGH THE YEAR WEAR INTO THE FIBER OF THEIR HOME LIFE AND WORK THEIR NEWLY GATHERED STRENGTH AND INSPIRATION. THIS WAS MY CONCEPTION OF A GATHERING IN THE GROVE, WHERE PEOPLE COULD GATHER WITHOUT LIMIT OF SPACE OR TIME.

It was only then, let it be remembered, that Lewis Miller had turned to his colleague and friend John H. Vincent as the one man he knew who could work such a scheme successfully. And the reluctant Reverend Vincent, let it be remembered, had to be forcefully sold on the idea "by persistent effort of two years" because he feared "a wild, impractical, visionary camp meeting."

Mary Valinda's "petty boys," John and Theodore, thrived at Yale after successful courses of study at St. Paul's, faring far better than their two Edison cousins. Joining the glee club, the Miller brothers stopped by often at Glenmont during the holiday season, gathering around the grand piano to sing Christmas carols. Handsome "Thede," the youngest Miller, and the great hope of the family—amateur actor, president of the debating society, violinist, wrestler—he could do no wrong. Little Madeleine revered him. In the days when the lower meadows at Glenmont were still rustic hayfields, she would sit on the fence and to her great delight watch Theodore ride bareback in prancing circles. His biographer characterized Theodore as "his father's Benjamin." Lewis was grooming his son for a perfect future. He sent Theodore, while the lad was still an undergraduate, texts to help him start his own Sunday school modeled after the Akron Plan, recommendations for supplemental religious studies, even a draft mission statement for potential religious school teachers. The entire family attended Theodore's commencement ceremonies in June 1897 and proudly wished him well as he went on to New York University law school.

On April 17, 1898, four days before the United States declared war on Spain, Theodore told his father he was ready to enlist at first call. Lewis, who regretted not having fought in "the Great Rebellion," still found it difficult to give his unqualified blessing to Theodore's wish, cautioning his son to seek out a clerking job, perhaps, or some sort of "official work" in the quartermaster's office, or safer guard duty on coastal defenses—anything but the pitched fray. "This war is an interloper," Lewis said, "and must be treated by you as such." And Mary Valinda "moaned" through the entire breakfast at which her son's letter arrived: "It is a dreadful thing this matter of war. So many never return." Despite the tearful entreaties of mother and sisters not to abandon his law studies, Theodore was unswayed: "What I want to do is get into the 'scrap,' and be able to do something worth doing. I may be too eager and ambitious, but that is what I want to do . . . I can't say anything in contradiction to your desire for me to stay. For I would stay if I thought it was my duty not to go. . . . If everyone excused himself for selfish reasons, we could have no army."

President McKinley's public call for 125,000 volunteers was answered by a million men. Theodore Miller would not opt for the safe path. While brother John joined the Navy as an engineer aboard the U.S.S. *Marblehead,* "Thede" signed with Theodore Roosevelt's Rough Riders, trumpeting, "There could be no better place than where I am going, for this regiment is made up of the finest fellows in the country and I have several friends with it." In the rush to join the company, assembling at San Antonio, Texas, then moving on to Tampa, Florida, for the invasion voyage to Cuba, there was no time to stop off for farewells at Akron. "Darling," Theodore wrote his mother, "I think I am doing my duty and trust that you will agree with me. My train leaves right away, so I must close, darling. Love beyond expression from a loving son. Goodbye."

Officially the First Volunteer Cavalry, although in fact they went into battle with only their officers on horseback, and under the direct command of Colonel Army Surgeon Leonard Wood, the Rough Riders were a motley crew with a solid core of Ivy Leaguers. Teddy Roosevelt, as Wood's number two, was particularly proud of his Easterners: "I really doubt there has ever been a regiment quite like this . . . I warned them that work that was merely irksome and disagreeable must be faced as readily as work that was danger-ous . . . Not a man of them backed out."

Theodore, disembarking from the transport *Yucatan,* was among the twelve thousand American troops landing at Daiquiri, Cuba, on June 22 and 23, 1898, in preparation for the advance through the dense and hilly jungles, trudging inland in relentless tropic heat to meet the Spanish troops at Santiago to the northwest. After several engagements during the ensuing week, the Rough Riders found themselves in the late morning of July 1 at the outskirts of the city, with the San Juan River at their backs, facing the San Juan Heights, final barrier before their goal.

Spanish long-range snipers—"the spit of the Mausers"—kept the men pinned down with what Roosevelt believed was random fire; there was no choice but to go forward and take Kettle Hill by direct assault: "I waved my hat and we went up the hill with a rush." The journalist Richard Harding Davis was there. Unlike the romantic images that would later come to char-acterize this ill-fated charge, "It seemed as if someone had made an awful and terrible mistake. One's instinct was to call to them to come back . . . The pity of it, the folly of such a sacrifice was what held you . . . Single lines of men, slipping and scrambling in the smooth grass, moving forward with difficulty, as though they were wading waist high through water." Five men fell to the ground almost at the same moment; Theodore was hit in the left shoulder, at first glance not a serious wound. He warned his comrades not to bother with him, for he felt little pain. But as Theodore lay quietly, he realized the bullet had passed straight through his upper body, exiting at the right shoulder. He could not move. He was having difficulty breathing. His spinal cord had been nearly severed, and he was paralyzed from the shoulders down. Through a summer monsoon turning roads to mud, Theodore was taken to the rear, twelve miles to the American Red Cross field hospital at Siboney, an improvised compound of abandoned buildings, overburdened by casualties.

"Dear Mama," he dictated, "a rather narrow escape but feel sure I will pull through all right. . . . They are doing everything that they can for me. I remain your most loving son and will be with you soon." Stoic, he persevered with the faith that his injuries were not mortal. Theodore spoke often from his hospital cot of his family, his home life, and especially of Mina— "Oh! You must know my sister. She is the dearest sister a man could have and is one of nature's noble women," he told the Negro orderly who brought him fresh water in a canteen every hour and gripped the wounded man's hand "seeming loath to let go."

On July 7, twenty-three-year-old Theodore fell into a stupor and never fully awakened, and died at noon the following day with the trace of a smile remaining on his face. He was buried near the charred ruins of a blockhouse in a simple grave on a low rise several hundred yards from the shore of the Caribbean Sea. Military bureaucracy slowed news of his passing.

Three days later, at Glenmont, Mina Edison gave birth to a baby boy, and then within hours, John Miller's brief telegram announcing his brother's death finally arrived. Mina, weeping softly, pulled the infant tightly to her breast. He would be christened Theodore.

CHAPTER

21

GRIEVING AND STRANDED ON THE U.S.S. *MARBLEHEAD* IN Guantanamo harbor, John Miller could not at first claim the body of his fallen brother; a strict military quarantine was in force at Siboney in the aftermath of a yellow fever outbreak. "O, Father darling, it is all a mighty sad and hard thing to think of. I try to keep my mind off from it as much as possible but this is very hard to. We have certainly suffered a very very great loss indeed. Dear old Thede. How gladly would I have given up my life for him . . . If only I could have gotten here a little sooner and seen him before he died. If Thede could have only lived, what a glorious homecoming we would have had," John wrote on July 17.

He yearned for some "war excitement" to break the "monotonous life" on the very day that General José Toral, commander of the Spanish forces, surrendered to General W. Rufus Shafter and the American troops at Santiago as the clock on the old cathedral in the city square struck twelve, and the Stars and Stripes were hoisted over the palace while the military band played "Hail, Columbia."

A month later, the war over, Theodore was finally brought home to rest at the Miller family plot in Akron's Glendale cemetery on the hillside near Oak Place. Offering balm to the tensions between the families, Bishop Vincent officiated and delivered a moving eulogy. According to the bishop's son, George, the Millers were greatly gratified. "I think the personal relationships have never within a dozen years been so satisfactory as they are to-day. Of course," he added, confiding in William Rainey Harper, "whether there is any element of permanency in it remains to be seen." George's abiding affection for Theodore seemed to win out over loyalty to his father on the perennial Chautauqua origins issue. George proposed to Lewis Miller the idea of assembling a Souvenir Book, a keepsake in tribute to Theodore, made up of incidents, anecdotes, and "little happenings" gathered from talking with childhood friends, freely illustrated with photographic snapshots from family albums.

Lewis, as was his habit, encouraged this kind of ardent activity as a way to help the newer generation heal themselves from the loss of one of their most stalwart members. However, he referred to himself as an "invalid," failing in health and under a doctor's care. And then, the sudden death from valvular heart disease in November of eldest daughter Jane, only forty-three—the ever steady Jennie who had been Mina's moral guide during young womanhood—hit the family even harder as they had barely regained their equilibrium after Thede's departure.

The Miller children mourned, striving to regroup and support one another, while the junior Edisons back east maintained their errant ways. Tom had managed to avoided enlisting in the Spanish war; instead, his bogus "Improved Incandescent Lamp Company" could no longer be found at any fixed address. During the summer of 1898 he took up residence in a flat on the West Fifties near Broadway in Manhattan with an eighteen-year-old actress. Referred to in the popular press as a "casino girl," Marie Louise Toohey was immensely cheerful and self-confident, "with shapely figure and pleasing voice . . . a mass of wavy golden hair and blue eyes."

Brother Billy moved in with the couple for the winter, and the ensuing months passed quietly enough, until the fateful night of February 16, 1899, when, in the midst of her stirring performance as Parthenia in *La Belle Hélène*, Marie broke into giggles on stage. As soon as the curtain came down, the company manager, with whom Marie had frequently tiffed in the past, dismissed her on the spot, saying, "You leave Saturday night!"

"You need not tell me when I am to leave," answered Miss Toohey with her sweetest manner. "I'll leave when I get good and ready; in fact, I think I'll leave now."

"You can give her costumes to Miss Miller," said Mr. Solomon, the stage manager, turning to the wardrobe woman.

"Pardon me," retorted Miss Toohey, more sweetly than ever, "I am not *'her.'* I have a name, and it is Mrs. Thomas A. Edison, Jr!" And with that shocking revelation, she swept off into the darkness, telling the throng of reporters catching wind of her "last exit" that yes, indeed, she and Tom, Jr., had been secretly married some three months before.

A somewhat more scandalous story emerged when the elder Edison received news of this latest embarrassment and sent his private investigators on the trail, whereupon they discovered the New York State marriage license certifying that the groom and bride had in fact been joined in marriage (as it were, after the fact) on February 19. The horror did not end there, as the new Mrs. Edison, intoxicated with her public spotlight in more ways than one, began to make the rounds of the unsavory Tenderloin district in the company of a disreputable and "vile" crowd, drunkenly boasting to one and all that she was "Edison's daughter-in-law and his favorite," the "Old Man" gave her money, and she was playing the family for "suckers."

"I never could get [Tom, Jr.] to go to school or work in the Laboratory," an exasperated Edison told his friend Thomas Commerford Martin, editor of the *Electrical Review*. "He is therefore absolutely illiterate scientifically or otherwise." Within a year, Tom and Marie had split up over her affair with one of his best friends, and Tom was out on the streets.

For Lewis Miller, the end, when it came, was prolonged, as his favorite son's had been. The sickness that had plagued him—variously described as a disorder of the kidneys, abdomen, or stomach, but most probably prostate—eventually reached the stage where his doctors decided immediate surgery was required, and he was sent to New York City's Post-Graduate Hospital for the procedure. Caught in a heavy blizzard, the express train from Akron became stranded in the Jersey meadows for two days before an ambulance arrived. The catheterization was performed on Monday, February 13, 1899. The shock was too much for Lewis. With Mina at his bedside, he hung on until Friday's early morning hours, when uremic poisoning took hold. He was seventy years old.

Fellow townspeople affectionately remembered him as a large man, "big in every way," outgoing in disposition, hospitable with his magnificent homestead, philanthropically generous with his money, tireless at the factory, and devoted in his attentions to the church, Sunday school, and the education of young people from primary through college years. We have seen that Thomas Edison identified with his father-in-law's pioneer roots, indefatigable energy, civic pride, and especially the way he applied innovative imagination to create "labor-saving devices." After all, Edison had come to know Lewis Miller as a colleague in the field several years before he met and married Mina.

But contrary to their affinities, the most interesting comment Thomas Edison made about Lewis Miller after his passing was contained in an outtake from the introduction Edison wrote to Ellwood Hendrick's 1925 biography of the inventor of the Buckeye Mower and Reaper. A paragraph excised from the final manuscript reveals more to us about the writer than his subject: "[Miller] was in a business all his life he didn't like," Edison asserted. "He should have been a great educator. He liked doing things like Chautauqua but he felt he always had to get the money. He wasn't like me. *I am always doing things I enjoy doing whether I make any money or not.*"

This statement deserves scrutiny as we approach the fin de siècle point in our story, by which time our hero had taken out a grand total of 765 patents, with 328 more yet to come over the next three decades. Ever the public man, and on the threshold of the most intensively litigious phase of his entire career, Edison vented his spleen on the subject of patents for a New York *Sun* reporter, "owing his fortune and his fame to some of them; [and having] lost greater fortune and perhaps greater fame because he was not able to protect his rights in others." To Edison, the value of a patent rested on a sliding scale: at one end were simply quantifiable *things*, like, say, a wrench, which had "real value," and could be manufactured replicatively and at a defined, if modest profit—a practical tool. At the other end of the spectrum, however, resided "*systems* which revolutionize things," machines with multiple and variable components representing new methods. Here, Edison railed on piously, the "industrious inventor" was bound to lose out even as he contributed to the general welfare of society.

Because of the cumbersome and drawn-out legal process surrounding patent registration, trade secrets were easily stolen, and poaching competitors with no respect for intellectual property would set up their own factories to produce knock-offs of legally protected machines. Since injunctions to cause "the other fellow" to cease manufacture immediately were not permitted in a democratic society, there was no other recourse, at least certainly not for Thomas Edison, but to sue, under burden of establishing proof of originality and therefore at far greater cost than that incurred by the defendant. Meanwhile, the adversary maintained his presence in the marketplace, and "you are manufacturing at a loss, while the enfringer [*sic*] is manufacturing at a profit."

As a consequence, "every good man," in Edison's opinion, was being driven out of business by fear of the "patent sharks," forcing Edison to resort to greater and greater detail in his own patent descriptions as a way to hedge against any conceivable imitation.

The brunt of Edison's wrath was reserved for the United States Circuit Court. It was no coincidence that immediately after this irate interview was published, Edison's attorney, Richard N. Dyer, brought a bill of complaint before the Circuit Court's Southern District against the American Mutoscope and Biograph Company in the matter of their infringement upon his beloved kinetographic camera.

Reading accounts in the press of his relentless legal maneuvering, Edison is frequently depicted as the biggest "shark" of them all—ruthless, determined to take no prisoners, sanctimoniously protecting the integrity of his products. But the transcripts of the Biograph case, especially the deposition in his own words (among the few instances when we know we are hearing Edison's unadulterated speech) serve as a revealing articulation of his philosophy of the deeper *nature of invention*, and help to illuminate the unquantifiable, emotional, and not exclusively financial, reasons why he felt compelled to defend himself.

For reasons still intensely debated by film scholars, Edison and W. K. L. Dickson had come to a parting of the ways in the spring of 1895. One theory is that Dickson did not get along with the newly appointed general manager of the West Orange laboratory, William Gilmore. As good an incentive as any for Dickson's departure might also have been the conflict between Edison's conviction that the peephole kinetoscope was the wave of the future—one viewer per machine, a private entertainment, even when enhanced by synchronous attachment to a phonograph—versus Dickson's intuitive technological sense (after all, he *was* a photographer by training) that magical images thrown upon a screen would offer irresistible allure to the public. And where did Dickson go? To the "other side," the American Mutoscope Company, which would become Edison's chief commercial rival and where, it was supposed, Dickson may already have been moonlighting. Changing allegiances was a cardinal sin in the Edison moral hierarchy.

Edison had no choice but to respond when Francis Jenkins and Thomas Armat of Washington, D. C., came forward with their "phantoscope," essentially a kinetoscope liberated from its coffin (which resembled, according to an incredulous account in *The Illustrated American,* "a cedar clothes-chest some five feet long and two feet wide, [with] a suggestive lens poking out of one end. Its weight is about 300 pounds"). Gilmore, the aggressive new member of the Edison team, saw that the two men possessed a sensational, albeit derivative idea and were undercapitalized. A deal was struck incorporating the Armat/Jenkins method of projection with the superimposition of Edison's name as a surefire marketing ploy, "so as to keep within the actual truth and yet get the benefit of his prestige," renaming the machine as the "Vitascope," to go head-to-head, and not a moment too soon, with such players as the "Kineopticon," the "Centograph," the Lumière brothers' "Cinematographe," the hyperbolically advertised LARGER, BRIGHTER, STEADIER AND MORE INTERESTING Biograph, as well as a virtual epidemic over the next several years of smaller independent exhibitors and distributors across the land.

Suddenly, screen projection was big business. By the time the Vitascope, complete with hand-tinted film of the Leigh Sisters smiling, kicking, and bowing their way through an umbrella dance, had its premiere before a "crowd [that] went wild" at Koster and Bials' Music Hall in April 1896, it was touted by agreement with Armat and Jenkins as *the latest invention of the Wizard Edison."* Front-row patrons cringed, seeming to fear they would be drenched while watching a film of waves at Dover Beach "rolling as if in a dream," then breaking on the seashore. Truly, "the magic wand of electricity . . . kept the audience spellbound in darkness."

Six years after it was first filed, Edison finally received approval for his fundamental motion picture camera patent at the end of August 1897, and was then able to engage in legal combat with many of the lesser competition, who backed down and resolved the matters: instead of contesting the suits they agreed to become subsumed as licensees under the Edison corporate aegis. Even as real guns began to blast in Cuba, these preliminary corporate skirmishes on home ground prepared Edison for his protracted war with Biograph, which would carry on into the first decade of the next century.

The brief presented, arguments constructed, and depositions taken on behalf of plaintiff Thomas Edison by his solicitors Richard Dyer, Frederick Fish, and Samuel Edmonds transcended the boundaries of conventional legal strategy for a patent case. Edison's point was simply that the American Mutoscope (later Biograph) Company had unlawfully violated his rights as the inventor of the kinetographic camera by placing its own derivative apparatus on the market, exploiting the increased popular appeal of motion pictures, presenting their own entertainments and thereby undercutting Edison's potential revenues. However, the commercial arena was really not where the most meaningful drama was played; rather, the defense of the sanctity of Edison's imaginative ideas and aesthetic goals was at issue: Where does the invention as a discrete thing end, and "invention" as a concept begin? And, by extension, to what degree can "invention," aside from the nuts and bolts of machinery, itself be claimed and protected?

Edison had the opportunity to test these issues in his deposition given at the West Orange lab on January 29, 1900. "I was trying to do for the eye what the phonograph does for the ear," he said, lapsing into the mantra-like definition of the kinetograph, at the same time reminding the court, and alerting posterity, that the moving picture, a filmstrip applied first to a cylinder, was the direct sensory descendant of moving sound, making Edison the father of them both. Retracing the steps taken to "perfect" the moving picture revealed no surprise information; but at one point, Edison emphasized that "kinetoscopes of various forms" were constructed in the early experimental years of 1888–89, and that he "had considerable difficulty in getting them into commercial shape."

Market concerns, Edison said, were of less pertinence to him than the motion picture as "*art*." He had no hesitation about admitting to having seen the work of Muybridge and Marey; however, for the first time publicly he revealed the aspect of their work that bothered him: their pictures were "very, very jerky," and did not produce the desired "*illusion*," to replicate movement with versimilitude: "What I mean to say by 'illusion,' " Edison explained, taking care to reach back again to his phonograph, "I could illustrate by taking a record of a song on a phonograph and cut away seventy-five per cent of the record. What would remain would, of course, give the tune, but it would be of very poor quality."

"You considered your invention to be in the apparatus which you had designed for producing the illusion of motion?" Mr. Parker Page, attorney for the defendants, asked Edison in cross-examination; to which the inventor, who would not be pinned down, repeated his refrain, once more emphasizing the big picture rather than minutiae: "I was trying to do for the eye what the phonograph was doing for the ear, and to make it commercial." As if reinforcing his more comfortable role as the imaginer, the artificer, rather than the mechanic, Edison was on much less secure ground, becoming downright rambling and prolix when asked to explain technicalities, making distinctions between the broader character of the motion picture camera with which he was concerned, and those production details that were the rightful purview of "*the photograph people*," (i.e., the unmentionable turncoat Dickson): "That was one of the things we got the photograph people to work on, to give us the maximum sensibility [sensitivity?], because we were very short of light, and as we were taking photographs at a very rapid rate, we had to have something extremely sensitive, and they made us a special film, different from the film we printed on, that is to say, different from the positives, and this special film was made extremely sensitive so that the smallest amount of light would produce great results, and we, at first, did not get them very sensitive, but they, by working on them, made them extremely sensitive, so that with the small amount of light we had, they were quite satisfactory after they had experimented on them . . . [We] tried to get [the emulsion] finer so as to get better definition, and I believe, if I remember right, they said that was a question of the fineness of the emulsion, and we finally got emulsions that were quite fine, but they have never been able to get them fine enough yet; they are not perfectly satisfactory yet."

Edison's claim to advancement over his predecessors was based upon his immediate conviction that photography of motion such as that practiced by Muybridge and Marey was "unnatural and machine-like," designed primarily for the documentary or clinical purpose of *analyzing* motion, rather than *re*-presenting it artfully as "*commercial, living picture art*" which would be "suitable" for the delight of the common spectator. Even though it was conceded that the design and components of Edison's apparatus as well as the method by which he embarked upon his motion picture taking did bear resemblance to prior mechanical developments by many others, the most crucial fact remained that Edison was the man who *created* the natural and faithful reproduction of motion, taken from a single, that is to say, humanly visual, point of view.

This courage required to make art, his adroit counsel argued, elevated Thomas Edison above mere preoccupations with "utility," beyond the zoetropic "toys" of the mid-nineteenth century, beyond experimental "trials" or "anticipations" that could not be measured with the same currency as results, and into the rarefied air of the inventor. "Perhaps [Thomas Edison] was a child of fortune," said Mr. Fish in his closing rhetorical flourish, "but

he grasped Fortune when she
came in his direction," and, again,
a true inventor was the man with
the prescience to recognize opportunity without sacrificing his
sense of pragmatism.

On July 15, 1901, U.S. Circuit Court Judge Hoyt Henry
Wheeler ruled in favor of Thomas
Edison and against Biograph,
which promptly appealed the decision to the higher court, and
won a reversal nine months
later—to which the doggedly persistent Edison countersued, effec-

EDISON AT FIFTY

tively forcing his opponent to shift toward more representative, newsreel-
type and panoramic fare, while the Edison Company strengthened its
position in the vaudevillian "entertainment" arena and built a new studio on
21st Street in New York City.

With Edwin S. Porter's *The Great Train Robbery*, Edison's feature films
abandoned the debate about the nature of motion pictures as the replication
of actual anatomy of movement through persistence of vision, and embarked
upon narrative stories, the full-fledged *cinema*. But we are jump-cutting too
far ahead.

The new century brought a new Thomas Edison, rising phoenix-like out of
the ashes of "Edison's Folly," his failed ore-milling plant, resurgent into
what Brooks Adams called "The New Empire." Now that the Western frontier, according to its revered chronicler Frederick Jackson Turner, was officially closed, American imperialism claimed another territory—applied science. "The labors of successive generations of scientific men have established
a control over nature which has enabled the United States to construct a new
industrial mechanism, with processes surprisingly perfect," Adams announced. "Nothing has ever equalled in economy and energy the administration of the great scientific corporations. These are the offspring of scientific thought."

Edison was the consummate recycler. Out of the Ogden plant's dismantled, skeletal gloom, and its considerable debt, which had to be paid off, coalesced an idea that according to Edison's notebooks had been gestating for at least ten years. He would adapt the same techniques used for ore-milling to enter the just emerging Portland cement business.

Following the pattern laid down with his landmark inventions, Edison grafted himself onto a preexisting tradition, convinced that his imagination would herald "an epoch" for the industry. In the case of Portland cement, the technology, while relatively recent in America, extended back to antiquity. Burned gypsum was mixed with sand for construction of the Egyptian pyramids, and volcanic ash (*pozzolana*) in Roman times, but it was not until 1824 that Joseph Aspdin, an English bricklayer, alighted upon the solution that gave the cement its prototypical modern ingredient: limestone. Reportedly fined for stealing the material from local roads after it had been crushed by passing wagon wheels, Aspdin combined limestone pebbles with clay, burned the mixture into lumps called clinker, then pulverized it into finished Portland cement, so-called because its hue resembled a limestone quarried on the British island of the same name.

During construction of the Erie Canal, natural, quarried cement was used, requiring little processing; it was simply limestone direct from the ground with random veins of clay deposit running through it. The results after baking in chimney-like, brick vertical kilns were uneven, and the actual chemical reaction that formed usable cement—calcination—was not calibrated precisely, so the product was porous and chalky.

The essential difference in the Portland cement method, which delivered a sturdier and more consistent product, was its much higher heating temperature, during which a specific "recipe" of raw materials was cooked: silica, alumina, magnesia, iron oxide, and sulfur trioxide added to the lime.

After the Civil War, especially with the increased burden on the sprawling American transportation infrastructure and the need for bigger and stronger factory buildings and warehouses, Portland cement was imported to America from England cheaply or at no cost, the heavy barrels used as ballast for freighters. It was only a matter of time before indigenous American competitive spirit took over. In the late 1860s, in the little Lehigh Valley town of Coplay, Pennsylvania, David O. Saylor began experimenting with local cement rock, mixing the clinker on his kitchen stove and then pulverizing it with a pestle improvised from a cart axle. In 1871, he filed the first U.S. patent for an improved cement "in every respect equal to the Portland cement made in England," and the domestic industry was born. The distinctive Schoefer vertical kilns built by Saylor—proud, tapering turrets rising out of the low hills—can still be toured today just outside the town of Coplay, five miles north of Allentown, part of a well-preserved industrial museum under the auspices of the Lehigh County Historical Society.

As the ever attentive Mr. Mallory tells it, he was dispatched to the New Jersey side of the Lehigh Valley "neighborhood" to assess the potential for ideal cement plant sites on home territory. At Stewartsville, a mere five miles northeast of Easton, Pennsylvania, and thus a hairbreadth across the border, a rich vein of cement rock was found, and Edison snapped up eight hundred acres of land there for his newest *magnum opus*. Again according to Mallory, who would become the vice president of the Edison Portland Cement Company, his chief entered another one of his renowned visionary fugue states, secluding himself at Glenmont over a weekend and emerging with a fifty-page handwritten proposal for the entire cement works, including a location sketch for all the buildings.

Edison described the commercially strategic location of the place—it was on the Delaware, Lackawanna and Western railroad line, as well as near the esteemed Morris Canal, which ran from adjacent Phillipsburg all the way to Jersey City—heralding a "new departure in cheap mining." The mine would be established in a narrow valley surrounded by mountains rich in gneiss rock coated with magnesium limestone. At the center of the valley was a core of cement rock with a layer of fossilized limestone above it, posing a distinct logistical advantage in that the mining "could be [more efficiently] along the lines of stratification." Edison estimated the potential yield of this tract as seventy-five million tons.

Many of the basic sequential principles (and same machines) employed at Ogden would be utilized in the Stewartsville crushing process, moving the raw rock to ever smaller forms toward granular powder, beginning with dynamite blasts to loosen the core, and then on to the venerable steam shovels, which had been improved in efficiency and "[could] load a seven-ton skip in one and a half minutes." Edison planned to add an ongoing assaying function to the production line, an automatic sampler which would draw out a small amount of the dried, crushed rock every thirty seconds and allow his chemists to keep the "intimate" cement rock–limestone ratio constant.

When it came time for the most important operation in the sequence—the formation of clinker which, when subsequently powdered, would result in the final cement product—Edison added his distinctive signature: a horizontal roasting kiln more than twice the length of any existing model. According to custom, he constructed a prototype of cast iron lined with brick at the West Orange laboratory, six feet in diameter and one hundred feet long (the actual kiln on-site would be one hundred fifty feet). His goal was to produce the maximum amount of quality clinker using the least amount of costly fuel and labor. He accomplished this by reasoning (it turned out, correctly) that the greater length of the kiln thereby extended the "combustion zone," the period during which the rock could be cooked to perfection gradually, over a longer stretch of time, using less extreme temperatures and less coal than employed in the shorter kilns. The limestone fed into the kiln was converted to lime by a blast of hot gas at the front end of the path, so that the carbon dioxide given off by this reaction would not interfere with the subsequent burning of coal.

Edison's long kiln turned out to be one of his most profitably sustained inventions. Into the 1920s he was receiving a royalty of one cent on every barrel of cement produced by other companies through this method.

Edison's goal for cement production was to reach ten thousand barrels per day, beyond which, he predicted, "there is danger of poor management" because "the technical staff would require to have higher abilities, and ability is scarce." The Stewartsville plant took three years of planning and labor, during which time, as might be expected, Edison oversaw every detail, to a level that would only be described as unbelievable if we did not possess the documentation to prove it. As he virtually masterminded the deconstruction and salvaging of viable parts of the "Ogden baby," reconfiguring them to build Stewartsville from the ground up, list upon numbered list headed "Jobs that can be done now," was generated, ranging in dimension from the modest to the intolerably complex, from "Steam pipe through mill number 1 leading to Mackintosh engine removed—If lagging can be saved, do it" to "Drive pulleys—frictions—Extension & steel outboard bearing holder & idler on 1st 36-Roll can be removed—Pulleys & friction go to Store room, Idler to pond. This idler we may want again but I think not, as it will be of a different shape" to "Angle gears on main line shaft & shaft pulleys & bearings of shaft driving conveying belt from bottom of Dryer removed & piers levelled to ground & debris removed."

While Edison worked on the design for the cement mill taking shape in erector-set fashion forty-five miles to the west, the West Orange factory had become overwhelmed with orders for the spring motor phonograph, a machine for the masses, "light, compact, and readily operated." The middle-man had been absorbed into the Edison corporate structure: There were no more subcontracted agents in what had become a million-dollar annual revenue stream. Phonographs were cranked out from West Orange and passed directly to the impatient public by Edison's own sales force. Production of this "amusement" model signaled the beginning of Edison's shift away from the office-machine market and toward domestic consumer entertainment.

Just as the vitascope motion picture machine demanded new films to feed it, so did the latest and cheapest model phonograph, the $10 Gem, require cylinders to play upon it. At the turn of the century, one thousand men and women labored in the Edison Phonograph Works, where molds for three thousand different musical and dramatic subjects were housed and available for reproduction. The first step in the manufacture of the records was to melt and purify the wax. The liquid was poured into molds made up of an iron core surrounded by a brass sleeve, mounted on a rotating table. Once hardened, the smooth forms were ready to be incised.

Three floors up, "specialists employed by the Phonograph works were kept steadily at work speaking, playing or singing into the recording machines" providing fodder for the releases. It must be remembered this was still before the age of the microphone; thus, the sound came directly from a violin, say, into the immense, cornucopia-like tube, down to the stylus of the recording phonograph. It was extremely interesting for the visitor from *Scientific American* permitted into the inner sanctum to observe "the endless stream of fine, hair-like turnings which falls from the little tool while the record [mold] is being made."

An instrumental band was employed full time by the Edison Company—in residence at the factory to keep up with the special demand for martial music, airs, and waltzes. Gilbert and Sullivan songs, brief overtures, vaudeville monologues, and sentimental folk tunes helped fill out the catalogue. Quality control was strict. Each record cast from a master cylinder was audited by a white-collar "corps of experts," discarding those with the slightest hint of imperfection. Ones that passed the test were wrapped in cotton, then in wax paper, then placed in titled, numbered cardboard boxes for ease in order fulfillment.

After nearly a quarter of a century Edison's work on this stage of the phonograph's evolution was less purely innovation than progress, in another area about which he always cared deeply, refining an ever better methodol-

SELLING PHONOGRAPH CYLINDERS

ogy—a *system* that worked. He came to cement-building chastened but resolute from the mistakes of his ore-milling crusade. He developed the phonograph with a sure feel toward profit. However, in the third chapter of Edison's industrial renaissance, the storage battery, he played the inventor's hand.

"All the gasoline motors which we have seen, belch forth from their exhaust pipe a continuous stream of partially unconsumed hydrocarbons in the form of a thick smoke with a highly noxious odor," wrote the electrical engineer Pedro Salom in the 1896 *Journal of the Franklin Institute.* "Imagine thousands of such vehicles on the streets, each offering up its columns of smell!"

Imagine, indeed, which is precisely what Thomas Edison did on a spring day in 1900 while stalled at dockside on his customary route into New York City via the Cortlandt Street ferry. For several hours he observed the "seething mass" of wagons, horses, men and goods making their cacophonic, impenetrable, malodorous way through lower Manhattan; never one to waste time in idle thought, Edison jotted in his pocket notebook, "Problem:—Narrow streets. Comparatively large street area covered by a horse drawn vehicle. Slow speed. Limited loads. Congestion . . . Solution:—Electrically driven trucks, covering one-half the street area, having twice the speed, with two or three times the carrying capacity."

Edison's ideas never seemed to die, but coexisted in tandem for decades, transmuting, growing, held in abeyance until the right moment for exploitation when a synapse fired and they became manifest. He had in fact been intrigued with chemical batteries since childhood, and there had arisen, more recently, the need for a way to power the electric phonograph motor, as well as the electric light on a local level. He wanted to move beyond the traditional and cumbersome lead battery, which could weigh anywhere between one hundred and two hundred pounds per horsepower output. The lead battery electrolyte, diluted sulphuric acid, was prone to corrosive leaks, as young Tom had experienced back in his telegraphic years, and the electrodes deteriorated with time. The lead battery itself was prone to short-circuiting and deterioration. Its recharging span, up to five hours, was too long, and its useful life too short.

A storage battery generates electricity through a chemical reaction that takes place between the electrolyte liquid and the plates, or elements. This reaction is triggered initially by passing an electric current through the battery, adding oxygen to the positive plate, and taking oxygen from the negative plate. (A contemporary writer charmingly characterized these transfer phenomena as "indefatigable little atoms of oxygen, called, with fine imagery, *ions*, or wanderers.") When the circuit is closed, by attaching the battery to a device that requires electrical power, such as a light bulb or an automobile engine, a reverse flow, or discharge begins, the positive plate giving up its oxygen and the negative plate becoming oxidized as a result.

For three years, Edison set his chemical research team at West Orange to work testing thousands of different permutations for the variety of elements within the structure of the battery, to find the right interaction of an alkaline (i.e., nonacidic) solution with a noncorroding plate, and to establish a constant chemical reaction that would not wear down the components, and could be replenished with plain water—the clear goals being longer life, durability, safety and a much more efficient weight-to-energy ratio. Whereas a lead battery weighing one thousand pounds could drive an automobile thirty miles, Edison's battery of the same weight should be capable of taking the distance to ninety miles.

Arthur E. Kennelly, along with the chemist Jonas W. Aylsworth, one of Edison's closest associates in the development and implementation stages of the battery, reported to the 18th Annual Meeting of the American Institute of Electrical Engineers at their meeting in New York City on May 21, 1901, about the correct electrochemical combination of ingredients found for the new battery. The negative pole, or positive element, was iron. The positive pole, or negative element, was a superoxide of nickel. The electrolyte was potash, in aqueous solution containing ideally 20 percent potassium hydroxide.

Unlike the lead battery, Edison's cell could be charged very quickly with no physical damage. The plates were thin sheets of steel, with grids "resembling window frames" cut out of them, into which the polar elements were inserted. The battery seams were soldered with a compound unaffected by the potash solution, a liquid that further demonstrated its viability by maintaining its chemical properties during the oxygen "channeling," whereas in the lead battery, the aqueous sulfuric acid invariably turned to plain water under the same conditions.

"OUR GREATEST LIVING AMERICAN" AND HIS WIFE
AT THE TURN OF THE CENTURY

Harper's for May 1901 ran a full-page portrait of Edison, "the famous inventor, who considers his new storage battery the most valuable of all his creations, and believes it will revolutionize the whole system of transportation."

Visitors to the laboratory settling in for a quiet chat in his library with the man recently called by the New York *World* "Our Greatest Living American, The Foremost Creative and Constructive Mind of This Country, Our True National Genius" were shocked to see valise-shaped objects plummeting outside past the arched windows as he tested the hardiness of this "valuable creation" by assigning a boy at the lab to drop prototype model batteries repeatedly from the third floor.

There was also an important underlying social motivation for a different battery, as Edison declared in his sixth and final article for the *North American Review*, "The Storage Battery and the Motor Car," appearing in the

same issue as a short story by a forty-year-old New Yorker named Edith
Wharton called "The Three Francescas." Miss Wharton's first published
work in 1897 had been *The Decoration of Houses*, co-authored with Ogden
Codman, but her physician advised her that a healthful way to stave off the
encroaching stress of her marriage would be to shift to fiction. By the time
of her current piece in the *Review*, she had brought out three volumes of fiction, and three years later published her first novel, *The House of Mirth*,
which became a bestseller. The *North American Review* went on as a monthly
until 1939, its one hundred twenty-fifth year.

In this august forum, Edison spoke with disdain of the critics' inclination to compare his newly constituted battery—not quite yet but soon to be
on the open market—with the performance of the ponderous, inefficient,
clearly inferior and imminently, he would hope, obsolete version of the secondary cell invented by the French physicist Gaston Planté in 1859. A *"real
storage battery,"* he said, in a telling equation, "must be reversible, like a
dynamo, which converts power into electricity and vice versa." The real advantages in transportation for his fellow residents of the verdant hills and
valleys surrounding the polluted metropolis would not merely be economic,
since it would be possible to fire "the chauffeur, the irresponsible instrument," but also sanitary, since, the disgruntled Edison reminded his readers,
there would be "no [more] horses to eat their heads off and no oats and hay
to buy." Furthermore, he said, look across the seas to our friends in France
and in Europe, where there are fewer automobile accidents, because the general public has been "educated up to the situation." The future belonged to
the electric car!

"At last we have a *real* storage battery," Edison concluded.

With the closing of the Ogden mine, the focus for Thomas Edison's industrial research and development activity shifted to West Orange, marking
his return to the home front after a decade in relative absentia. Life at Glenmont under Mina's auspices had become the model of "conspicuous consumption . . . [It] is not sufficient merely to possess wealth or power,"
Thorstein Veblen wrote in his classic *The Theory of the Leisure Class* (1899).
"The wealth or power must be put in evidence, for esteem is awarded only
on evidence."

At Madeleine's thirteenth birthday party in the early summer of 1901,
the forty children fortunate enough to be invited were received in the grand

entrance hall by the birthday girl, then sent into the dining room to be greeted by Mrs. Edison. There, the table and rug had been removed and chairs set up for the game of Going to Jerusalem, followed immediately by Spinning the Plate. Amusements continued outside on the lawn with a potato race, until all had worked up an appetite. Seating partners were arranged by drawing odd verses of Mother Goose rhymes attached to favors from a bag and matching them up. The birthday cake was decorated with American flags, and at each place setting was a flag stuck into a dinner roll representing a different foreign nation. (Not to be outdone, at little Theodore's third birthday the following month, he was blessed with a giant, cream-filled cake in the shape of a locomotive.)

Holiday menus on Thanksgiving and Christmas were similarly ornate, beginning with fresh oysters (a perennial favorite of Mina's), moving on to consomme, hard-shell crabs, turkey, potato croquettes, mushroom patties, cucumber salad, plum pudding, lemon ice, and nuts and figs, and at each setting was placed a miniature Santa Claus figure filled with candy. The massive tree in the downstairs den was adorned with every conceivable decoration from the conventional to the fanciful: evergreen roping, cranberries, pink poinsettias, a miniature hansom cab, icicles, piano, banjo, horseshoes, butterflies, silver dogs, and cornucopias. Any Christmas tree put up by Thomas Edison would have to be electrified—in this case with blue, pink, yellow, orange, green, violet, and fuchsia lights. Under the tree might be found a stamp collection or a train set for Charles, a dollhouse and sideboard with tea set for Madeleine, and toy hussars and stone building blocks for Theodore.

After dinner, all the immediate family and assorted Miller cousins with college friends in tow would gather for caroling around the organ, or take parts in plays they read aloud, then adjourn upstairs to the favored room above the porte cochere for a round of parlor games, going far beyond the ubiquitous Parcheesi, which could be played only according to Edison's house rules, to dominoes, jackstraws, and anagrams, or the more esoteric Ho Ko and Faga Baga.

For brother Charles, these were years when he became accustomed to ways of privilege. Glenmont was surrounded by eight hundred acres of "royal" estates; his playmates were the sons of other captains of industry—Henry Colgate of soap fame, Jerome Franks, whose father worked for Andrew Carnegie; and Lloyd Fulton of the great Newark ironworks dynasty. The Edisons' closest neighbor during their occasional summers at Deal

Beach on the Jersey shore was Robert Lincoln, son of the late president. Charles ran with a rough gang of lads who took pleasure in kicking a medicine ball around Glenmont's improvised gymnasium just off the kitchen, or shooting barn rats with slingshots, pursuing each other through bramble patches brandishing buggy whips stolen from family coachmen, and throwing rocks, or playing bicycle polo on the back lawn with croquet mallets and balls, knocking out spokes and nearly decapitating themselves in the process. By the time he was twelve, Charles was driving the family's electric Studebaker to and from Carteret Academy.

Madeleine, two years older, recalled with deep relief her father's greater presence on the scene as she entered her teens, but there was no doubt in her mind that he was "The King" of the domain, and the entire family, especially Mina, "bowed and catered to his wishes. We tried to please Father in every way. Everybody sat up when he spoke, and danced to his tune. He was next to God. Anything Father said was all right and we were supposed to abide by it. I think [Mother] was as much under his thumb as any of us, and enjoyed it . . . Anything Father wanted, she would do." Edison seemed always to be trying to teach his children a lesson of

PIANO AND READING LESSONS AT HOME FOR
CHARLES AND MADELEINE

some kind. Every day, before leaving for the laboratory, Papa came into each of their rooms to bid them good morning. On one such occasion, Madeleine was lounging back in her chair while the maid—one of five in attendance inside the house, along with cook, waitress, laundress, and nurse—buttoned her high-top shoes. Edison went storming downstairs to Mina and told her in no uncertain terms that no child of his should *ever* be treated to this luxury.

If Madeleine and her mother dressed to go out wearing a necklace *and* a pin *and* a ring, Edison would naggingly remind them of the Fiji Islands native woman who weighed one hundred pounds and wore two hundred pounds of jewelry. Papa did not approve of bracelets at all, and it was not until after her father's death that Madeleine felt comfortable wearing one. Red dresses were also forbidden.

At breakfast over habitual meat and potatoes—between spontaneous, pop-quiz questions based upon the previous day's homework and vocabulary checklists maintained by Mina—Edison might dip his spoon in hot coffee, then touch it to Madeleine's hand when she looked the other way, a teasing test of her endurance. Research and diligent reading were enforced from earliest times. Even Marion recalled having been required to memorize ten pages of the encyclopedia every day when her father took her out of school to be his traveling companion after Mary Edison died. In the evening, when Madeleine and Charles were heading for bed, baby Theodore sat quietly with

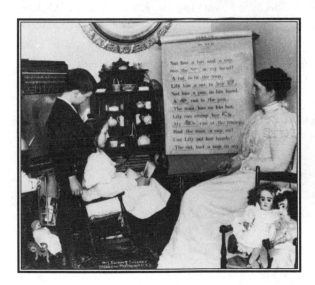

his governess, and Mina was looking forward to her knitting, crocheting, reading a bit of Shelley or Ruskin, or the latest issue of *Lady's Pictorial*, *Gentlewoman*, or *Country Life*, Papa would enlist them to help pore through volume upon volume of science reference books in the upstairs living room digging out citations and inserting slips of paper in the right places as markers to help him in his laboratory work. After his family finally retired for the night, Edison would remain at his "thought bench" until four or five o'clock in the morning, briefing himself for the next day's project, and still arose punctually at 7:30.

There were also the five newspapers to read each and every day, while Shakespeare, Tennyson, Charles Lamb, and the Old Testament were *de rigueur* and considered good for one's moral development. Mina read selections from such classic works to the children for an hour before they dined. In secret, Madeleine took her beloved copy of *Little Women* and other romantic novels to her room and devoured them time and again.

Even nature walks on springtime Sunday afternoons had a pedagogical theme. Edison would dispatch the children through the surrounding woods to collect varieties of every form of vegetation they could find, then match all their cuttings at the end of the stroll with the flowers he had collected, to see who had the most. As in Parcheesi, Edison liked to win, and he most often emerged victorious from these outings.

"Two Years Rest for Edison, He Is Tired Out and Will Stop Inventing for a While," announced the Philadelphia *North American*, having found the great man of the century "in a stain-spotted suit and an old slouch hat drawn well down on his forehead," working away at a "dilapidated little table over which he had been bending. [He] threw a stub of a pencil down on a yellow pad of paper, and settled back in an armchair. 'At last,' he said, 'I've finished work on my storage battery and now I'm going to take a rest!' "

Two years rest was, as might be expected, something of an exaggeration. However, on February 28, 1901, just after his wedding anniversary, Thomas Edison and family finally returned to Fort Myers after a fourteen-year hiatus—the illustrious resident reassuring the townspeople he was back for real this time, and that the place was "finer than ever, the prettiest in Florida, and sooner or later visitors to the East Coast will find it out . . . For a winter residence, nothing equals it, and the fishing is unlimited and includes every known variety." As if to prove his point with a vengeance, Edison spent the next month indulging in a nonstop succession of voyages, "entering into the sport of the day with the enthusiasm of a boy . . . playing

EDISON FISHING ON HIS FORT MYERS PIER

hookey from school." Down the Caloosahatchie to Fort Thompson and over to Sanibel Island with his friends, he hauled in jack, snook, Spanish mackerel, lady fish, tarpon ("the mighty Silver King leaped into the air to free himself from the hook of the electric wonder of the age"), black bass, sea trout, red fish, and snapper.

A new steamer to ply the serpentine Caloosahatchie between Fort Myers and Fort Thompson was built at Apalachicola over the coming year and christened the *Thomas A. Edison* in honor of Florida's celebrated prodigal son. A sleek sternwheeler heralded as "clean and beautiful both fore and aft" by all who watched her glide past banks of orange groves and leaning cabbage palms, in full burn at ten miles per hour capable of outrunning any ship in sight, she was eighty feet long, with two decks, ten staterooms and a nine-foot wide dining room—a fittingly magisterial tribute to the inventor on the cusp of his comeback.

CHAPTER

22

WHILE MINA WAS IN CHARGE OF BIRTHDAY PARTIES, THANKS-
giving, Christmas, and Easter, when it came to the Fourth of July, the most
American of all holidays, Thomas Edison was the undisputed impresario.
Everyone had to arise at the crack of dawn. Father insisted the children go
barefoot and meet him outdoors on the curved driveway under the porte
cochere, where echoes were especially loud. Although he laid in a huge supply of every known species of fireworks for the evening display on the front
lawn—musical candles, jeweled jets, rosettes, pinwheels, geysers, double
triangles, saucissons, floral fountains, and dragon's nests—Edison's favorite
variety were the firecrackers he invented and constructed himself, unbeknownst to Mina and the children, spiced with just a pinch of TNT. These
he enjoyed lighting and then, ever the prankster, playfully tossing at the
feet of his children, making them jump, dance, and scream with feigned
fright. After a short nap and another round of incendiary torpedoes—this
time Madeleine, Charles, and Theodore turning the tables on their Papa and
watching him skip about—watermelon and ice cream would be served.

Golden-haired Theodore, or Baba, as he was called, was from his earliest days a tribute to the adage that the apple does not fall far from the tree.
He certainly had his father's curiosity. While playing "camping out" with
his big brother, Theodore watched Charles take up a hatchet to cut kindling
wood, and tried to do the same, succeeding in slicing his index finger al-

THEODORE EDISON AS "BABA"

most to the bone. By age six, he was collecting pets, a number of them exotic. He returned from a trip to Fort Myers with five baby alligators, which he kept in a tub of water under a low spreading tree near the house. One summer afternoon, while Mina was serving tea in the conservatory to her lady friends, Theodore decided to play show and tell, but was unable to contain all the "wrigglers" in his arms, and they slithered out and under the table, much to the alarm of the guests. Far more submissive were Star, the gray rabbit who craved the nuts Theodore foraged for him in the forest and endured his master's delight in picking him up by the ears, and Snowy, the spaniel puppy. With the addition of Charles's hatching baby chicks, a diverse Glenmont menagerie took shape.

During his time at Fort Myers, Charles was drawn into another of father's obsessive activities, fishing. Upriver on the family's electric launch, *Reliance*, Edison hooked a forty-pound tarpon, whereupon Charles landed a one-hundred-pound monster.

After becoming accustomed to the gentleness of the Edisons' beloved companion, Lucy Bogue—nicknamed Boguelette by wordsmith Charles, she lived with the family for twenty years and taught all three children how to play piano—Theodore found it difficult adjusting to Miss Henning's kindergarten school, which Madeleine and Charles had also attended. The boy was always close by his mother, and Mina was ambivalent about letting her youngest out of her company, even as her own mother had wished long ago that dear Mina had "staid at home." His first week at school, she spent the better part of three whole mornings in the classroom, finally reminding Baba that she "had to leave some very pressing work to get to school today." He replied, "Well, why *did* you come, Mamma? I am not going to die today—there will be ever so many other days that you can come down to see me."

While Madeleine, forbidden to ride horseback, worked on her French, and Theodore, forbidden to chop wood, cared for his pets, Charles earned the nickname Toughie and became enlisted as his father's chauffeur and general automobile mechanic. During the years after the turn of the century, when work on the storage battery took up so much of the Old Man's energies, Edison acquired several different electric cars so that he could experiment with various types of batteries, and his barely teenage son would take him out on the road in a happy marriage of business and pleasure. One by one, they ran through the Mark III Columbia, the Baker, the Woods, and the Studebaker, proud and now forgotten names. Out for a Sunday drive near Westfield, in Union County countryside about ten miles south of West Orange, humming along at ten miles per hour, Charles was struck in the eye by a bee, and he had to pull over, in a rare act of necessity relinquishing the "tiller" control lever to father, who promptly careened from one side of the road to the other before mounting the embankment and cracking the front axle. During the three-hour wait before a horse and buggy came by to pick up the dusty pair, Edison simply curled up in the grass and slept.

In the summer of 1905, with the next acquisitions in the Edison stable—a pair of White steamer cars promptly christened Discord and Disaster—*automobiling* in the popular sense of the day became the most serious pursuit for Edison family leisure time. No more perfunctory day trips for Thomas Edison. A veritable entourage set forth for a jaunt through New England: Thomas and Mina, Madeleine, Charles, Mina's sister Grace, brother John, and niece Margaret Miller. (Ira's daughter, at eighteen, had become a close pal of Madeleine's and spent her Easter holidays and summers at Glenmont.)

August 15 was a sultry (and dark) Tuesday morning, and jolly Uncle Thomas, resplendent in midcalf-length duster, peaked cap, and goggles, awoke everyone at his scheduled hour of five; perhaps in excitement and anticipation he had not even been to bed the night before. A limited amount of technological progress had been made in the design of the automobile, to the extent that isinglass windshields had entered the picture, although windshield wipers had not yet made their appearance. Glycerine was liberally applied to the glass to prevent rain from obscuring the view.

Northward they headed, through the upper reaches of New Jersey and into Nyack, where the party rolled onto a ferry for the choppy trip across the Hudson to Tarrytown, then on to Peekskill, Poughkeepsie, and Millbrook in New York before forging on into the Berkshires—Stockbridge and Lenox—and east toward Springfield and Worcester, Massachusetts. Veering north again, four days later, the famed White Mountains of New Hampshire loomed into view, and through Franconia Notch the caravan flew, ar-

riving at an inn called Crawford House, where Mr. Edison was naturally recognized, and the party was paraded around the parlor like celebrities,
dressed in tuxedos and gowns for dinner. The next morning, a trip up
Mount Washington by train and cog railway culminated in dinner at Tip
Top House, then back fifteen miles to their inn via six-horse stagecoach,
Margaret noting that "the driver kept the horses on the jump by pelting
them with stones." The reveling continued with dancing until the early
morning hours, such that although Edison himself was up by sunrise, he
could not rouse the others until the late hour of eight o'clock.

The journey back home down the coast via Portsmouth to Boston was
held up slightly by the bane of all automobilers, a puncture. Charles was
pressed into service to pry the tire off with an iron, pinching the inner tube
in the process and having to start all over again. Then the two cars became
separated near Salem when Edison, in the lead and insisting upon making
up lost time, nearly burned up his engine. Always the history buff, Edison
picked up the trail retracing Paul Revere's route, stopping at Lexington to
view the statue of the Minute Men and then the Common where the first
battle of the Revolution had been fought. At Concord, the group stopped at
the homes of Nathaniel Hawthorne, Louisa May Alcott, much to
Madeleine's delight, and, needless to say, Ralph Waldo Emerson's cottage.

Through Connecticut and approaching Danbury, Discord (or it may
have been Disaster) broke down in *its* turn, and Charles was dispatched by
horse to survey the local hotels available. The following day, "the boys"
spent six grease-smeared hours working on the stalled machine, finally finding the problem to be a broken boiler, which had to be sent up by special
courier from New York City. Edison then adventurously insisted upon traveling straight into the metropolis, reaching Harlem quite lost after midnight, fortunately running into a friendly streetcar motorman who steered
them to the Holland House, where they arrived past two in the morning
and left with daybreak for the final dash home, at August's end breathlessly
concluding a two-week adventure.

As Edison set forth the strictly constrained but moderately affectionate
ground rules whereby he tried to accommodate his wife and three children
at home into the industrial priorities of the rest of his life, he at the same
time entered into ever more convoluted patterns of domination and dependence with the three children in exile.

Tom, Jr., who addressed his father as "Dear Sir" and wrote to him only through Johnny Randolph, endured an exceedingly bad patch after his breakup with the unfaithful Marie. Tom blamed his indigent situation on his father, claiming that prospective employers inevitably decided against him by saying "If you was [*sic*] any good at all your father would employ you and if you are not good enough for him you are not the man we want." For about a year, suffering from an undefined illness that turned out to be anxiety fueled by excessive drink, Tom boarded with uncle Charles Stilwell (mother Mary's elder brother) in Newark and finally succumbed to relentless headaches and attacks of dizziness, committing himself to a sanitarium for two months at the end of 1903. Edison castigated his son for "passing bad checks and [his] use of liquor which is known to everyone."

Reconciliation driven by self-interest and a modicum of pity took significant shape after Tom entered into an especially amateurish fraudulent business deal, "selling" the use of his famous name to a homeopathic medicine company. His father became the brunt of ridicule and reached the limit of tolerance. He offered to send Tom a regular allowance of $50.00 weekly, and help him fulfill his dream—to establish and fully equip a mushroom farm in the rural reaches of Burlington, New Jersey, on the Delaware River outside Philadelphia—if his son would once and for all just drop the name Edison altogether. The trade-off was designed to remove the threat of further mortification by financial and material means. Above and beyond the check came funds for a team of horses with harness and plow, a farm wagon, seeds and spores, even money to build and paint a farmhouse and outbuildings.

THOMAS A. EDISON, JR., AT HIS MUSHROOM FARM

Thus did Tom, Jr., and his second wife, Beatrice Heyzer of Louisville, Kentucky, become metamorphosed into "Mr. and Mrs. Thomas Willard."

William Leslie had also tied the matrimonial knot, with a young woman named Blanche Travers. Despite the fact that his bride's father, according to Will, was "one of the biggest gentleman farmers and wholesale produce men in Baltimore," Mina was opposed to the marriage, and in a highly unusual move, she ceased writing to him. This gave Will a convenient outlet for his wrath. He characterized himself as "the forlorn son of a great man, without a home." He accused his stepmother of "disinheriting" him because she wanted to "get rid of [her] husband's first children for [her] own sake."

Attempting to establish yet another company, this time to distribute automobile parts, Will ran into the same problem as Tom—a father furious at the tarnishing of his good name. It was bad public relations: "You are being used for your name like Tom," he telegraphed, "and as you seem to be a hopeless case I now notify you that hereafter you can go your own way and take care of yourself . . . I warned you not to do it. Your action makes it hopeless for me to do anything again—I am through—E." At this point, even Blanche jumped into the fray, letting her own anger spill forth at what seemed to her to be the most abhorrent of all affronts, a father's excommunication of his sons: "Do you realize that we [*sic*] are the children of '*The Greatest Man of the Century*'—and for them to live, as they should, upon an income of forty dollars per week, takes very much more ability than I can display. Surely you would not have us . . . live in a cheap boarding house with plain people."

Arguments from this new quarter caused father to stand his ground even more resolutely. Now he had *two* recalcitrant people to deal with in the same household. "I see no reason whatever why I should support my son," came Edison's immediate riposte. "He has done me no honor and has brought the blush of shame to my cheeks many times."

The young couple led a wandering life, surfacing over a seven-year span in Staten Island, Washington, D.C., Yonkers, Philadelphia, and Pittsburgh, as Will, by his lights "paddling my own canoe" and living in "Hell holes" and "rotten joints," took low-paying jobs as a mechanic and factory worker between stints on public assistance. Eventually he followed in the footsteps of brother Tom and established a game and poultry farm in Waterview, Virginia, where he and Blanche bred and sold turkeys, pheasant, homing pigeons, quail, ducks, and leghorns.

And what of Marion? "Ten years [since her wedding] is a long time and brings many changes," she wrote to Mina from her aptly named *Villa Mon Repos* in Dresden, still happily married to Oscar, although childless, and consumed more and more with memories of home. Dot's hardships lay far ahead, in the still unimaginable shadows of war.

A Saturday afternoon in January 1903. Thomas A. Edison, "the happy-hooligan light shining out of his gray eyes," made his introductory speech to M. A. Rosanoff, a nervous new man in the chemical research department of the phonograph factory at West Orange who "approached him in a humble spirit: 'Mr. Edison, please tell me what laboratory rules you want me to observe.' And right then and there I got my first surprise. He spat in the middle of the floor and yelled out, 'Hell! There *ain't* no rules around here! We are tryin' to accomplish somep'n. . . . Do you believe in luck?' " The trembling Rosanoff replied, "Yes and no. My reasoning mind revolts against the superstition of luck, my savage soul clings to it." "For my part," said the Old Man, launching into a volatile subject, "I do not believe in luck at all. And if there is such a thing as luck, then *I* must be the most unlucky fellow in the world. I've never once made a lucky strike in all my life. When I get after something that I need, I start finding everything in the world that I *don't* need—one damn thing after another. I find ninety-nine things that I don't need, and then comes number one hundred, and that—at the very last—turns out to be just what I had been looking for. It's got to be so that if I find something in a hurry, I git to doubting whether it's the real thing; so I go over it carefully and it generally turns out to be wrong. Wouldn't you call that hard luck? But I'm tellin' you, I don't believe in luck—good or bad. Most fellows try a few things and then quit. *I* never quit until I git what I'm after. That's the only difference between me, that's supposed to be lucky, and the fellows that think they are unlucky.

"Then again," he went on, "a lot of people think that I have done things because of some 'genius' that I've got. That too is not true. Any other bright-minded fellow can accomplish just as much if he will stick like hell and remember that nothing that's any good works by itself, just to please you; you got to *make* the damn thing work. You may have heard people repeat what I have said, 'Genius is one per cent inspiration, ninety-nine per cent perspiration.' Yes, sir, it's mostly *hard work*."

The Old Man himself said it best: The nature of his *creative* relationship to each of the diverse companies that had sprung up like seedlings and characterized his businesses at this frenetic time was directly a function of the kind of labor—just plain work—required to gratify him. When a new inventive or industrial path was chosen, a new company was established, and Edison dipped his hand in to varying degrees. Thomas C. Martin of *Electrical World* compared his friend to "a Japanese juggler, balancing half a dozen little affairs in the air, giving them the deft spin or kick at the moment when they might drop."

In the movie business, for example, Edison was never interested in becoming involved with producing and directing, not when he had the talented Edwin S. Porter at the helm for eleven years, assisted by William J. Gilroy, to develop a constant succession of "story" (fiction) films that promoted Edison's indestructible, patriotic belief in the importance of glorifying great American tales for the entertainment of the common man and the reinforcement of his values. As long as he was assured that this informing principle remained in place, aesthetic standards were secure in the capable hands of others, and Edison devoted his inspirations to technical problems such as color film and an affordable home kinetoscope.

The years from 1903 to 1908 were commercially sucessful ones for Edison films, sparked by the sensation of the ten-minute *The Great Train Robbery* with its exciting narrative line, triumph of justice over evil, and brazen theatricality. The Western saga was homegrown, shot in New Jersey's Essex County and Orange Mountain parkland, and on the Deleware, Lackawanna and Western Railroad line along the Passaic River, within ten miles of New York City. Interiors, including daring matte shots, were filmed at the West 21st Street studio in Manhattan. *The Life of an American Policeman* and *The Life of an American Fireman* (filmed in Newark) were hymns to upstanding citizens devoted to safeguarding hard-earned virtues.

Even after the Motion Picture Patent Company was formed in September 1908—a huge corporate entity establishing a general licensing agreement for all the competing movie moguls and film stock producers in recognition of the necessity to combine forces or be in peril of killing one another off—Thomas Edison personally preserved the moral prerogative to bring out "good, clean pictures."

At the cement plant, unlike the movie business, he opted from the outset to get himself dirty (or, as a reporter for the New York *American* found him, "greasy and white-spotted") up to the elbows and beyond. Among the twenty-three patents applied for in 1903 was a "Means for Operating Motors in Dust Laden Atmospheres," an ingenious improvisation on Edison's endless concern with protecting the bearings and moving parts of the machinery required to grind the rock. At the Ogden

mine, he had already perfected a "circulatory system" for
oil to move via narrow pipes
through the complex from
a central "heart" pump. He
had found that hermetically
closing in the moving parts
of the electric dynamos—the
armature, commutator, and
brushes—caused them to heat
up excessively, leading to loss
of efficiency. Hence Edison's
idea for a permeable, double-
walled gunny-cloth textile,
acting as a "dust-sieve," per-

A MODEL OF THE CEMENT HOUSE

mitting clean air alone to enter through the meshes while letting heat out.
In an opening of the machine casing he placed a small, five-horsepower fan
so that the dust was further inhibited from interfering with moving parts.
Within two years of this innovation, turbine-driven electric drive-shafts had
replaced the common line shaft to which each machine had been connected
by a belt-driven pulley, and the cement plant at Stewartsville was on-line,
generating three thousand barrels a day—far short of Edison's high-flown
early estimate of ten thousand barrels daily, but respectable nevertheless
against great odds. From the temporary shack he referred to as "The
Monastery" just outside Stewartsville, Edison exercised impatient command
over his successful domain.

Edison sent his movies forth to provide edification for a certain stra-
tum of American. Likewise in the forefront of his philosophy about cement,
beyond the primary challenge of manufacture, was the professed desire to
"evolve something to make the struggle for existence easier for the wage
earner."That intention became realized in one of his most quintessential inventions: the poured cement house. The germ for this inspiration is traceable to dicta published as early as 1901, when Edison donned what we shall
call his "populist Wizard" hat to promise that once he had removed his designs off the drawing board for the purest cement ever made, he would
channel that material "within the reach of the builder of the humblest cottage . . . to stop the ruthless devastation of pine forests" and "place homes
within the reach of thousands." He would devote his resources to granting

the dream of every American, the cheap one-family home. "There's a cry from millions of mothers and children who grope about in the human beehive tenements of the great cities, craving a breath of air, a glimpse of blue sky, a few blades of grass," sang the *American Carpenter and Builder* with Whitmanesque longing. "Swallowed up and submerged as they are in the dark back rooms, the Genius of Invention is to be their salvation."

It took eight long years for Edison's earnest vow to become a reality. Keeping with the tried-and-true genesis of all his important works, Edison hired two able young engineer-architects, George E. Small and Henry J. Harms, to help him get up drawings based upon his concept for a three-story house that had to come in at under $1,200 complete, including plumbing, heating, and lighting. The unit cost of Edison's house was able to be kept low because it was designed to be mass-produced in Levittown-like planned communities—idealized small towns from this smalltown boy—and because all of its parts, including sides, roofs, partitions, stairs, floors, dormers, even cupboards, bathtubs, fireplace mantels, and ornamental ceilings, were poured into a single mold. This house was to be made *from the top downward.*

A 25- by 30- by 40-foot-high modular "house mold" with two-foot-thick double walls was built first out of five hundred cast iron sections bolted together and placed upon a concrete footing. At the very top of the roof mold was a funnel-like opening, with tributary pipes and open troughs leading to other, smaller apertures around the periphery of the roof for ease in distribution. A quartz-strengthened and clay-tempered mixture of quick-hardening cement, fluid and tensile, especially adapted for this purpose, was raised in buckets to the top of the mold from an agitator-mixer connected to a continuous vertical conveyor belt. Over a six-hour period, the mold, punctuated by air vents to prevent internal bubbling, was filled to the brim, cement continuously kept in circulation. Four days after the pouring, the molds were dismantled. Six more days were allowed for hardening the cement. Only windows and doors then remained to be added.

Variety was made possible by baroque ornamental designs cast into the molds, and a broad palette of waterproof iron oxide tints was developed by the Edison laboratory, colors that could either be blended into the cement or applied after construction was complete. One of the first structures built to test the properties of structural concrete was a garage on Edison's Llewellyn Park property which became necessary as his assortment of automobiles kept expanding. He then went on to put up a cluster of eleven concrete houses for his em-

ployees on Ingersoll Terrace in Union, New Jersey, ten of them still standing today as tribute to one of the first modern efforts at a low-cost housing
development in America.

Always with an eye for potential spin-offs, Edison went on to produce
cement cabinets for the phonograph, and seriously considered building a
concrete piano.

Carrying the storage battery out of the laboratory and into the marketplace
was, as we have seen, a painfully slow process. Not until May 1901 was an
assembly plant ready in an old brass factory in nearby Glen Ridge. And even
then, Mina saw her husband's restlessness returning, felt the "Japanese juggler" in him poised for another shift, as he made "progress with the cement
works so I am hoping that he will soon be through with them however then
he starts on the battery works which is going to keep him busy for some
time to come. Poor dear, he does not get through with one before another
problem is before him." She wrote these words to her mother at 8:00 p.m.
on a Sunday night, in a hastily scribbled note while she waited at the laboratory for Edison to wrap up loose ends. Having brought her "Dearie" down
from Stewartsville the previous day, Mina was all too soon in melancholy
anticipation of his departure early the following morning.

Again, lofty production goals were set—"five hundred cells daily,"
said the notebook entry—for a "pioneer plant" factory complex that would
run to about $385,000 in startup costs, with twenty-one departments,
arranged in a cavernous shell four hundred feet long, fifty feet wide, and
three stories tall. The batteries would be made cheaply in a classic assembly
line, at a unit cost of ten dollars, to be sold for fifteen. In Edison's rudimentary plans for the factory, he took great care to tie the particular varieties
of cells by amperage into the different types of electric automobiles and
"delivery-wagons" (trucks) he had field-tested, as well as to a standing order
from the Pennsylvania Railroad for cells to illuminate passenger cars. "We
are on to every defect and all can be cured perfectly," he wrote with a burst
of early excitement to Herman Dick, calculating that in New York City
alone in 1902 there were sixteen thousand licensed trucks, each and every
one a prospect for his new Type E cell, "practical, every-day, hard-working,
and low-priced."

As something new from the "New-World Alchemist," the battery enjoyed an immediate popularity that lasted about a year before troubles sur-

faced, and Edison ruefully told Mina that "he [was] leading a slave's life again, and he hopes there will be no batteries in heaven." Containers leaked and power losses under the stress of repeated charging ran as high as 30 percent. To counter the structural durability problem, Edison's men went back to the drawing boards and developed a new welding machine to supersede the old soldering method. No longer dropped out of windows, six cells at a time were mounted on a hand truck and then slammed against a brick wall five hundred times at a speed of fifteen miles per hour. The chemical problem, it turned out, was caused by expansion and swelling of the electrodes during charging and discharging. This was remedied by introducing microscopically thin nickel "flake" with a volume of four pounds per cubic foot—mined locally at Silver Lake, New Jersey—within the positive plate, increasing the watt-hour capacity per pound of battery by a dramatic 40 percent. The electrolyte was even further purified by adding a small amount of lithium hydrate to the potash solution. The advanced, hermetically sealed, high-energy, sleek Type A battery was ready.

It was 1908. Enter, spotlit at center stage, Henry Ford. *I will build a car for the great multitude,*" he said of the Model T. "It will be large enough for the family but small enough for the individual to run and care for. It will be constructed of the best materials, by the best men to be hired, after the simplest designs that modern engineering can devise. But it will be so low in price that no man making a good salary will be unable to own one—and enjoy with his family the blessing of hours of pleasure in God's great open spaces."

The author of this "consumer vision of the world" (as the late cultural historian Warren Susman so aptly defined it) was born on July 30, 1863, on a prosperous farm near Dearborn, in Wayne County, southeastern Michigan, the second of eight children, the grandson of an English tenant farmer who had come to America from Ireland during the famine of 1847. Henry Ford's mother, Mary Litogot, died when he was thirteen years old, and Henry went to work helping his well-meaning although occasionally alcoholic father, William, work the land, tend the sheep, and raise the younger siblings. From childhood he resented schoolwork, although his mother introduced him early to *McGuffey's Reader*, and hated the drudgery of farm chores. He gravitated instead toward collecting "pockets full of trinkets— nuts, washers, and odds and ends of machinery," and spent his spare time at his workbench repairing any broken watch he could find. Henry took great pride in his handiness with all things mechanical and thought nothing of walking into Detroit—which took an entire day—to visit hardware stores.

Biding his time until the right moment for escape, and tired of bickering with his father, at seventeen Henry signed up at the Michigan Car Company in Detroit, then with James Flower and Brothers, a small machine shop, and finally as an apprentice machinist at the Detroit Dry Dock Engine Works, the largest shipbuilding firm in the city. On his lunch hour, Henry read machine parts catalogues and delved into Faraday's works on the steam engine. He was known to be a wanderer about the shop, always more curious about what the other men were doing rather than his own assignments.

Returning to the family farm after his grandfather died for three years in the early 1880s, Henry applied his interest in steam traction to building a car-locomotive with a kerosene-heated boiler, enrolled in shorthand and accountancy classes at Goldsmith Business College, worked part-time for the Westinghouse Engine Company, and took dancing lessons. Self-improvement was his raison d'être, and it paid off.

At a harvest moon dance, Henry met Clara Jane Bryant, the daughter of a local farmer in Greenfield township, and told his sister Margaret that within thirty seconds he knew this was the girl of his dreams. In the spring of 1888 they were married, and settled on eighty acres of land where Henry operated a sawmill (steam-driven, of course) and sold lumber to make a living while continuing to pursue experiments with "horseless plowing."

As Ford tells it, bluntly and unromantically, within two years "the timber had all been cut," and a $45-per-month engineering job came available at, of all places, the Edison Illuminating Company, leading producer of electrical power in Detroit. Clara and Henry moved into a house a few blocks from the Edison substation, where he started on the twelve-hour night shift, six p.m. to six a.m. Ambitious and driven, he rose to the position of chief engineer soon after the birth of his son, Edsel, and set up a workshop in a shed behind his home at 58 Bagley Avenue. Here he began intensive experimentation building upon his earlier labors with a four-cycle gasoline engine—one-inch bore, three-inch stroke—moving to a double-cylinder model that he had originally conceived as attaching to the rear wheel of a bicycle. These prototypes would become the basis for Ford's horseless carriage.

We must emphasize that the idea of an internal combustion engine was far from new, as Ford was always the first to admit. Among many pioneers, in 1860, Jean Joseph Étienne Lenoir, a Belgian living in Paris, made the first commercially successful model. In 1877, Nicolaus August Otto of Deutz, Germany, built the first "silent" four-stroke engine, which was exhibited at the 1878 Paris Exposition. Beginning in 1882, the Otto was manufactured in the United States, and Ford repaired one in 1885 at Detroit's Eagle Iron Works, with ample opportunity to dissect and analyze it. In 1879, George Selden, an attorney in Rochester, New York, applied for a patent (witnessed by George Eastman) for a three-cylinder gasoline motor to drive a "road engine." This initial application, which became the bane of Ford's existence, was amended and enlarged more than one hundred times over the next sixteen years. Karl Benz's internal combustion tricycle was built in Germany in 1885, and a Benz car was exhibited at the Chicago World's Fair in 1893, the same year that Charles and Frank Duryea of Springfield, Massachusetts, built the first American gasoline car—which virtually "bewitched" Ford into inspired action. Like Edison, he knowingly entered territory well occupied by predecessors—and with the same confidence that he could make the thing better, and bring it out to a wider audience.

Ford's gasoline "quadricycle," a buggy frame mounted on four twenty-eight-inch bicycle wheels, with a two-cylinder, four-horsepower, air-cooled motor handmade by "Crazy Henry" from the exhaust pipe of a steam engine, lumbered out onto the cobblestones of Bagley Avenue in the rainy predawn hours of June 4, 1896. It weighed five hundred pounds, had a horn made from a doorbell, and a three-gallon gas tank, no reverse gear, and ran at two speeds, ten and twenty miles per hour.

Two months later, at the concluding banquet of the annual convention of members of the national Association of Edison Illuminating Companies in the ballroom of the Oriental Hotel in Manhattan Beach, Long Island, the wiry, shy, nervous young "tinkerer"-engineer met the legendary Thomas Edison in the flesh. They were introduced with a bemused smile by Henry's boss, Detroit Edison Company superintendent Alexander Dow: "Here's a young man who has made a *gas* car!" At Edison's encouragement, leaning close to him, Henry described and sketched the components of his creation. The Old Man pounded his fist onto the table and heartily urged the starstruck Ford to "keep at it!"

From that moment onward Ford's admiration ripened into hero worship, "like a planet that had adopted Edison for its sun." The two would not meet again for sixteen years, during which time Edison ironically turned away in disgust from the pollution and noise of gasoline engines and invested his time and money on the future of the electric car.

By the time Edison's A cell was in its penultimate year of refinement, Henry Ford had traveled a long and very different path than his idol. He had resigned from Detroit Edison and, with no personal funds of his own, entered an industry "strewn with the wreckage of failures," determined to create a market for a gas car—a commodity generally perceived at the turn of the century as being little more than a recreational frill. That notion had to be subverted. Capitalized at $150,000, Ford first founded the Detroit Automobile Company, which manufactured twenty-five vehicles before collapsing a year later. He became obsessed with automobile racing, and this glamorous attention-grabber helped gather twelve shareholders among the wealthy Detroit social set willing to give Henry another chance in a new incarnation as The Ford Motor Company. What raised the profile of this endeavor above the more than one hundred fifty other automobile manufacturers then active throughout the United States? The first Model A was lighter, less expensive, less mechanically daunting, than the competition. "Anybody can drive a Ford," said Henry, and indeed, it was consciously dedicated to *utility* rather than *luxury*, "designed for every day wear and tear. . . . Built to save you time and consequent money," read the initial advertisements, "providing service—never a 'sporting car.'" The strategy worked. Despite a nationwide economic panic in 1906–07, by lowering prices, and increasing as well as varying production into and then immediately out of the luxury sector with eight successively different models, the Ford car was a countercyclical phenomenon.

Ford's most critical decision was to take the success won by diversification, consolidate his gains, and bring down the focus to a single, simple, egalitarian, and uniform car, manufactured as much alike as "pins or matches . . . a car that the farmer could afford." It was the Model T, riding high off the road, slight yet strong, forged from vanadium steel, and deep black—only black—the mass-produced, assembly-line "Tin Lizzie," Ford's version of democracy on wheels. It was faster and had a wider range than the electric car; hand-cranked, it did not need a storage battery.

In 1905 Henry Adams, with his contrarian frame of mind, yet again captured the pulse of the period seizing men like Edison and Ford at the turn of the new century, when "the typical American man had his hand on a lever and his eye on a curve in his road; his living depended on keeping up an average speed of forty miles an hour, tending to become sixty, eighty, or a hundred, and he could not admit emotions or anxieties or subconscious distractions. . . . The new American—the child of incalculable coal-power, chemical power, electric power, and radiating energy, as well as of new forces yet undetermined—must be a sort of God compared with any former creation of nature."

A few years earlier, William Graham Sumner looked back with less trepidation to meditate upon what the nineteenth century had bequeathed to the twentieth, aside from "new powers and devices which are just in their infancy." He saw as its most significant legacy a vast fund of theoretical, *pure* knowledge and process born and raised in scientific laboratories, which could now be *applied* by industry to the general betterment of mankind and to "material and economic welfare."

Both men agreed—Adams for ill, Sumner for better—that industrial organizations and institutions now unambiguously held the accumulated authority, the *force,* formerly deemed sacred in isolated individuals. Even as the Enlightenment had refined notions of natural rights and social compact, Sumner reasoned, the nineteenth century willed populism to the twentieth. Adams viewed this dynamic much less rosily as "bowing one's neck to railways, banks, corporations, and trusts . . . the multiplicity of unity," in a worrisome attitude of submission.

Thomas Edison had been a rambunctious child of the nineteenth century. He entered the twentieth as an irascible man in his early sixties plagued more and more by hearing loss—he had two more ear operations by 1909—yet if anything, perhaps in compensation, more attuned and attentive than ever to the routine and solace of his daily work.

Those in his circle who knew him well realized that Thomas Edison would never consciously relinquish control of the tiller (or steering wheel) of his scattered and complex enterprises. Therefore, when his attorney, Frank Dyer, came to Edison with the plan for a necessary "reorganization under one corporate name . . . to bring home more firmly to the public the fact that this is Mr. Edison's personal business and that his personality stands behind it," he had the public relations savvy to suggest *Thomas A. Edison, Inc.* Let the middle managers and department heads and factory foremen and "muckers" and errand boys do their jobs within an actual hierarchy written down on an organizational chart, and let the dotted-line commands continue to flow in a ceaseless stream from the top down, as they undoubtedly would, but at least the structure of Thomas A. Edison, Inc. acknowledged the direction the rest of American business had taken.

CHAPTER 23

EAGER FOR GLIMPSES OF THE EDISONS *EN FAMILLE*, **WE HAVE** been fortunate for the existence of the memoirs and interviews, published and unpublished, left behind by Madeleine, Charles, and Theodore; they are the stuff of biographers' dreams, the flavor of distant times. Charles, especially, was a student of the human condition from an early age. His handwritten, finely honed, humorous, and loving drama in one act, *Prayer Meeting Night*, has survived among Mina's letter books, and parts the veil on the smiles of a summer night nine long decades ago.

The scene is the dinner table at Glenmont. The cast of characters is "Mother – Father – Lynn [Madeleine] – Theo – Unc [John Miller, Mina's

THE EDISONS AT HOME

younger brother; after the Spanish-American War, he graduated from Cornell University with a degree in engineering, came to live with his sister's family, and worked for Thomas Edison] – Miss B [Lucy Bogue, the piano teacher] – and C. E. [Charles]." The fifteen-year-old author's spelling remains unaltered.

THEO *(banging gently on the table with his jelly-covered soup spoon and kicking vigerously at the table leg)* Did you know they wern't sending any telegrams to Washington?

CHORUS No, why?

THEO *(putting his elbow forcibly in his soup)* Why, Washington's dead! *(Loud laugh from* PETER [the butler], *groans from the rest.)*

LIN [*sic*] *(suddenly)* Ahoooo! Oh Snow [*the family dog*] get off my feet!

THEO *(sweatly)* [sweetly?] Here Snow! *(dumps his meat on the carpet)* *(then)* Hey bring me some more syrup.

MOTHER Dearie, wait, Peter's out drinking his coffee, we mustn't disturb him.

THEO *plaintively* "Boggy," [Miss Bogue] you go get me some.

BOGGY No dearie I haven't finished my meat yet.

THEO REACHING OVER AND GRABING THE MEAT, There, now go, you haven't got no meat left.

MOTHER *with a poorly controlled smile,* Why Theodore, not haven't got *no* meat, but haven't got *any* meat.

THEO Oh well what's the difference? *(Squeezes the meat tightly and watches gravy run down his trouser leg.)*

MOTHER TO UNC Are you going to—down Snow, you know he really hurts me when he scratches that way—I saw the prettiest bird yesterday on the way home from the Woman's Exchange. It had a red breast and slate—

FATHER *suddenly* to UNC How much of that nickle [battery] mix did Ralph send up in that last batch? It wasn't the same as the old stuff. What's he done to it?

UNC Why he made it by exactly the same process.

FATHER No he didn't. He's done something different, hasn't dried it the same or something. *(To the rest)* You can't expect a man to do the same thing twice, it takes too much brains. Freddy Ott is just the same way but he hasn't any brains at all. *(Retreats behind his piece of pie and leaves the table.)*

3/4 of an hour afterwards.

FATHER   *(returns to the Dining Room)* What, still at it? It's so necessary to eat so as to be sure to keep warm? It's only 81 degrees out. *(Points to* MOTHER*)* She'll get at her eats and don't take any exercise. *(Sits down in rocking chair and reads)*.

MOTHER   Well that's right we shouldn't eat so much.

5 min. later

FATHER   *(looking at the clock meekly)* Prock? [Proctor's Vaudeville Theatre on Park Place in Newark, Edison's favorite Tuesday night haunt until it closed in 1908] *(Absolute stillness, each looking at the other with a sheepish "Gee, I'd like to go only I don't dare to say so" smile.)*

MOTHER   *tiredly.* I don't think I care to go. You others go and I'll stay home with Theodore.

FATHER   *breaking the record stillness.* Well, will you go. You've got to hurry or I won't go.

CHARLES   *jumps up and runs to the door.* How many seats shall I order?

UNC   Five.

MOTHER   No, 4 I don't believe I'll go.

MISS B.   Now go ahead, I'll stay with Theodore. I don't want to go anyway.

MUD   [Charles' pet name for Mina] Yes you do—now—

CHAS.   *impatiently hanging on the portiere* [archway between the dining room and the foyer] Oh hurry up and decide who's going.

MOTHER   Well you three boys go on. Father doesn't want me anyhow.

CHAS   Oh that makes me sick, why don't you come? I'll order five seats. *(Goes to telephone)* 1040 Newark – 1 – 0 – 4 – 0 – Newark! Allright Newark 1040. Busy? Yes, call me please. *(Ten minutes wait)*. 1040 I mean Newark 1040. Have you five seats in the first row? [Because of his hearing loss, Edison always had to sit in the very front of the orchestra]. Nothing nearer than the 8th? These are for Mr. Edison. Yes, Edison – E – D – I – S – O – N in Llewellyn Park. You've got them—Oh, alright. Goodbye.

CHAS   *bursts into Dining Room.* Come on, Unc. We'll go down and get the car.

UNC   No, I don't think I'll go to Proctor's, it's such a waste of time.

Exit CHAS.—*Exit* UNC.

Ten minutes later.

Honk – Honk

FATHER   All aboard!

More honks

VOICE FROM ABOVE   All right!

LIN *comes down stairs with Miss Bogue and a book.*

VOICE UP STAIRS   Good night, dearie.

MUD *comes half way down stairs.*

THEO   *from bed.* Say, Mahahuh.

MUD   Yes, Theodore.

THEO   Come up here a minute.

FATHER   *sweetly.* Aw hurry up!

MOTHER   No dearie not now. What? Allright I will. You go to sleep now. Yes, Theo—Ernest [another butler] is just down in the den. Allright, good night. *(Comes down and gets to the step of the auto.)*

THEO   *upstairs.* Oh Muthur!

MOTHER   Oh dear, that child. Yes, Theodore?

THEO   Come here a minute.

LIN   Oh for heaven's sake let's go if we're going.

MOTHER *goes back up stairs.*

Machine [auto] *goes to the barn after robes* [lap-robes for protection against the wind] *(Temperature is 82.)*

FATHER   *takes off his coat.* Well, it's too late now I won't go.

Machine comes back and nobody goes out. UNC *enters.* What's the matter why don't you come?

LIN   Father says it's too late now.

UNC   Oh pshaw come on.

MOTHER *comes down smiling.*

Enter CHAS. Say, Mud, where have you been—Why don't you come.

MOTHER   Why it's the machine what's been the matter with it—you just this minute came around. I've been ready for a long time.

LIN   Well, father won't go now.

FATHER   All right I'll go this time cause I said I would but this is the last time. *Puts on his coat and goes out. All get seated in auto.*

MOTHER   You know, that child got to talking about perpetual motion and he's got the cutest idea for—*Exit Motor Car.*

Such comedies of manners became less regular when, under Mina's influence, Madeleine (who from the age of ten had been forewarned she would be attending college) went off to Bryn Mawr in the fall of 1906, and Charles started his two-year stint at Hotchkiss the following year. Neither child was especially accomplished in the academic side of education. Charles applied himself with diligence and struggled with virtually every subject from geometry and German (which he habitually failed) to Bible studies and rhetoric, but his grades averaged from the low 30s to the high 70s. He spent most of his time counting the days before his *"vacance,"* or "luxuriously lolling on soft pillow cushions" and spraying talcum powder into the air while listening to phonograph records in his room, or juggling Indian clubs in the gymnasium.

Madeleine dreaded classes in math and science, a realm in which her heart was not at all engaged. Impatient, she feared she "would walk with a cane" before she escaped from the tedium of academic life: "This afternoon from two to four I endeavoured to find out how many centimeters there are in an inch. That may sound simple but I had all the instructors in the laboratory trying to explain it to me. Tomorrow afternoon I go to the lab. again," she wrote to her mother, "gnashing my teeth with terror and weeping copiously. It simply isn't thinkable that I can ever like that stuff. Why, there is actually trigonometry in it—for me—who had a tutor in arithmetic! . . . Psychology is also too much for my feeble brain and fainting soul."

"The only relief" for poor Madeleine were the Glee Club, and her English and drama classes, shared with two fellow freshmen, Mary Worthington, the niece of Martha Carey Thomas, the president of Bryn Mawr, and Margaret Mary James, daughter of William James, whose texts, especially *The Principles of Psychology*, were an integral part of the undergraduate syllabus. Her Uncle Henry had spoken at the college the previous year on Balzac, then delivered the commencement address.

Like her father, Peggy James suffered from periodic bouts of depression. In the second semester of her first year, she endured a breakdown and had to be sent home for a month; by the spring of 1908, it was clear that Peggy would not be returning to Bryn Mawr at all, Mrs. James writing to President Thomas that "while she can have [the company of] her father and her brothers, it seems fitting that she should be sharing their very interesting life."

Although Mina had hoped her daughter would complete the entire course of study, the stubborn Madeleine said she had never intended to stay at college for four years, and followed Peggy's example, returning at the conclusion of her sophomore year to Glenmont. In mid June Mina threw a debut afternoon reception for Madeleine, her formal presentation to New York and New Jersey society. Music was provided by Victor Herbert's orchestra—Herbert acted as "musical advisor" at Edison's Phonograph Works—and hundreds of guests filed by a receiving line under a bower of palms and pink roses in the Glenmont entrance hall. Later in the week, Madeleine held a luncheon, including all the Akron cousins and out of town friends staying at the house. Papa Edison took time off from the laboratory to join the young women, taking his place at the head of the table, welcoming all, and announcing that since he did not hear well, he would read a book. While the ladies feasted on fruit cup; broth and crackers with toasted cheese; molded, steamed salmon with fancy muffins; salad; and a six-inch-high four-layer cake drenched in chocolate sauce, Edison nibbled slowly at his single lamb chop, two tablespoons of fresh peas, and dish of applesauce. Over the hour-and-a-half meal, he remained at the table engrossed in reading, looking up every so often to survey the scene quietly. When the after-dinner coffee accompanied by mixed dried fruits and mints had been served, he stood up slowly and asked, "Well, ladies, who is ahead now? I, who fed my body the spare nutrition it needed, and then fed my mind—or you, who have been digging your graves with your teeth?"

Little Theodore first left home for any kind of extended stay on his own, at the tender age of nine, to attend the exclusive Camp Pasquaney in Bridgewater, New Hampshire. He loved the outdoors and was rarely homesick; rather, "Your loving sun, Baba's" main concerns were the health of dog Snowy ("Give him a good patting for me"), whether he could buy a camp "Baner," which cost $1.25—but he would have to know "omedeatly"—and in the literary area, after reading his way half way through the "*Mistory of the Yellow Room*," wondering who "realy did the crimb in the end." Unlike Charles, Theo did not attend boarding school. He studied at Montclair Academy, just up the road from West Orange, where his marks were good, but his behavior criticized by Headmaster John G. MacVicar, as "inattentive and silly."

For a change of scene after her daughter's break from schooling, Mina took Madeleine and Theodore to visit their Aunt Grace Miller on rugged, fogbound Monhegan Island, a slab of gray granite with the highest seaward cliffs and one of the the deepest inner harbors on the New England coast, ten miles off the Maine shore. First discovered by Captain John Smith in 1614, the island became the last safe port of call on the direct route to Europe, home to Maine Indians, and a pirate camp later in the eighteenth century, with not even a lighthouse erected until 1824. Monhegan's hills bristled with evergreens beneath the Great Atlantic Flyway, path of vast bird migrations. After the last farming holdouts departed, the "Ash-Can" School of Painters in New York City—Robert Henri et al.—discovered the raw beauty of Monhegan as a summer retreat. Rockwell Kent, Andrew Wyeth, and Clifford Odets followed in search of secluded inspiration.

Only thirty cottages had been built on the rudimentarily developed northwestern side of the island when the Edisons arrived after a ferry ride from Boothbay. They took rooms at the Old Monhegan House Inn. At Grace's suggestion, Madeleine brought her oil colors and block-printing kit. Theodore ecstatically tramped the trails; although he did not return until after he was married, the place made an indelible impression.

Since leaving college, Madeleine had become involved with a gadabout crowd of "forty or so" (by her estimation) local chums who danced and partied and had "a grand time" at gala balls until the wee morning hours, and motored to football games in Hoboken together—including her supposed boyfriend, John Eyre Sloane of nearby South Orange. She leaped at the invitation to spend the waning days of the following summer with Peggy James at her family compound in Chocorua, New Hampshire. Father William—despite his seemingly endless series of illnesses and indispositions, including breathing difficulties and "violent cardiac pain triggered by exertion or any mental hesitation, trepidation or flurry"—was hard at work reading proofs for his collection of fifteen essays, *The Meaning of Truth: A Sequel to "Pragmatism"* forthcoming from Longman, Green & Company that fall. Shortly after Madeleine's arrival, William left the country retreat to meet Freud, Jung, and Ernest Jones at a conference at Clark University in Worcester, Massachusetts.

Having a "fine and frivolous time" but in the same breath reassuring her parents "not to lose any sleep over me and my 'gentlemen friends,'" Madeleine nonetheless could not help becoming intrigued by Peggy's two unmarried elder brothers, thirty-year-old Henry and twenty-seven-year-old William—Harry and Billy, as they were universally known. "The James brothers are most attractive," she told Mina, even though (or because) she had been warned that they "detest being flirted with." Billy had attended Harvard as an undergraduate for two years where he made his mark on the crew team, then returned for a year of medical school, but tired of it and gravitated to studying painting. "I could go into aesthetic and soulful raptures about colors and cloud effects," Madeleine reported, hoping to find common ground for conversation, "but he seems not to take to that kind of thing." Harry had been editor of the *Crimson* at Harvard, spent some years working in Washington at the department of the Interior, and then graduated with distinction from law school. "Harry I think I could get along with," said Madeleine, flitting her head in the other direction, "if I had the nerve, but certain marks in his physiognomical character rather trouble me. He has eyes with the colored part set up too high so that lots of white shows." One afternoon, after an idyllic picnic at the top of a rocky hill, scrambling down in a rush along an unmarked trail, Madeleine caught the lining of her skirt on a jagged edge, tearing yards of braid which she had to hold together on both sides or else reveal more than she felt appropriate.

The girls kept in touch by letter over the coming months. According to James family biographer R. W. B. Lewis, "bereft (it seems) both of family and of private purpose," Peggy drifted into another depression when her beloved father died of heart failure late the following August, precipitating the disintegration of the close household. Reciprocating for past hospitality and hoping to bolster her friend's spirits, Madeleine invited Peggy to join her and two other friends for a month of rest and relaxation during the winter of 1912 at Fort Myers. The great patron Thomas Edison had made yet another contribution to the welfare and beautification of his adopted town, planting two rows of stately Royal Palm trees—more than two thousand of them, twenty feet apart, imported from Cuba—along the main thoroughfare, Riverside Avenue, subsequently renamed McGregor Boulevard, leading from Manuel's Branch Creek, out of the village, and past his estate.

Peggy was delighted, her spirits lifted when she encountered the surprising luxury of Seminole Lodge, its white walls, red roof, wide porches, and French windows giving out over lush vegetation and an uncountable spectrum of bold flowers that, she told her mother, "almost destroy one's reason." The Edisons had purchased and completely redecorated Ezra Gilliland's former house, and connected the two adjoining structures with a shaded walkway.

The group went for a fishing cruise up the Caloosahatchie on the steamer *Suwanee*, and toured Lake Okeechobee and onward forty miles to the Everglades. Peggy marveled at the huge alligators that nudged up to the side of the boat, the tropical colors, the great orchids close enough to touch. At night, drifting silently on the water, through the porthole of her cabin she heard the "whippoorwills and herons and bitterns and owls calling."

His undistinguished stint at Hotchkiss concluded, Charles entered MIT in September 1909, to study general science, at his father's suggestion, rather than focus upon any specialized area. The goal was to groom him for a well-rounded managerial role in Thomas A. Edison, Inc. Charles joined Delta Psi fraternity and lived in a town house at Number 6 Louisburg Square near Beacon Hill in Boston, "very quiet, very aristocratic, very respectable . . . oh very dignified!" He felt constrained by the structure of his academic schedule and preferred the elective route, allowing him to "pick and choose." He apologized constantly and contritely to his parents for the mediocrity of his grades, but never made excuses. Extracurricular activities—not to mention "West End slumming parties"—and pursuits more toward the language and rhetoric area continued to distract Charles. He joined an informal reading and study group called The Symposium, which met every other Sunday night over coffee, beer, and crackers in the rooms of a young English professor and playwright. At one such session, Charles wrote excitedly to his mother, Lincoln Steffens's assistant told the young men about "all the muck raking things that he ferretted out."

By the end of his second year at MIT, an exhausted Charles Edison went off for a camping trip in the mountains of Arkansas to rest. And Madeleine, visiting Aunt Grace and Grandmother Mary Valinda at Oak Place, seemed to be "nervous, restless" and "on a high tension," for good reason. She had been spending more and more time of late in the company of John Sloane and his family, much to Mina's displeasure. Mother did not approve of their growing attention. Theodore, as usual in some kind of mischief, had been dangling off the running board of one of the Edison automobiles when his foot slipped and the right rear wheel ran over his stomach. He emerged bruised and shaken, but fortunately none the worse. Amid all this hubbub, Papa—unbelievably—decided that he, too, needed some respite from the strain of the laboratory.

In early June Edison announced that he was taking the entire family to Europe. Marion had not seen her father for seventeen years, and was the most excited. "Our happiness has risen somewhere about fifty degrees. Please do not let anything prevent your coming," she wrote from her summer bungalow in Eichwald-on-the-Rhine. "Oscar has begun to take English lessons so that he can be able to understand you when you come. We have never spoken English together."

According to Charles, Edison set forth apparently determined to keep to smaller cities and avoid publicity by traveling unannounced. Mina, Theodore, and Madeleine (who resisted the very idea of the journey, but was spirited away for her own good) were sent on one month ahead for a tour of England, Belgium, and Holland. Charles and Papa, sailing on the *Mauretania* at the end of July, planned to join them in Boulogne.

One day, lounging in a steamer chair, Charles noticed his father and two other men standing by the railing deep in conversation and laughter, folding paper darts and throwing them overboard against the wind. The banter was first about aerodynamics, and then as they cast their gazes downward to contemplate the path of froth skimming by, turned to a discussion of "the skin friction along the side of the ship" and how it could be reduced. Papa introduced the new acquaintances to his son. One was Peter Cooper Hewitt, the inventor and scientist, and the other was Henry James. The author had stayed on with his late brother William's family since the tragic events of the previous summer. He had recently received an honorary degree from Harvard, visited with William Dean Howells while in New York, and was on his way back home to Lamb House in Rye. "The great bland simple deaf street-boy-faced Edison" was on board, James wrote his sister-in-law. "He has asked very sympathetically about Peggy."

Edison and Charles landed at Liverpool, then drove down to London, stopping at Coventry. "It was sort of like going through a green trough most of the way," said Charles of his first impressions of England. Their escort was Sir George Croydon Marks, MP, who took his guests to Parliament, where they witnessed the historic debate and vote abolishing the veto power of the House of Lords over the House of Commons.

A ferry trip took them from Folkestone to Boulogne, and the family was reunited. They drove south in a Daimler along the Normandy coast to Saint-Malo and Mont-Saint-Michel, stopped at selected chateaux, then moved back up through the Loire Valley. Despite his earlier protestations, Edison agreed that a short stop in Paris was acceptable—for he had asked Marion to bring her husband there, rather than have to make a detour to his daughter's home later on. Aside from Dot's protestations at being "miserable from nights and nights of sleeplessness" because she had to travel such a distance and stay in a hotel, the stay in Paris turned out also to be the moment of the trip that generated media attention. The happily-reunited family was discovered there by a sleuthing correspondent for the New York *World* named E. A. Valentine, who, Madeleine remembered, "buttonholed

Charles" and pleaded for access to the Great Man, sternly denied. Off again on the road to Aix-les-Bains, the Edisons soon realized they were being followed. Edison resorted to his old tricks, not releasing the itinerary until everyone was packed and in the car at the break of day.

On through Switzerland to Geneva and Lucerne. On into the mountains of the Austrian Tyrol, with Mr. Valentine in dogged pursuit, until he suddenly became gravely ill. Suffering from a heart condition, he had taken an overdose of strychnine. Again, he begged Charles to allow a few words with Edison so that he could file his story. The inventor relented, and spent some time with the grateful reporter. Charles was the amanuensis and cabled the piece in deadline time. Valentine recovered and was "adopted as part of the family" by the Edisons for the rest of their tour. From Klagenfurt in Austria, they moved to Vienna, and then Budapest, where they had an emotional reunion with Francis Jehl, who had been one of Edison's most trusted assistants at Menlo Park. Edison went out for an afternoon stroll, was immediately recognized, and was practically mobbed. Striving to stay informal, the Edisons were told that dinner would be in an intimate gypsy tea-room with private serenading. However, the hotel management had other ideas. A dais was set up in the main dining room, and there was an elaborate "surprise" reception.

"Did he ever give you any indication what he thought his greatest triumph as an inventor was?" Madeleine Edison Sloane was asked in an interview given in her eighty-fourth year. "I think he liked the phonograph best," she replied, "but that was just, he liked it. I don't think he thought it was his *greatest*. I don't know that he ever had time to consider it much."

In the years approaching the Great War, Americans were purchasing more than five hundred thousand phonographs annually. As in the movie business, Edison found himself inescapably caught up in the swirl of competition. It was no longer a given that "Thomas A. Edison [was] the Alpha and Omega of the Phonograph," as his corporate publications once declared with pride.

The most difficult technical transition for Edison was actually sentimental—from his beloved cylinder to the inevitable disc. Although the cylinder represented to him the very archetype of the phonograph, as we have seen, ever since the earliest days of 1877 Edison had flirted occasionally with the idea of "a revolving *plate* (having a volute spiral cut both on its

THE NEW CYLINDER PHONOGRAPH

upper and lower surfaces)." The template for the disc was a plaster cast taken from flat instead of cylindrical tin-foil impressions. At one point early on Edison even made one of his characteristically off-the-cuff announcements that he would "right away—abolish this whole cylinder and supersede it with a flat steel plate about as big as a dinner plate . . . reamed with a fine groove running around itself beginning at the centre and ending in the circumference." When he returned with a vengeance to the phonograph a decade after its birth, Edison nevertheless remained focused for most of his research and production on the cylinder, continuing to conduct informal, abandoned experiments on the disc through the turn of century.

When the Victor Talking Machine Company and the Columbia Company invaded the market with their three-minute ten- and twelve-inch discs beginning in 1908, Edison's counterthrust was to assign Jonas Aylsworth and his men to develop an improved four-minute *cylinder*, as well as a double-sided disc. He would fight the opponent's advantage of time with a product vaunting even more recorded time, rather than modify the sacred shape of the medium. The key to longer recordings resided in the thermoplastic used for the coating of the cylinder, which as it became more and more refined allowed a greater number of grooves. He called the new cylinder the "Amberol." But this did not still the outcry from Edison's dealers and jobbers in the field—this was no longer what their "Victor-crazed" public craved! "When is the new *disc* coming out?" Paul Cromelin, the general manager of Edison's phonograph interests in London pleaded to Frank Dyer in the West Orange home office; to which came the diplomatic reply, "It is difficult always to reconcile Edison's views with principles of commercial expediency. At the same time he is a man of such great experience and so tremendously more intelligent than any of us that when his mind is fixed on a proposition, I am not disposed to believe he is making a mistake."

According to the exhaustive "Diamond Disc" scholarship of George L. Frow, Ronald Dethlefson, and Martin Bryan, it was ultimately the insistent pressures of Edison's trusted colleague Dyer that caused the Old Man to acknowledge, slowly and painfully, that "the cylinder's tide was going out fast." At the end of 1912, three prototype model Edison Disc Phonographs coupled with Edison Disc Records were released—and yet, even then, out of Edison's deep-seated ambivalence, the cylinders remained on the market, and the inventor would not allow a cessation of advertising for them.

The first Edison discs were one-quarter of an inch thick and weighed ten ounces, made of compressed wood flour coated with condensite varnish. Disc Phonographs (in lieu of earlier proposed brand names such as Edisonola and Edis-co-phone) were spring-wound and, because of the energy required to keep the supporting turntable spinning under the weight of the accompanying records, were incapable of cycling all the way through a twelve-inch disc without slackening from 80 r.p.m. and sounding rather lugubrious. It would be another three years before this apparatus posed a credible alternative to Victor's.

Edison's patent files for the years 1906–1915 are crammed with plans: finer phonographic recording styli working up from sapphire to diamonds; feed mechanisms for the phonograph; enhanced "talking machine" diaphragms (layers of Japanese rice-paper .001 inch thick, lacquered, dried, and compressed); sound modifiers; more sophisticated compositions for the inner body of the record itself; better mold-making methods; and more exotic and handsome Circassian walnut furniture designs to go against the eponymous, revolutionary Victrola inward-facing horn concealed in the top of its elegant mahogany cabinet.

And technology was only one part of the story. Since the turn of the century, the Victor Talking Machine Company and the Columbia Company also gradually brought to the market more highbrow music, opera and classical, while Edison at first stuck with old chestnuts, marching songs, and ballads of yore. Victor's Red Seal recordings of Enrico Caruso and other stars of the Metropolitan Opera became must-haves for every self-respecting American middle-class home. Victor's *Book of the Opera* dictated what records should be purchased for the ultimate in music appreciation.

"Since my return [from Europe in late summer of 1911]," Edison advised Thomas Graf, managing director of the Edison phonograph division in Berlin, "I have taken up the direction of the musical end of the new disc. . . .

None of the men we have can recognize a good from a bad opera singer; they think that if they sing at the Metropolitan Opera House they are fine, whereas it is notorious that many of them are press agent singers or are great on dramatic parts, their singing being indifferent. They select and record opera which is merely recititive [sic], *without tune or connection when put on the phonograph*." During the ensuing years of product development, the near-deaf Edison took the dictatorial task upon himself day in and day out to acquire an entirely revised catalogue of recording artists. The policy was clear: "We intend to rely entirely upon the tune and the high quality of the voices, and not on the name of the artists." Henceforth, Edison, alone, would approve each and every artist and selection, and to drive the point home he would not even allow the names of the performers to appear on record labels until 1915.

For the handicapped Edison, music had always been a sensory rather than an aesthetic challenge. Now, he paid singleminded attention to artists who would be "phonograph friendly" as opposed to merely "high-priced opera people." As we have seen in his relationship to all the arts, it was not sufficient for Edison to assert his standard of taste. He was in a position to promote and purvey it to the rest of society, and was not shy about exercising that prerogative.

"You know music in one way, and I know it another," he scolded John Philip Sousa. "I know nothing about musical notation and have never tried to learn. I am glad that I don't know. I try to form my own opinions."

Edison had a deep antipathy to what he called "complicated" music. Even as he continued to work on a recording stylus that would incise twice as many threads per inch with greater precision than its predecessors, he was in the business to find the best *mechanical* way to "establish music on a *scientific* basis." Along with a more finely honed consumer product, he likewise sought to eliminate the intervention of the singer's dramatic personality, expressed through vocal devices like vibrato; he believed these "false notes" were symptomatic of "interpretative" music and its concomitantly misleading and superfluous nuances and subtleties—which, of course, he could not truly discern or appreciate.

Instead, Edison sought clear diction, what he called "straight" tone : just plain, unadorned notes. Samuel Gardner was a young man of twenty when he came to the West Orange laboratory in 1911 in response to Edison's need for a violinist-in-residence, more commonly called a house-artist, so that he could make sound tests on the spot. One day, the Old Man asked Gardner to listen to recordings of *Ave Maria* by Albert Spalding and Carl

Flesch, two very respected musicians. "Awful sound," Edison said when they had finished. "Those people have a very shaky bow. They don't know how to draw the bow. . . . Those people don't know how to draw a *straight sound*. Draw a straight sound for me." Gardner caught on immediately. He "drew a dead sound, the worst kind possible," utterly devoid of vibrato. "That's great! That's great!" Edison said. "He didn't like chromatic runs because they distracted from the melody," recalled Ernest L. Stevens, Edison's personal pianist and arranger in the early 1920s. "If things didn't go right for him scientifically, he would experiment. So therefore, if he didn't like a tune, *there was something missing in that tune* that could have been perfected, according to his ideas."

The genealogy for these aesthetic ideas can be traced back to Edison's initial engagement with the writings of Hermann von Helmholtz (1821–1894), the revered "scientist-sage" whose volumes on thermodynamics, conservation of energy, and—most important to Edison—perception dominated the last quarter of the nineteenth century.

Helmholtz's influence in Thomas Edison's career began in the spring of 1875. The young inventor, working on acoustic resonance in the telephone and the telegraph, and on the brink of his earliest breakthrough toward the phonograph, purchased a copy of the English edition of Helmholtz's pioneering work first published in Germany in 1863, *Die Lehre von den Tonempfindungen als physiologische Grundlage fur die Theorie der Musik* (*Sensations of Tone As a Physiological Basis for the Theory of Music*). This precious book is now in the MIT Archives, donated by Mina Edison. Francis Upton joined with Edison at Menlo Park fresh from his studies with Helmholtz at the University of Berlin. Edison himself finally met Helmholtz in Berlin just after the 1889 Paris Exhibition.

In *Sensations of Tone*, Edison came upon a landmark theory: the "science of the beautiful [*Wissenschaft (des Schönen)*]," a domain that could—and should—be inhabited by the empirical experimenter. In the particular case of critical evaluation of music, Helmholtz posited, a "discrete, pure sensation" was transmitted by each cochlear nerve fiber from the middle ear directly to the brain. Edison, working first as a hearing-impaired scientist and later as a self-styled musical critic, was reassured by Helmholtz's insistence that pleasurable musical perception resulted from "regularity, connection, and order" among discrete tones. As a trained listener seeking this inherently classical harmony, Helmholtz declared that the rigorous scientist had the authority to discover and define what was rational and utilitarian (in other words, beautiful and *civilized*) in music. Is it therefore any wonder that Edison, the proponent of the "straight tone," despised what he called the "musical jumbles" and "interrupted conversations" of Debussy's music in the same way that he ridiculed the new invasion of Cubist art after the 1913 Armory Show, which appeared to him "as though someone had accidentally upset a pot of paint on the canvas"? Edison greatly preferred Beethoven's funeral march over Chopin's, in the same manner as he lauded the "craftsmanship" of Verdi and Johann Strauss while condescendingly assessing the inadequate "melodic inventions" (read subtle complexities) of Mozart.

How did Thomas Edison "listen" to his recording artists? Daughter Madeleine recalled the era of Edison's decision to audition proposed numbers for the Diamond Disc. During the winter of 1912, what seemed to Madeleine like every night, a pianist would "pound out" waltzes in the downstairs den. Sometimes her papa would put his teeth on the piano—literally bite it—so that the vibrations resonated through his skull bones. The noted Italian educator and physician Maria Montessori was lecturing in America at this same time to coincide with the publication of her book *The Montessori Method*—espousing self-motivated learning through disciplined structure, a philosophy Mrs. Edison shared. One evening, Madame Montessori was a dinner guest at Glenmont while the waltzes were being auditioned, and the great lady huddled in the corner of the den weeping because Edison could not hear, and was putting his teeth in the side of the grand piano. She thought it was excruciatingly pathetic. Edison's personal Disc Phonograph, preserved at the Laboratory, also shows teeth marks on its soft wood framework. Photographs of the period reveal Edison hunched over the phonograph, right hand cupped to his ear, white mane of hair virtually tucked inside the cornucopia horn, as the machine blasts full volume.

By the time he was granted a reunion audience with Thomas Edison on January 9, 1912, Henry Ford had also attained household name status, but not without his own brand of struggle. For the past decade, reaching back to before the mythical Model T took over the country, Ford beat back a succession of lawsuits based upon assertions of infringement of George Selden's 1879 gasoline-driven automobile patent. The rights to this original patent had been purchased by a series of competitors, most recently the Association of Licensed Automobile Manufacturers, a consortium that refused membership to the young upstart, fearing Ford's potential dominance. Confident of victory, Ford taunted the Association by scornfully offering to pay them $1,000 to initiate the suit. "I am trying to democratize the automobile," he said in testimony. "Progress happens when all the factors that make for it are ready." But his bravado was to no avail. The first decision, reached by the New York court of appeals, found in favor of the ALAM's charge that Ford was pilfering principles of the Selden engine. If Ford wanted to continue with the Model T, he would therefore have to pay royalties to the Association.

Ford's rejoinder was immediate, and this time the result was in his favor. Judge Walter Chadwick Noyes of the appeals court retraced the genealogy of the motor vehicle gas engine with precision, stating that while he had made improvements upon prior incarnations, *"no invention"* was involved in Selden's effort; he had had many predecessors. Reinforcing Ford's allusions to democracy, Judge Noyes said in his decision that the essential gasoline engine had evolved into a "social invention" that one might consider to be in the public domain, belonging equally to all manufacturers. Selden's company and its successors collapsed within the year, and Ford was seen by an adoring public as the David who had taken on the Goliath of the big conglomerates and emerged victorious.

On the site of a sixty-acre former racetrack in the Highland Park suburb of Detroit, and with the guiding design genius of architect Otto Kahn, Ford built a new, four-story glass and steel factory where a Model T could be produced on the moving assembly line every two hours and thirty-five minutes. Each man was assigned one task, stood in one place, and conveyor belts took the parts to him.

Complete automation of every manufacturing process was inevitable. Ford wanted Edison to share in the prosperity engendered by his swift system, and asked him to design a battery, starting motor, and generator for the Model T. Ford was prepared to reconfigure the engine compartment of the car for this purpose. He guaranteed a minimum of $4 million in annual orders for these parts. Toward the end of 1912, Ford lent Edison $100,000 as an advance against eventual delivery. In the ensuing two years, he advanced a further $1 million toward research and development—to no avail. Edison could not find the integrating mechanism to make the right fit.

Their inability to forge a business partnership did not diminish the intensity of Ford's reverence toward Edison: "His knowledge is almost universal," Ford wrote of his mentor. "He is interested in every conceivable subject and he recognizes no limitations. He believes that all things are possible. At

the same time he keeps his feet on the ground. He goes forward step by step. . . .
He knows that as we amass knowledge we build the power to overcome the
impossible."

As he approached the symbolic age of sixty-five, Thomas Edison's public
image began to burnish noticeably. The glow of a mellowing patina could
be discerned that over the next two decades would evolve into Edison's
greatest invention of all—his final persona. Ford's tribute, unabashedly encouraged by its subject, was echoed in countless other forms by statesmen
and journalists alike. However, although his specific inventive triumphs
may have come fewer and farther between, and with less sensation and fanfare than in the 1870s and '80s, Edison's imagination and his inclination to
invent ideas and opinions remained undiminished. As we shall see, he was
an unrepentant fount of late-life pronouncements.

And a chronic night owl. Like his "muckers," Thomas Edison proudly
punched in and out at the Laboratory every day. His time clock card for the
week ending September 10, 1912, registered 111 hours, 48 minutes. "Say,
we are working all night tonight," telephoned Edison's sidekick Fred Ott—
nicknamed "Santcho Pantcho" by his self-styled Don Quixote boss—to
young chemist M. A. Rosanoff, "The Old Man says to ask you if you want to
come up to the lab." Rosanoff groaned under his breath, got out of bed, and
dressed hurriedly. He found the factory brightly lit, Edison with son
Charles at his side, and a group of assistants in attendance. "Say," Edison
hailed the chemist, "let's you and I go to work on your damned problem
[formulation of Aylsworth's hardened coating for the cylinders] to-night
and make a resolution not to go to sleep until we have solved it!"

"Mr. Edison," Rosanoff pleaded, "you know I have been at my problem
for months; I have tried every reasonable thing I could think of, and no result, not even a lead!"

"That's just where your trouble has been," the Old Man interrupted.
"You have tried only reasonable things. Reasonable things never work.
Thank God you can't think up any more reasonable things, so you'll have to
begin thinking up *un*reasonable things to try, and now you'll hit the solution in no time. After that, you can take a nap," he added reassuringly.

"I work 18 hours daily—have been doing this for 45 years," Edison
wrote in typical response to many fan letters he received by this time from
the public at large, asking him for the secrets to his success. "This is double

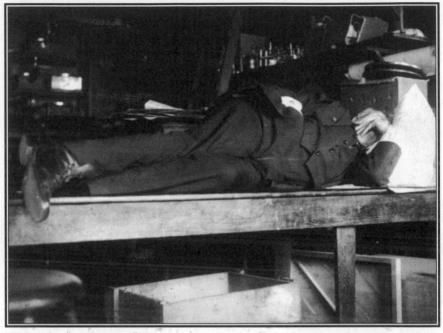

A QUICK NAP AT THE LAB

the usual amount men do. This makes me 90 years old. Add twenty years for youth and you have 110 years as my age. I am still working 18 hours and expect to keep it up for 20 years more. This makes me 150 years old. The reason I can work 18 hours is because I eat very little, sleep very little, and wear clothes that do not pinch the blood veins in the slightest."

The image of the indestructible Methuselah was codified in a 65th-birthday *Harper's* interview Edison granted to a physician, Richard Cole Newton, "How Can a Man Keep Well and Grow Old? Thomas A. Edison Tells Why He Is Never Sick." Dr. Newton noted with delight how the great inventor came bounding down two steps at a time from his personal laboratory on the second floor of the West Orange Laboratory building to receive him in the vestibule outside the library; then, when they were settled by the fireplace for their conversation, the inventor realized he had forgotten something, and went running back up the stairs "like a schoolboy, not at all out of breath, jolly, gay and busy."

Yes, Edison admitted to occasional headaches and slight colds. His middle-ear infections were shrugged off as inconsequential; no mention was made of lanced abscesses. Rather, Edison drew the extremely apt (and refreshingly modernist) analogy, *"The body is only a piece of machinery,* and every

EDISON AT 65 WITH THEODORE, MADELEINE AND MINA

practical man knows that to get good work out of a machine and keep it in repair at the same time one must know how to take care of it."

Here sat the ideal machine man for the machine age! Edison came across as comfortable in his era, conversant with its gospel of progress, moving easily within the world of speed and efficiency. Like a well-oiled, streamlined engine, Edison did not consume excess fuel: a small mutton chop, some toast, and one cup of coffee for breakfast; two fried sardines or a single anchovy on toast, perhaps an apple, and tea for lunch. Pie (apple, of course) or pastry for dessert was his only culinary vice, and one of long standing, almost affectionately tolerated by now. The secret to his eating, again like the good combustion engine providing the proper mixture of fuel and oxygen, was "careful mastication." Edison thoroughly chewed every particle of starchy or vegetable food so as to extract the full nourishment and energy from it.

And when he was not feeling well, Mr. Edison told Dr. Newton, he did not eat at all—why waste fuel on an inoperative machine? He recalled his late beloved father Sam going without food for three or four days at a time when he felt under the weather: "Unassimilated food is only a drag on our digestive systems."

At 5 feet 9 inches, Edison weighed within the range of 175 to 178, but he was proud that if necessary—if the pace of work required—he could get himself down to a wiry 165.

Finally, Mr. Edison discussed another key to his success: He simply did not allow himself to worry. In self-fulfilling logic, he explained, the best way to achieve a worry-free life was to "keep thoroughly occupied and follow a regimen" so that one did not have time or psychic room to indulge in unproductive and superfluous thought.

"Unquestionably," Dr. Newton concluded, in praise that would echo through ensuing decades, "Mr. Edison's *almost complete control of his environment, including his own body and mind,* has done much to make him one of the greatest men of all time."

CHAPTER

24

THROUGH THE CRITICAL WRITINGS OF NATURALIST JOHN BUR-
roughs (1837–1921) Henry Ford discovered and was immediately seduced
by the transcendental voices of Emerson, Thoreau, and Whitman. Their
siren song provided relief from the assembly-line din and clamor and guided
Henry to solitary woodland pilgrimages on which he jumped fences, strayed
willfully from the beaten path, and listened to birdcalls. But Burroughs,
whom Henry James dubbed "the child of the woods," did not share Ford's
rabid enthusiasm for the Tin Lizzie, and in various magazine articles condemned its Manichean, "befouling" incursions. Ford, distressed, shipped a
Model T for direct delivery to Burroughs's Catskills fruit farm in West
Park, on the Hudson River across from Poughkeepsie. Within months, Burroughs was gamboling around Ulster County at high speed, his grand-
daughters clinging to each other and squealing as they bounced in the back
seat. In June 1913 the admiring Ford brought the ambivalent but grateful
Burroughs to Detroit for the unveiling of a bird fountain at Fair Lane made
from stones gathered at the Burroughs farm. The talk at their first meeting
was of birds—to Ford, "the best of companions"—and Emerson. Ford proposed a camping trip to Concord, Massachusetts, that coming fall to visit
the great philosopher's home, with stops at Walden Pond and Sleepy Hollow Cemetery; it would be a great honor if Burroughs would accompany
him on this journey of the spirit. They went back to nature in name only.

The caravan included a field kitchen and dining tent, cameramen from Ford's publicity department, and laundry service with starched collars daily.

The following winter, Ford took Burroughs to Fort Myers, reminiscent to the conservationist of Jamaica and Honolulu, and introduced him to friend Edison. The three men sallied forth into the Everglades for a few days of communing. Thus began a decade of shared annual excursions, deliberate exercises in rough-hewn living.

What a dramatic and patriarchal figure the white-bearded John Burroughs presented at nearly eighty years of age standing flanked by the slightly stooped but solid Thomas Edison, and Henry Ford in shirtsleeves seeming even thinner than usual. Burroughs was deservedly the spiritual center of this odd trinity. Born in Roxbury, New York, by the East Branch of the Delaware River, John grew up on a farm, the seventh of ten children of Chauncey and Amy Kelly Burroughs. As a lad, he tended cows in the spring and collected maple syrup in winter, and one day, reclining under a tree during a break from cutting hay, spotted his first Black-Throated Blue Warbler.

A retiring, religious, and bookish youth, John dabbled in local journalism and held an assortment of teaching jobs, living for a while with his wife, Ursula North, in East Orange, New Jersey, during the late 1850s. In the Civil War years, Burroughs met and talked at West Point with Emerson, who resembled "an alert, inquisitive farmer . . . [with] a serene, unflinching look." Working in the Treasury department in Washington, he hiked with Walt Whitman in the Rock Creek woods—"Walt is as great [as Emerson]," he wrote to a friend, "though after a different type. . . . It is as if Nature herself had spoken."

After the war, he established a home constructed of native stone ninety miles north of New York City, named it Riverby, and settled seriously into writing. Burroughs' essays are models in the romantic, pastoral tradition. The personal influence of Emerson and Whitman is spiritual, rather than rhetorical. Burroughs does not linger on metaphysics, become entangled in abstract concepts, or try to find the godhead; nor does he dive headfirst into the overtly sensual and let his ego take the lead. We can see, rather, how Burroughs's self-styled "home instinct" strikes the tone that appealed—*and made useful sense*—to Ford and Edison. He exalts the local sights and particular sounds at his doorstep. There is a comforting, moderate stance in Burroughs's affection for the natural world and the simple pleasures of rural life. In *The Heart of the Southern Catskills*, he wrote of Slide Mountain, "a summons and a challenge to me for many years. . . . In a tangle of primitive woods, the very bigness of the mountain baffles one. . . . The eye is of little service; one must be sure of his bearings and push boldly on and up. One is not unlike a flea upon a great shaggy beast, looking for the animal's head." In another early essay, he observed: "Look up at the miracle of the falling snow, the air a dizzy maze of whirling, eddying flakes, noiselessly transforming the world, the exquisite crystals dropping in ditch and gutter, and disguising in the same suit of spotless livery all objects upon which they fall. How novel and fine the first drifts! The old, dilapidated fence is suddenly set off with the most fantastic ruffles, scalloped and fluted after an unheard-of fashion!"

Or thus, in a different season, "If I were a trout, I should ascend every stream till I found the Rondout. It is the ideal brook. What homes these fish have, what retreats under the rocks, what paved and flagged courts and areas, what crystal depths where no net or snare can reach them!" John Burroughs would produce a corpus of thirty books over four decades of labor.

By the time he met Thomas Edison in Florida, Burroughs had attained his own degree of celebrity, in a different sphere. He was the literary darling of the New York social crowd, hosting Oscar Wilde on his lecture tour of the States, and lecturing himself at the Authors' Club, hobnobbing with Matthew Arnold. He had sailed to Alaska with Edward H. Harriman, gathered around the campfire with Theodore Roosevelt at Yellowstone Park, trekked into the Grand Canyon with John Muir, and dined with Andrew Carnegie. Of late, ranging less far afield, Burroughs had spent more time tending his muck swamp celery plot and taking meditative constitutionals through the hemlock forest. He wrote daily in his journal, sitting in a favorite rocking chair on the porch where he received visitors at his cabin two miles inland from his big house overlooking the Hudson. Built in the summer of 1895 by hand from birch and cedar with the help of his son, Julian, Slabsides became John Burroughs's retreat for a quarter-century. It has been maintained and can still be toured in the Town of Esopus, New York, off what is now Route 9-W.

Because the participants (except Edison, as the most resolutely "on vacation" of all) each recorded their thoughts during and after the outings, there is a *Rashomon*-like quality to the accounts that survive of the camping trips, which fortuitously shed the most light on our protagonist.

EDISON CAMPING WITH
(FROM LEFT) HARVEY
FIRESTONE, JR., JOHN
BURROUGHS, HENRY
FORD, HARVEY
FIRESTONE, AND
R. J. H. DE LOACH
(SEATED)

The group covered vast territory. In late October 1915 the Edisons and the Fords were joined by Mr. and Mrs. Harvey Firestone (the Akron tire and rubber magnate), and Mina's sister, Grace, for a West Coast sortie after Edison Day at the San Francisco Pan-Pacific Exposition, which included a trip to visit Luther Burbank in Santa Rosa. The following summer, the entourage, sans wives, traveled through the Adirondacks and the Green Mountains of Vermont, accompanied by R. J. H. De Loach, a professor at Georgia Teachers College in Statesboro, and a friend of Burroughs. War intervened and prevented a reunion in 1917, but the following August saw the three comrades, again with De Loach, in the Great Smoky Mountains, on into West Virginia, Tennessee, North Carolina, and Virginia. In 1919, the focus was "the trail of the Minute Men in Historic Old New England," a route Edison knew well from his 1905 steam car journey. In 1920, John Burroughs hosted husbands and wives for brigand steak and bacon cooked outdoors on sharpened green wooden skewers at Riverby; the following summer they were joined by none other than Warren G. Harding (Edison pronounced the President "all right," because he preferred to chew tobacco "like a man," rather than smoke cigars "like a weakling"), Edsel Ford, and Harvey Firestone, Jr., for a caravan that began at the Firestone home in Columbiana, Ohio, and then moved on to Licking Creek, near Hagerstown, Maryland; West Virginia, and finally Uniontown, Pennsylvania.

After Burroughs's death, Ford, Edison, and Firestone ventured down through Michigan's Upper Peninsula, including a voyage on Ford's three-hundred-foot yacht, the *Sialia*; and Edison's nostalgic visit to the old Homer Page Farm, then into nearby Milan and the house of his birth, when the whole town turned out to sing "Hail! Hail! The Gang's All Here." During the summer of 1924 they stopped at Ford's Wayside Inn, a 1686 tavern that had inspired Longfellow's *Tales of a Wayside Inn*, and the surrounding five thousand acres which Ford had recently purchased near South Sudbury, Massachusetts, followed by an old-time picnic and fiddle contest and a so-journ with President Coolidge and his family in Plymouth, Vermont. Here, Coolidge presented Ford with an heirloom, sixteen-quart wooden sap bucket that belonged to his great-great-grandfather and had been in his family since the late eighteenth century, and then all toured a cooperative cheese factory nearby, Edison sampling the product and coming down with a bad stomach ache. "Will Coolidge be elected?" the crowd of reporters asked Edison. "Sure," the pundit replied. "If he doesn't talk too much!" The President doubled up with laughter.

Each of the four central players gradually assumed a consistent role in the annual camping team headed for "Nature's primal sanities." Harvey Firestone, the youngest of the group, had hoped for "simple, gipsy-like fortnights," though they inevitably later degenerated into "a kind of travelling circus" as the media got wind of their plans, and the men were endlessly photographed and filmed. Firestone's responsibility was to load up his Packard and a supplementary, rearguard truck with provisions and hire the best cooks. There were usually five helpers, including drivers and tent-pitchers. Firestone was also the amateur poet of the group, moved from time to time to set down doggerel verse to immortalize their experiences: "Down in the Vermont country / Where the air is bracing / and the spirits bound, / In a field by the roadside / With a flowered carpet found." Compass in hand, Thomas Edison was the navigator, proposing itineraries, then changing them at the last minute, always in the lead car, his four-cylinder Simplex, avoiding large towns in favor of little hamlets, diverting from well-traveled highways onto rutted, dirt roads in search of new routes. Edison liked to patronize the local farmers, purchase fresh chickens and pressed cider, and hand out chocolate bars to children who recognized him ("Mr. Graphophone!" they would cry, to his amusement), waving as he passed. Henry Ford was the first to alight from the car, stripping off his jacket and shirt to search for the nearest stream to bathe in the icy water. While Edison curled up *under* a tree for his afternoon nap, Ford would be more inclined to *climb* it. He chopped and split wood with demonic fury, building a fire immediately no matter how warm it was in the pine or maple grove they had chosen. He organized rifle-shooting contests, dance festivals (when the ladies were present), and running races. He was the mechanic of the team. When a spring shackle bolt sheared on the highway, it was Henry Ford who spied a thresher in the middle of a wheat field, dashed out, "borrowed" a bolt, and grafted it into the afflicted engine. When the Packard fan punctured the radiator, Henry stitched the blades back into place with copper wire. And John Burroughs was the "Guest of Honor," sometimes a bit cranky and reluctant to "romp and prance around like a boy," the late-rising eminence grise with a tendency to snore and keep the others awake. Burroughs's esteemed presence conferred the stamp of credibility on the journey. He conducted flower identification searches at Edison's side, and individual birdcall seminars for Henry Ford. A little boy spotted Burroughs stopping once to buy some cold drinks at a general store in the hamlet of Johnson City, North Carolina, and called out, "There goes Rip Van Winkle!"

Since he first encountered "the good playfellow," Burroughs had become a student of Thomas Edison's every move. In Burroughs's eyes, the industrialist Ford soon took second place to Edison the modest philosopher, the man who shunned crowds and needless palaver. To the old naturalist, the inventor was without a doubt "the centre" of the group, encyclopedically aware, his "meditative and introspective mind . . . a storehouse of facts and figures on all subjects . . . crude and uncultured . . . a big-brained man & genial & good-natured [with] a big fund of all kinds of stories, humorous or nasty." He watched Edison read magazines in his tent at sunrise, oblivious to the hustle and bustle around him. He watched Edison subvert his supposed dietary restrictions, consume fried eggs and ham for breakfast, gulp down a bottle of soda pop and "eat pie by the yard," then without a pause "turn vagabond very easily," and with hair unbrushed and clothes wrinkled, wander off into the woods to spend hours examining the cut stalks of plants for milky sap in search of organic rubber.

Even in the roughest weather, or when heavy frost was predicted and the other men decided to find a hotel room and a hot bath, or board with friends in the nearest town rather than brave the chill—or even to use a civilized bathroom for a night—Edison resolutely stayed behind in camp, unshaven and disdainful of the "dudes and tenderfoots." Reveling in the unkemptness of it all, he rigged up a lighting system with a storage battery so that he could continue reading the newspaper and, past midnight, when the skies had cleared, survey the stars.

Once the evening meal had been consumed and the tables emptied, the men hunkered down around the fire for some serious talk, Edison and Ford, indefatigable as always, taking the lead, while Firestone, Burroughs, and De Loach, mindful they were "present at the creation," interjected occasionally, and took notes. Firestone was proud of the one important time when the conversation turned to literary matters, and he simply had to take issue with Edison. The Old Man insisted, not surprisingly to those who knew him well, that *Evangeline* and *Les Misérables* were the greatest works of poetry and fiction of all time, Firestone opting instead for Shakespeare. Edison wondered (in much the same way that he preferred the "straight" violin notes) whether Shakespeare could be reproduced in "common, everyday speech," as relief from flowery, Elizabethan rhetoric.

During the later war years and just after, the men were preoccupied with the conflict and its repercussions. "The Huns came in for their share of censure and cussing at regular periods," De Loach recalled. Edison lashed out with his harshest criticism, scorning the Germans as "boneheads . . . not efficient. . . . The pure German is as dull as a cow," he said. "*We must win*," agreed Ford, "and to do it, we shall have to throw away alot [*sic*] of our resources. It is all waste, but it seems necessary, and we are ready to pay the price." Edison ruminated on the chemistry of explosives and how they could best be deployed from submarines. A bald eagle swooped by overhead. Burroughs looked up: "Sail, O Bird of Freedom!" he exclaimed. "Show the Germans how we are to win this freedom for the world!" And while they were on the subject, as for German music, Edison scoffed, "Mozart was no good at all." All concurred solemnly that Woodrow Wilson was an honest man, even a "Great Man for the Ages."

On the 1919 "Historic New England" trip, however, Ford strayed off the safe, jingoistic path and attacked the Jews, "attributing all evil to the Jewish capitalists," wrote Burroughs in his private pocket notebook, mightily offended. "[Ford said] the Jews caused the War, the Jews caused the outbreak of thieving and robbery all over the country, the Jews caused the inefficiency of the Navy of which Edison talked last night." When Ford lashed out at Jay Gould as a prime example of the avarice he abhorred, Burroughs argued that the industrialist, who had been his childhood wrestling playmate, was a tried-and-true Presbyterian. The satisfaction was momentary, for this was an entrenched issue, a vicious, persistent stain on Ford's psyche.

Stunned and embarrassed by the depth of this aggressive anti-Semitism, Burroughs did not transcribe this aspect of the fireside debate into the published version of his camping story, "A Strenuous Holiday," an otherwise affectionate reminiscence that appeared in his 1921 book, *Under the Maples*. Nor did he ever mention Ford's bias in any other related essay, merely demurring that "not much of the talk that night around the camp fire can be repeated."

Burroughs sensed, correctly, that Edison was empathic to Ford's prejudices, although in a far less strident, more stereotypical way. The Jewish businessmen in Germany, Edison declared, according to John Burroughs's jottings on that same night in August 1919, were "keen and alert . . . efficient" in contrast to the lowbrow German industrial and military leadership wrongfully taking possessive pride in the prosperity of their nation. Five years earlier, indignant at a similar manifestation of Edison's typecasting published in an interview in the Detroit *Free Press*, the banker Jacob H. Schiff at Kuhn, Loeb & Company dashed off an angry letter asking Edison to deny his "flighty . . . assertion[s]." Edison drafted a conciliatory response insisting he was only trying to give credit where credit was due; if Mr. Schiff wanted to discover the secret of the "enormous industry of modern Germany," he merely had to "dig up a Jew who furnished the ability and that made them a success."

A year later, what we might call Edison's "Shylock complex" resurfaced in transparent clarity. "The Jews are certainly a remarkable people," he conceded in a letter to one Isaac Markins, "as strange to me in their isolation from all the rest of mankind as those mysterious people called Gypsies. While there are some 'terrible examples' in mercantile pursuits, the moment they get into art, music and science, and literature, the Jew is fine." The next and most telling sentence in the draft was excised with Edison's rapid, pencil-stub scribble: "I wish they would all quit making money."

Edison went on with his thesis regarding the accumulated causes for this inherently competitive behavior: "The trouble with the Jew is that he has been persecuted for centuries by ignorant malignant bigots and forced into his present characteristics and he has acquired a 6th sense which gives him an almost unerring judgement in trade affairs." To Edison, it was this "racial" characteristic, this "natural talent" and "natural advantage" at becoming rich, that had caused the Jew to be "disliked" in Europe. However—he concluded with optimism and encouragement—within American democracy, because the Jew is "free," therefore "in time he will cease to be so clannish."

But the common ground between Ford and Edison had a more positive foundation than any straw men masquerading as Jewish robber barons. Edison's affection for Ford does not require subtle analysis. Edison warmly perceived Ford as "a natural-born mechanic," a plain, unadorned fellow from humble roots who shared his pleasure in the joys of ordinary, applied thinking. And Edison trusted the younger man (his successful, surrogate son) because Ford did not think of himself as an inventor, and so they were never in direct competition—the death knell for several Edison family relationships and business collaborations. Until Mr. Ford appeared on the scene, trust was a feeling daughter Marion feared her Papa had abandoned long before: "What a good and faithful friend [Mr. Ford] has been to you, Father dear! A good and faithful friend is a precious possession," she wrote. "After your experience with Mr. Gilliland it is a wonder that you ever opened your heart again to another friend." Ford's unabashed admiration evolved into near-chronic worship with the advancing years.

Ultimately, Ford just wanted to be *near Edison* as much as possible, to bask in the Old Man's reflected glory. When the luxurious $3\frac{1}{2}$-acre estate adorned with over one hundred grapefruit trees and fifty orange trees fronting the Caloosahatchie Bay just next door to Edison in Fort Myers came available in the spring of 1916, Ford snapped it up for $20,000 in cash. Built in 1911 by New York City financier Robert W. Smith, the two-story, fourteen-room home known as The Mangoes came complete with a huge garage to store successive Ford automobile models; expansive, fern-adorned verandas upstairs and downstairs with wicker rocking chairs; hardwood floors perfect for Henry and Clara's other passion, square-dancing; a study for Mr. Ford's ruminations; and a master bedroom with French doors opening onto the water view through palm trees. Of course, the place was completely electrified.

At last, a new neighbor "Damon" had settled in for the next fifteen years to accompany Edison's "Pythias."

At 5:17 p.m. on the evening of Wednesday, December 9, 1914, an explosion and fire broke out in the film inspection department of the Edison laboratory and factory complex in West Orange. Immediately knocking out all power in the compound, it spread quickly to the large adjoining cabinetry and carpentry shop, and leapt from there to devour the wooden window frames of the Phonograph Works, then to huge Building 24, where the new disc records were cast. Edison was not present at the time the blaze started, but he soon appeared, and stayed steadfast at the scene for the next twenty-four hours, directing efforts to save the main laboratory on the corner of Valley Road and its perpendicular, outlying buildings, as well as the six-story kinetophone, electrical, and storage battery works just across Lakeside Avenue, once he realized that the phonograph manufacturing structures were lost. Damage was estimated to be from $3 to $5 million.

While the flames roared, Edison was mindful of the need for immediate damage control. Approached by a reporter from the Newark *Evening News*, he handed the man a laconic, handwritten statement, which appeared the next day under the headline EDISON'S MOBILIZING PLAN, as if it were a dispatch for action issued by a five-star general in the field: "Am pretty well burned out but tomorrow there will be some rapid mobilizing when I find out where I am at." Holding his head high, to another journalist, Edison said, "Although I am 67 years old, I will start all over again tomorrow. I will begin to clear out the debris if they are cool enough, and I will go right to work rebuilding the plant." True to his word, after a catnap on his corner cot, Edison set up a much-publicized command post in the library—large charts, photographs, and site plans arranged on easels around his desk so he could survey the recovery day-by-day.

"It's a goner," Edison said with a smile to general manager Charles H. Wilson, Jr., "but we'll build up bigger and better than ever. Why should I be downhearted? I can't take any of this away when I die, anyway." A round-the-clock salvage brigade went to work immediately. "Arc, search and incandescent booming all night," exclaimed Edison with glee. "It's like the old days to have *something real* to buck up against!" By the end of the first week after the blaze, forty-four carloads of iron and steel wreckage had been carted off, besides hundreds of wagonloads of bricks; the men and surviving machinery of the Phonograph Works were temporarily housed in the storage battery plant.

Edison moved swiftly to reassure his cylinder and disc jobbers across the land. While structures were leveled, most manufacturing equipment had in fact been saved. Outside machine shops were immediately subcontracted to keep up with the demand for phonographs. Nearly all of the master molds for cylinders were intact. The recording laboratory in New York

City continued full steam ahead. Blue Amberols would be back on line in one month.

After of a banner growth year, demand for the phonograph was soaring as never before. Lavish new Edison display showrooms had recently opened in Milwaukee, San Francisco, and New York City. Playing time per disc side had been increased to five minutes. Thirty thousand Diamond Discs per week were flying out of the Edison warehouses in 1914. An electric turntable motor was on the verge of reaching the market. Upscale Art Model phonographs in deluxe Chippendale-design consoles were all the rage. Thomas Edison stoked this demand with one of the most ingenious modern marketing ploys he ever concocted: The Tone Test. As early as 1908, the Victor Company had published photographs of their ubiquitous star Enrico Caruso standing next to his disc of "Celeste Aida," with the caption, "Both are Caruso." But until Edison, no phonograph manufacturer had dared to take the dramatic, theatrical chance of placing a live opera star beside a functioning phonograph, performing before an audience, and challenging them to distinguish to whom or what they were listening.

Soprano Anna Case was Edison's choice to take the leading role in the earliest Tone Tests in spring 1915 after the great fire. And for Edison, she was the sentimentally right choice, a blacksmith's child, native daughter of New Jersey, born October 29, 1889, growing up in the hamlet of South Branch. As Anna liked to say, hers was "the little village of one hundred souls." She sold soap, ran a hack service with her pony and cart for twenty-five cents a ride, scrubbed floors and cooked for her neighbors for fifty cents a day to be able to afford singing lessons in New York City every Monday. Coming back in the dark on the train, she helped her sickly mother care for three younger siblings. At fifteen, Anna played the organ in church for $12 a month, and began to teach music around town, driving miles out into the countryside to meet with pupils, carrying a revolver for protection along the dark roads. She found a position in a church choir in Plainfield, but her deeply religious Dutch Reform father considered travel on Sundays to be sinful, so Anna went on Saturday, scraping together the money to pay for room and board that night.

At twenty, Anna moved to Philadelphia and took a job singing three afternoons a week as tea-time entertainment in the Bellevue-Stratford Hotel. After the applause died down from her very first concert, she was approached by a member of the audience overwhelmed by her elegance and beauty. He was Andreas Dippel, general manager of the Metropolitan Opera Company, where Anna Case made her debut as the critics' darling, the rags-to-riches "Miracle Girl of the Opera," a truly all-American phenomenon, the first soprano star at the Met who had never studied abroad.

Anna claimed most of the leading lyric soprano roles over the coming eight years, including Marguerite in *Faust*, Sophie in *Der Rosenkavalier*, Olympia in *The Tales of Hoffmann*, Micaela in *Carmen*, Nedda in *Pagliacci*, Gilda in *Rigoletto*, and Mimi in *La Bohème*. Not an opera aficionado, Thomas Edison was nonetheless on the rampage for talent. He first heard Anna Case sing not a classic aria, but rather the showpiece "Charmant Oiseau" from *The Pearl of Brazil*, and was astonished: "She has the most perfect scale of any singer." Moreover, Anna Case was exquisite to behold, long-limbed in white satin, her auburn hair piled high, wide brown eyes half-focused, curved fingers of her left hand loosely pressed against her pale neck, in deep decolletage partially shrouded by a lace mantilla. As America moved closer to war, Anna had gone on

ANNA CASE

the road to raise money for the Liberty Loan drive, appearing on stage draped sinuously in an American flag, her hair unfettered so it tumbled thickly to her waist. She pointed to a poster boldly asking, "Think—Have YOU Bought Your Limit?" After her country entered the conflict, Anna volunteered to sing "The Star-Spangled Banner" before troops preparing to ship out. To Edison, the soprano's patriotic streak only enhanced her appeal, and even he became nervous in her radiant presence. "I hope the tremolo will stay out of your voice as long as the tremble has stayed out of my hand," he said to Anna on first meeting, giving her a signed photograph of himself. In an extraordinary exception to his rigid musical rules, Edison allowed Anna to choose the arias *she* preferred to record with her longtime piano accompanist, Charles Gilbert Spross.

Edison's Tone Tests followed a regular ritual. When the concert hall had filled, Anna Case—or any of the other stars recruited by the company, names such as Elizabeth Miller, Marie Rappold, Alice Verlet—would assume a position in evening dress next to the machine on the otherwise bare stage, and the two would commence together. From time to time, the artist would pause, and allow the machine to carry on alone, or the house lights were dimmed, so that the audience was left to wonder which voice they were hearing. The artist might even leave the stage at this juncture to add to the mystery. One memorable evening at Carnegie Hall, after a diverse Test repertoire that included Charpentier, Gounod, Thayer, and Nevins, Anna sang harmony with her recorded voice. Wrote the admiring critic for the New York *Telegraph,* "The audience showed no desire to leave the hall, but remained literally at the feet of the radiant young American."

Remembering those heady times, Anna Case described herself self-deprecatingly as a "copy cat." But the revenue streams that resulted from the Tone Tests before more than 200,000 people nationwide in 1915 and 1916 proved that they sold Diamond Discs—which was Edison's purpose, after all, to meld the pretense of music appreciation with blatant commercialism.

Measuring the precipitous decline of the motion picture side of Thomas A. Edison, Inc., during the same period of soaring energies in the phonograph business confirms the extent to which Edison's personal and intellectual attention conditioned the failure or success of his enterprises. Film historian Charles Musser, the authority on this pivotal era, pinpoints the beginning of the end of Edison as film mogul with his firing of Edwin S. Porter and William Gilroy in 1909. Although Porter's production values—one-track scripts, harsh lighting, zealously applied makeup, stilted acting—came under criticism, the underlying reasons for his dismissal resided more in the corporate realm as Edison, Frank Dyer, and the T.A.E., Inc., management team diminished Porter's directorial authority and he ceased to enjoy the creative freedom of the auteur.

Porter moved on to work with Adolph Zukor and profit from the trend toward full-scale feature films, while Edison remained entrenched in the narrow world of vaudevillian one-reelers, preachy vignettes, and didactic "home library" subjects. He turned his technical resources to the imaginative fantasy of "presenting the illusion of scenes in colors," and to projected film in synchronization with phonograph records. This expensive and unwieldy exercise in novelty, the home phono-Kinetograph, while a technical advance, ultimately proved too difficult for the average American to operate properly in his parlor.

As early as 1914, Thomas Edison was already talking about his motion picture business in the past tense. His heart was just not in it: "Well, I was not so anxious to make money," he said wearily. "I only wanted to break even . . . at first we used to turn out films of an average length of two hundred feet and the subjects were all scenic. *The idea to utilize the invention for the purposes of drama came much later.*" Drama was an afterthought. For all the renowned capacity of his teeming brain, Edison was never able to conceptualize the film medium as a vehicle with the potential for sustained narrative function. Adolph Zukor, D. W. Griffith, Cecil B. DeMille, and others transcended the limits of film as momentary entertainment and created a veritable movie industry. Edison was much more comfortable pushing the limits of what film could *do*, rather than exploring what it could express.

As soon as he was able to shift gears in a desultory interview with *Motion Picture World*, Edison switched the topic to his beloved phonograph— an "invention which will give cheap opera to all the people." With the phonograph, ambiguity vanished: the medium truly *was* the message.

CHAPTER
25

IN THE BEGINNING, THE GREAT WAR WAS EUROPE'S PROBLEM,
a distant distraction. Even President Wilson and Theodore Roosevelt agreed
silent sympathy was the only response—until May 7, 1915, when the un-
armed Cunard passenger ship *Lusitania* was torpedoed in Irish waters, German U-boats fracturing the facade of American neutrality. Over the next
two years before the threat of rampant submarine warfare finally drew the
United States into the conflict, this singular trauma at sea precipitated di-
vergent and telling responses from Henry Ford and Thomas Edison.

The wellspring for Henry Ford's initial pacifism has been traced to the
childhood moral standard set in the pages of the McGuffey lessons he first
read as a lad of eight years old in the Scotch Settlement School a mile and a
half down the road from his father's Dearborn farm. This primer was laced
with antiwar sentiments in passages such as "Things By Their Right
Name," a tract characterizing soldiers as murderers. However, this same
McGuffey Reader also reinforced the cliché image of the Jew as an ethically
suspect, avaricious Fagin. During Henry's teenage years, he was aware of the
midwestern Populist ethos blaming America's post-bellum economic col-
lapse on European Jewish banking syndicates. During the formative phase
of his struggling automobile company, Ford, as a self-styled "upright businessman" witnessed a dramatic rise in the Jewish population of Detroit, part
of a nationwide urban immigration wave he xenophobically considered a

pernicious threat to the indigenous purity of the "Anglo-Saxon Celtic Race." As the first shots were fired in Europe, Ford was off the starting blocks with his euphemistic announcement that "the quick hunger for dollars" provided the igniting spark for this "purely manufactured evil . . . this orgy of money."

During the anxious summer of 1915 in the wake of the *Lusitania* tragedy, Ford proclaimed his fears that Wall Street's "absentee owners and parasites" would drive the common man of America toward an irrelevant, useless, and faraway battle. He grabbed headlines with his shrill commitment to peace. Ford's voice caught the ear of Rosika Schwimmer, a Hungarian feminist working with the American Peace Foundation toward the goal of "continuous mediation"—a congress of representatives from neutral countries who would present their terms to the adversaries. In November she managed to arrange an interview with Henry Ford at his home, where she sold him over luncheon on the idea of underwriting a delegation of prominent private citizens to Christiania, Norway, on a personal diplomacy mission. At the meeting with Mme. Schwimmer, Ford insisted he possessed incontrovertible proof that the "International Jew" had instigated the War. In a retrospective memoir, *My Forty Years with Ford*, his righthand associate, Charles "Cast-Iron Charlie" Sorensen, cast aspersions on the "clever" Mme. Schwimmer, implying she exploited his boss's noble intentions.

Rebuffed at the White House by the resolutely noncommittal Woodrow Wilson, Ford took the responsibility of shutting down the warmongering capitalists into his own hands: "Out of the trenches before Christmas, and never go back!" he declared, setting up headquarters at the Belmont Hotel in New York City, where a phone and telegraph blitz was mounted to convince other prominent patriots to join the crusade against the Kaiser and set sail with Henry Ford aboard the chartered steamship *Oskar II*.

Instead, Ford was subjected to a torrent of ridicule, as William Howard Taft, William Dean Howells, Ida Tarbell, Lincoln Steffens, Julius Rosenwald, Jane Addams, and John Wanamaker, along with every college president and forty-seven out of forty-eight state governors—and yes, even his friends John Burroughs and Thomas Edison—conveyed regrets. Clara pleaded with her husband not to embark, but he was resolute, having invested nearly half a million dollars in the mission, and was not about to call it off, although demoralization and bitterness had already set in.

Amidst a chaotic boatload of college students, eccentric "peace-nut pilgrims," clerks, and journalists, Ford took ill on the chill December journey from Hoboken. Charles Sorensen said that Ford knew he was enmeshed in an impossible situation even before the *Oskar II* reached Norway. Ford deserted the conclave virtually upon arrival and went silently back to Dearborn. Despite the disappearance of her supporter, the determined Mme. Schwimmer did manage to inspire and draft an eloquent "Appeal to the Belligerents" four months later.

"I do not regret the attempt," the quixotic Ford said. "The mere fact that [the Peace Ship] failed is not, to me, conclusive proof that it was not worth trying." He believed it to be his duty on principle to stay resolutely away from a war in which his country was not involved, to oppose the conflict "up until the time of its declaration"—after which, as a faithful American first and foremost he turned one hundred eighty degrees and dedicated every facility of the Ford industry to the government, cranking out six-ton trucks, aero-cylinders, caissons, listening devices, helmets, armor plates, body armor, and antisubmarine Eagle Boats.

Whereas Henry Ford's dialectic of war and peace was black and white, good against evil, purported altruist opposing venal capitalist, Thomas Edison's sentiments inhabited a gray zone. "My solution for war is not Peace Congresses alone," Edison wrote in 1922. "It is *preparation*. Preparation, not provocation, and this preparation or preparedness may one day involve the discovery of some terrific force, some engine of war the employment of which would mean annihilation for the opposing forces." Seventy years ago, Thomas Edison anticipated the military-industrial buildup mentality that led to the Cuban missile crisis. "The way to make war impossible," he predicted, "is for the nations to go on experimenting, and to keep up to date with their inventions, so that war will be unthinkable, and therefore impossible. . . . So far as atomic energy is concerned, [it] could be turned into electricity and projected not only across the Atlantic, but flung from any part of the world to any other part. Neither the Atlantic nor anything else could interpose an obstacle. The force residing in such a power is gigantic and illimitable."

Preparedness—how Henry Ford despised that compromised ground! Yet what his friend derided as equivocating became Edison's watchword, and indeed, Woodrow Wilson's moderating slogan. A creature of his time, Edison likewise began from a posture of distance, placing the Allies and Germany *over there* at arm's length, regarding their battle as "civilization's surgical operation," an inevitable consequence of retrograde nationalism.

The most immediate result for Thomas A. Edison, Inc., was a drastic short-age in carbolic acid, an essential ingredient in phonograph discs, at the very time when demand was cresting. "This war should teach us to depend upon ourselves . . . and to take care of our own needs," Edison said, setting Jonas Aylsworth and colleagues to remedy the situation as a challenge to their chemical ingenuity and a way to break free from the strictures of "commer-cial paralysis. . . . Now is the time for the United States to go ahead. We can manufacture [our own goods] cheaper today than in many years to come." From the waste gas generated out of the coking ovens of rural Pennsylvania, Edison's laboratory muckers condensed much-needed benzol, cut off from importation by the blockade of Germany.

Two long months before the sinking of the *Lusitania*, Edison preached the virtues of a submarine navy, a vast fleet that would patrol the entire eastern coastline as "peace insurance. By all means!" he told a visiting corre-spondent from the Brooklyn *Citizen*, beating the "preparedness" drum again. "A man carries insurance, why shouldn't a nation?" Two weeks after the *Lusitania* "horror," Edison told Edward Marshall of the Detroit *Free Press* he was pained over the loss of 128 American lives, but stopped short of ad-vocating war, saying yet again that as a "deeply injured" nation, we must concentrate on becoming "ready" psychologically and materially for a cata-clysm that, if and when it came, would be caused by others, not ourselves. The next day, Secretary of the Navy Josephus Daniels—who had recently stopped at the West Orange factory to sign a lucrative submarine and battleship battery contract with Edison rumored to be in excess of $15 million—spoke to reporter Marshall in favor of eventual disarmament, even as he admitted that the only secure road to peace was one paved with in-creased military expenditures, overproduction, and then storage of com-modities and materiel to aid in "the readiness to defend ourselves." The idea that a soldier "should love war is a misconception," Daniels observed. "He should not be bloodthirsty, but thoughtful."

Between the inventor and the enlightened bureaucrat there had been a meeting of the minds, a philosophical veering away from the abject waste-fulness of a great standing army ("eating its head off," said Edison) and to-ward a defensive posture on native soil. By its very nature, Edison believed, America never belonged to the community of "predatory" nations. Its safety should be enhanced by powerful armaments and long-range backup artillery in place at strategic domestic harbors, and our nation's foremost imperative should be to *rule the seas* with viable "war craft," extremely rapid vessels of the highest power.

After offering these sensible strategies on how best to guard his beloved land, Edison made it clear to Josephus Daniels that if called upon to serve, he would respond with patriotic alacrity. On July 7, 1915, after consultation with President Wilson, the secretary of the navy formally approached Edison with the request that he command "a department of invention and development" to be established within the navy as a clearinghouse to appraise technological ideas and suggestions from civilians and military inventors—a way to "mobilize" the best minds of the nation, in anticipation of wartime. To Daniels, the special emphasis of this Naval Consulting Board, the successor to a similar braintrust set up during the Civil War, must be upon "the new and terrible engine of warfare in the submarine," an area marking the convergence of Edison's longtime preoccupations on the one hand, and Germany's subversive and dangerous forte on the other.

It was a powerful group coming together in Washington, D.C., that autumn for its first meeting. The young assistant secretary of the navy, Franklin D. Roosevelt, was in attendance, as were Frank Sprague, L. H. Baekeland, Peter Cooper Hewitt, and Elmer Sperry. Edison immediately relinquished the office of chief executive, and took the honorary title of president; he did not want to sit behind a desk all day. The informing theory behind the Consulting Board (to some degree a public-relations outreach initiative on the part of the Navy) was that original ideas were just as likely to emanate from amateurs inspired by their country's potential peril as from professionals. However, it soon became abundantly evident after the first one hundred thousand concepts poured in over the transom that (according to the wise words of Robert S. Woodward, chairman of the Carnegie Institution of Washington) "revolutionary advances and inventions do not arise suddenly or in necromantic fashion." Overwhelmed by a mass of useless if well-meaning information, Edison and his colleagues transformed their mission into individualized consultation and supervised research projects within specialized Navy departments. They recommended that the Navy move to institutionalize research and development, resulting in the establishment of the first Naval Research Laboratory.

In early 1917 Germany declared unlimited submarine aggression toward all ocean shipping. After the United States entered the War, Edison launched his naval work full time, leaving son Charles to mind the store in West Orange. A contingent of Secret Service men was assigned to Edison while he lived in a Washington hotel and worked sixteen-hour days for four months, a tribute to the national-security value placed upon his imagination by the government. He did not disappoint the powers that be, showing off his sea legs experimenting on no less than six different converted yachts—in Long Island Sound accompanied by Mina during the summer; and with son Theodore as an able-bodied assistant aboard navy boats off Key West in the wintertime. Over an eighteen-month intensive period in the field, Edison concocted and masterminded forty-eight documented projects, ranging from the esoteric to the global: modified periscopes with rotating, spray-free viewfinders to function in low-visibility weather conditions; submarine sonar rangefinders and sounding cartridges based upon telephone diaphragm signals; a new nocturnal travel route plan for merchant ships to hug the shoreline in seven-fathom water and avoid submarine peril; refined shore-to-sea "under-water projectiles" (torpedoes) fitted with turbine heads and capable of penetrating ⅛-inch thick sheet iron; a rapid-shutter Venetian-blind searchlight signal with the facility to transmit forty words per minute in code; a hydrogen-detecting alarm to avert the danger of undersea explosions; smudge decoys mounted on a buoy imitating the smoke pattern of a steamer "so that submarines would be on wild goose chases"; vaseline and zinc antirust coating for submarine guns; a lightweight silicate of soda fire extinguisher which snuffed out the oxygen supply to "incandescent bodies"; an antiroll platform for ships' guns to ensure accuracy in rough seas; a "collision mat" device for patching canvas over damage holes in hulls caused by torpedoes; a mine detector wire linking two subchasers six hundred feet apart cruising one-half mile ahead of a cargo boat; a quick ship-turning device made of strong conical canvas bags deployed underwater, adding instantly to the vessel's resistance; an extension ladder capable of adding eighty-seven feet to the top of a mast; and on and on.

Edison's nautical gusto comes through in the scribbled dispatches he sent home daily to Mina when he writes of having "been out [off the coast near Annapolis] for two whole days experimenting with the towed phone. . . . We are now firing smoke shells from a three-inch gun and they work well. . . . Another Naval man is making my mine detector and a warboat will take it to Guantanamo Bay, our big Naval station in Cuba. . . . I am feeling fine . . . from your lover as ever steady, reliable and unchangeable."

He loved the clutter, camaraderie, and rituals of sea life. But while the salt air seemed to invigorate him, by the end of the war, Edison's official rhetoric had become riddled with criticism of the impenetrable myopia and frustrating red tape that ultimately prevented any of his four dozen ideas from being implemented by the navy beyond the prototype stage.

For Marion Edison Oeser, war was undeniably hell. Stationed in a garrison in the remote Thuringian town of Mühlhausen, Oscar Oeser spent most of his off-duty time carousing with comrades-in-arms until the first daylight hours, when he would bring a staggering group of night owls back to his home on the Modenheimerstrasse, pounding on the front door until Marion came to open it. Although the couple had long since abandoned the possibility of having children, Oscar remained an ardent lover, thrilling his wife with the pricklings of his "scrub-brush of a moustache" on the back of her neck. Not until a decade later did Marion discover to her disgust that Oscar's chronic philandering had begun even before the war. His string of affairs continued during the time he was mobilized, and Marion lived virtually alone in Neuenberg, thirty miles east of Stuttgart. The cannon bursts rattled her window panes and turned the glass red. One day, she watched in horror as a procession of more than two hundred half-famished French prisoners filed past her parlor window: "If we have a loving father in Heaven why does he let millions of his innocent children suffer so and the guilty ones go unpunished? . . . I wish I could run away and get away from it all," she wrote to Mina. "Of course, it would be no use, you cannot get away from your own soul. . . . I feel as if my nerves were on the outside of my skin." Prone for weeks at a time to bouts of nervous tension and suicidal fantasies, Marion tried a bromine cure, leaving her weak in the knees, prone to memory lapses, and short of breath. Girlhood fears of the dark returned, and in the middle of the night she went to her housemaid's room to share her bed.

In June 1917, continuing her southward path, Marion escaped successfully to Basel, and lived in the Hotel Schweizerhof until the end of the war. She begged her papa to send simple amenities unavailable in Europe; requests for several pairs of shoes went unanswered because in a flash of sibling rivalry, Madeleine warned Mina that "America and Germany are at war and [Marion], having married a German, is technically at least an enemy." Cut off from any dependable communication with her family in America, she cried constantly and at the slightest provocation. Her husband had a nervous breakdown and was hospitalized. Visiting him in a Neuenberg sanitarium after the armistice, she was shocked to see his hair had turned completely white. Their marriage was doomed; at forty-seven, Marion began protracted divorce proceedings against the unfaithful Oscar, who feared most of all the dire economic consequences of not being married to an Edison, and the pittance of an army pension he would be forced to live on.

Down at the Edison/"Willard" farm in Burlington, New Jersey, the furnace failed, the roof leaked, and winter winds whistled through chinks in the deteriorating stucco walls of the century-old homestead. Try as he might, Tom, Jr., could not stay away from the bottle; drink made him morose and gloomy and led to doggedly recurrent splitting headaches, in turn causing him to strike out against the devoted Beatrice. Drifting out of these sudden attacks he euphemistically called "neuralgia," Tom pledged time and again to start his life anew, but he had become inert, lodged in a morass of unceasing contrition. For her part, Beatrice had nowhere else to turn except God. At Thanksgiving, 1917 she stoically told her mother-in-law that "Tom is far from well, but the weight of my Cross is Love, and with my faith strengthening at every heartache, I patiently look for Victory." It was not to come.

William Leslie and Blanche Edison fared better breeding pheasants and milk-fed turkeys in Salisbury, New Jersey, until Will was drafted and shipped overseas to France, where he served his country with trepidation and misery as a sergeant in the salvage and repair department of the Army Tank Corps, "a mere foreman over a bunch of mechanics." Will's entreaties that Father use his influence and pull strings to obtain an early discharge were ignored.

Over her parents' protestations, Madeleine finally married John Eyre Sloane, her Catholic South Orange beau, in 1914. They lived for a year in a fourth-floor apartment on Eighth Street in New York City, across from the Brevoort Hotel, famed Bohemian gathering-place, and then left the city for Plainfield, New Jersey, where John's family biplane factory did a thriving business. Thomas Edison Sloane, called "Teddy," the first grandchild of the family, was born in 1916; and John Edison Sloane, called "Baby Sish," the second grandson, two years later. Old Man Edison could not keep his hands off the little ones, holding and jiggling them to his heart's content, but the coolness toward Madeleine's husband remained and was a source of tension and resentment on both sides. As the children grew older, Madeleine rebuffed Mina's invitations to spend summer weeks in Chautauqua or Quogue, preferring to stay at home in Washington where her husband was stationed during the war.

Charles's restless spirit finally overcame any vestigial commitment he may have felt to running the full course at MIT. Facing his senior year without a clue as to what branch of engineering he should adopt, he decided to quit college and work for the Boston Edison Company, much to his father's delight and his mother's chagrin. For $15 a week, he became a "floater" in the business, moving from one department to another—accounting, sales, turbine-testing, even security—as a jack-of-all-trades way to obtain a general indoctrination. In the summer of 1913, Charles the wanderer convinced his parents to let him take a motoring trip out West before (he assured them) settling down. He rendezvoused with a college friend, Robert Cox, in Denver, and they struck off into the Rockies, where they proceeded to get lost in the Argentine Pass before reaching Colorado Springs, where they attended their first rodeo. Onward by train to the coast, the comrades stopped off to view the Grand Canyon where Bob, running low on funds, sold his silver belt buckle to a Pullman porter. In San Francisco, they ran into Francis and Curtis Upton, sons of the Menlo Park veteran. The four young men shared an apartment on Bush Street and spent some delightful days at the beach before it was time for Charles to face the reality of further employment. On November 19, 1913, in accordance with a long-held family understanding, Charles joined the payroll of Thomas A. Edison, Inc.—at a whopping $10 per week higher than he had been earning as a novice in Boston. His first assignment was in the accounting department, to develop revised cost projections for the rapidly surging disc phonograph records.

But Charles Edison's other side, especially his demonstrated talent for writing and (despite the deafness plaguing him for most of his life) his love for piano improvisation, could not be suppressed. After the five o'clock whistle, he would hustle into New York City to visit his newlywed sister and brother-in-law in their Village digs. "The poet or whatever there is in me reacted favorably to that Bohemian type of atmosphere," Charles recalled half-a-century later. "Here [was] an undiscovered section of New York" that exercised immense appeal upon Charles the incorrigible explorer.

Down the street from John and Madeleine's apartment was a five-story brownstone at Ten Fifth Avenue; originally purchased by the Edison corporate interests as an adjunct office, it was now in Mina's name and stood essentially unused. Initially thinking he would install a performance space on the second floor for use as a demonstration site for phonograph Tone Tests, Charles slowly changed his mind as he spent more and more time in Village bistros, saloons, and chess parlors such as Romany Marie's and Jimmy Kelly's, and fell in with a crowd of struggling artists and literati. Dressed in a wide-brimmed felt hat and smoking a pipe, Charles Edison became a familiar figure ambling through the narrow Village streets. He decided to start a theater company.

In an old farmhouse cum studio at the corner of Washington Square

South and Thompson Street, Charles met Guido Bruno, a tall, swarthy, and charismatic Serbian, multilingual impresario of a notorious poetry reading series. The unlikely pair joined forces to establish the hundred-seat Thimble Theatre at Ten Fifth Avenue. Opening night was sometime during the spring of 1915; Marcel Duchamp had just arrived from France, and Walter and Louise Arensberg's modernist salons were at their height. Francis Picabia was in town; as were the cubist Albert Gleizes; composer Edgard Varèse, who held court in the Brevoort bar; actress Beatrice Wood; painters Charles Sheeler and Charles Demuth; and poets Wallace Stevens, Edna St. Vincent Millay, and William Carlos Williams (who practiced medicine just up the road from Charles Edison in Rutherford, New Jersey). It was indeed an intoxicating milieu for anyone in the arts experimenting in New York.

The young producers sent tickets to a list of critics and prominent socialites. Standing-room-only crowds became customary. The title of the vehicle for their inaugural performance has been lost, but scraps of other memories survive. Strindberg's plays seem to have been popular with the Thimble set, perhaps because they did not put too much strain on the budget, requiring only two or three actors. *Miss Julia* [*sic*] and *The Stronger* were offered in tandem, Monday, Tuesday, and Wednesday at 8:45 p.m., as well as a Saturday matinee at 3:00 p.m. Admission was one dollar. After the weekday performances, Charles would catch the Delaware, Lackawanna & Western "owl train" from Hoboken to Orange at 3:00 in the morning, reporting for duty punctually at the factory five hours later.

Perhaps this nocturnal routine explains why Charles wrote and published poetry under the pseudonym "Tom Sleeper" in *Bruno's Weekly*—the short-lived periodical he founded and published with his friend for two years after the theater had taken hold. Starting as a small, folio Village magazine, it grew to twenty-two pages with a subscription and newsstand list of four thousand readers paying $2.50 each. Charles's verse had a doggerel style to it. A representative effort, ten pages of quatrains called "The Mexican Border," was brought out as a self-contained pamphlet "By Guido Bruno in his Garret on Washington Square":

> *She was pretty enough and I liked her*
> *But her dad was a queer sort of ass*
> *Taught Greek to the kids of the ranchers*
> *In the mountains near Donegan's Pass.*

I flattered her some and I got her.
She was small—just a slip of a thing,
And her eyes had a queer fascination,
So I bought her a wedding ring.

We moved to the Mexican border.
Rough place, but we built us a shack,
She hated the plains, I the mountains,
And I swore that I'd never go back.

As we have noted, when war came, his father became consumed with work for the Naval Consulting Board, and Charles Edison reluctantly but understandably had to pull away from the Village crowd to take on more responsibility for the West Orange plant. Rejected from active military service because of his hearing loss, Charles personally discussed with Josephus Daniels as well as Newton D. Baker, secretary of war, the importance of providing secure and stable control at the Edison corporate headquarters while the Old Man was away.

Traveling to the office of a notary public in New York City to document papers that required witnessing, Charles stood before the clerk's desk as he read along without comment until coming upon the name at the bottom of the page. "You're not Thomas Edison's son, are you?" he asked, looking up in surprise. When Charles replied that he was, the man changed his demeanor completely. "Well," he laughed, "I guess *you* don't have to worry much! It's pretty soft for you fellows that can work for your old man."

Charles just smiled, thinking to himself that he wished the clerk could put in a few trial days working for Thomas A. Edison. "I'd like to hear his opinion of 'softness' of the job when he got through," Charles thought, "and it wouldn't make any difference whether he went in as my father's son, either."

On the contrary, the commonsensical young executive realized that "the curse of the soft snap" (as he regarded it) had ruined plenty of able men in their fathers' employ—but this fate would not befall Charles, because his father went out of his way *not* to bend or let his son stray, knowing the kind of contempt such patronism would create. Instead of becoming resentful of Charles when he entered the family business, the men at the plant actually pitied him, knowing that sooner or later he was going to receive the brunt of the Old Man's demands.

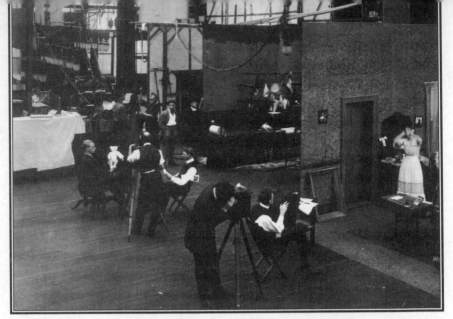

"I have heard experienced business men say of Charles Edison that he has the oldest business head of any man of his age whom they know," wrote a reporter from *The Literary Digest* who came to the factory to profile this self-effacing twenty-eight-year-old "whose tastes lie in the direction of business, with literature as a diversion." He was an inventor in verse, but not in the laboratory. Rather, Charles's first impulse in his father's absence was to try to bring some civilizing touches to the plant. Starting gradually, he initiated improved medical services, a dispensary, and other employee benefits. His next managerial step was structural. When he took up his duties, Charles presided (albeit de facto) over a motion picture business with a studio in the Bronx that would close with the end of the war; a nickel-iron-alkaline battery company; a phonograph company; a dictating machine business; a cement mill; woodworking plants; and even the manufacture of domestic appliances. "It was a mixed-up type of business," he wrote. "I decided that we weren't making enough money because responsibility was too far removed from the actual products." In response Charles advocated decentralized management authority, a divisional system for T.A.E., Inc., while maintaining central support services (what we would call now "back office") at West Orange. This approach was meant to trim overhead.

Mindful of not wanting to burden his father while the Old Man was in the service of his country, Charles concentrated upon dispatching positive bottom-line figures to him, especially salesmen's reports from across the land that showed how healthily the phonograph was progressing. For Charles's first five years with the company he seemed to be sailing smoothly. From his Fort Myers retreat, Father Edison even went so far as to praise his son's management style in the press: "I should not be going away to Florida if my industries were not in ship-shape condition," Edison said benevolently. "My son, Charles, managed the business during the [time] I was engaged in war work, and when I got back I was very much pleased to see how well he had handled things." Anyone who knew Thomas Edison's psyche, however, could have predicted that such a laissez-faire romance would never last.

In those honeymoon times, Charles married his college sweetheart, Carolyn Hawkins, on three days' notice after six years of courtship. The petite brunette, a humble secretary, was the youngest of three daughters of Dr. Horatio Gates Hawkins and Ada Jane Woodruff of Cambridge, Massachusetts, by way of Vermont. Her morphine-distracted father had committed suicide by drowning himself in the Connecticut River when Carolyn was a girl; the widowed Mrs. Hawkins devoted herself utterly to war relief work. The nuptials approaching, Father Edison and "first mate" Theodore were out at sea embroiled in naval researches.

Papa sent a laconic wire to Charles when he heard of the imminent ceremony: "The sooner the better. Anyway, it won't be worse than life in the front-line trenches." Madeleine and John, in Washington, could not make the trip on such short notice. Madeleine, particularly mindful of the sacredness of the Edison name, was put off by the impromptu nature of the announcement: "I was so afraid that there would be stories [in the newspapers] like those that came out when William was married or that it would be called an elopement, etc. . . . Wasn't anybody present but you and the fish?" she asked her mother.

It was a rather intimate ceremony set for three o'clock on March 27, 1918, on a balmy afternoon at Seminole Lodge. A red carpet was rolled out over the green lawn to an improvised altar with two candles burning in front of a camphor tree, a cinnamon tree, and a Japanese umbrella tree. The Reverend F. A. Shore of St. Luke's Episcopal Church in Fort Myers officiated. In the absence of Carolyn's mother, Mina gave the bride away, witnessed by longtime, faithful friend Lucy Bogue, and the family butler.

A week later, everyone left for Key West, then proceeded to New Jersey by steamer. The couple rented a furnished apartment on Park Avenue in East Orange and parked their Model T outside, soon thereafter moving to a proper cottage in Llewellyn Park so that Charles would be near his father, his boss, when the Old Man came home in November and tried to take over the tiller once more for the choppy seas ahead.

CHAPTER

26

UNDER CHARLES EDISON'S CAPABLE STEWARDSHIP DURING the war, and for two brief years immediately following, a well-organized Thomas A. Edison, Inc., showed a healthy although uneven balance sheet. On the income side of the ledger, gross cash sales marked an unbroken succession of increases, more than doubling from 1915 through 1920. Lucrative government contracts at the front end of this period preserved a niche for the batteries, and eleven phonograph models—ranging from the modest Chalet table version at $95 to the upscale XVIII Century English at $450— were available to consumers. An Army and Navy model accompanied by a tough carrying-case for records was a desirable item offered to the boys overseas at the cutrate price of $60, and did much to enhance Edison's patriotic image. More than a quarter of a million disc phonographs were sold in 1917 alone.

On the expense side, the picture was less consistent. The increasing cost of raw materials and scarce chemicals during wartime, and the constant pressure to align wages with rising inflation led to stress among middle-management, rapid turnover and occasional strikes among the nonunionized workers. In response to these vicissitudes, T.A.E., Inc., profit margins swung wildly from year to year.

Nevertheless, after shutting down the movie business the corporation displayed a positive cash flow, and the Old Man exercised a commensurately

THE OLD MAN
CONFERS WITH
AN ASSISTANT

light touch and benign presence. Coming down the mountain from Glenmont, arriving between 8:00 and 9:30 in the morning by open touring car,
he dutifully punched in at the time clock in the dark corridor just outside
the library, next to the sign prominently displaying the Joshua Reynolds
aphorism Edison had adapted as his motto, "There is no expedient to which
a man will not resort to avoid the real labor of thinking." An absorbed,
Dickensian figure, Edison shed his slouched black felt hat and overcoat.
Then, pale of complexion and blue of eye, he held court in the library's magnificent ambience, an ancient lion in his lair. In a rare silent film, *A Day
with Thomas Edison,* we watch the patriarch beckon to his white-collared,
dark-suited lieutenants, who approach bearing memoranda they lay upon
his huge desk as if priests with offerings at the sacred altar. A privileged few
with questions that cannot wait are beckoned closer. Edison cups a hand to
the side of his head, and a young executive with slicked-back hair and a
ready smile leans so close to the Old Man it appears as if he is about to kiss
him; but no, he yells into Edison's ear, and the Old Man knits his brow with
absolute attention. His response is quick—a few well-chosen words, a nod
of the head, and the fellow withdraws to make room for the next supplicant.

It is evident from contemporary accounts by those fortunate enough to
have been hired during this boom time that Thomas Edison acted as the

guardian at the gates of the company; after all, it was still *his* company, regardless of who was running it from day to day. To some degree his role was
symbolic—since there was a personnel department—but Edison lived by
rigid faith in his own instincts at "finding the right man." The scenario was
remarkably the same for all comers. Fresh out of the army, electrochemical
engineering student Paul Kasakove returned to Cornell to complete his final
year, and as the summer of 1920 approached, began scanning help-wanted
advertisements in the New York *Times*, wherein he spied a notice placed by
the Edison Company. He leaped at the chance to work for a living legend.

Arriving at West Orange, young Paul was interviewed extensively by
both the superintendent of the plant and the personnel manager, who cautioned the applicant that when he met Mr. Edison he should not be offended
by anything the inventor might say: "He was a genius, and as such must be
forgiven if he made any very blunt remarks." The trio walked across the yard
next to the main building and stopped before the door of the outlying brick
chemistry lab. The personnel manager turned to the superintendent and
said, "You take him in." The superintendent replied, "Nothing doing,
you're the personnel man, you take him in."

After considerable argument, Paul was shoved across the threshold,
and saw a white-haired man in a rumpled linen suit hunched over a cluttered table in the middle of the room, peering obliviously through a microscope. The superintendent strode forward, presented Paul's employment application, and shouted, "Mr. Edison, I think we have a man for you!" Edison
looked up, brushed a recalcitrant lock of hair from his brow, scanned the
form quickly, looked Paul up and down, and called out, "Come in Monday
morning!" Then he went back to his microscope. To Paul, the decision
seemed arbitrary; in fact, as he learned later on, Edison had already been
through a succession of young men sent to him by friends of the family and
employees at the plant, and had reached the end of his tether insofar as in-
bred connections were concerned. Fed up with the "pin-heads" he had seen,
Edison wanted someone fresh, unknown, from the outside; Paul was in the
right place at the right time.

A former Edison phonograph salesman writing anonymously in *The
Literary Digest* remembered a similar interview conversation with the Old
Man in armchairs side by side in the library. "I understand that you're the
whole thing with So and so," Edison began, disarmingly referring to the
firm where the young fellow currently worked. "No, I'm not the *whole thing,*
Mr. Edison," he responded, "I'm merely a cog-wheel in the machine." Edi-

son smiled approval, chomped pensively for a moment on his tobacco plug, and aimed a well-placed spurt into the white ceramic spittoon on the floor next to him. The superintendent, sitting in, interjected that this young man did not want to sign a contract for a specific term of employment. Mr. Edison reflected for a split second, and "with a princely disregard for [the applicant's] presence," replied, "He's got the right idea. If he makes good, he doesn't need a contract; if he doesn't make good he won't want to stay. Go ahead." The interview ended, and the fellow was to learn only later that before the meeting, Edison had already conducted an extensive investigation into his background and experience, character and ability. The interview had merely been a "once-over."

Thomas Edison vehemently resisted the invasion of radio, dismissing it out of hand as a "craze" and a "fad." Buckminster Fuller understood radio's traumatic implications for Edison, pinpointing that moment at the aftermath of World War I when "industry suddenly went from the visible to the invisible base, from the track to the trackless, from the wire to the wireless." But Edison pressed on, continuing into the twenties to make acoustic, mechanically reproduced phonograph recordings.

"The old masters," as Fuller referred to Edison's generation, "were *sensorialists* . . . [but] the big thing about World War I is that *man* went off the sensorial spectrum forever as the prime criterion of accrediting innovations." Holding on for dear life to this biological, "sensorial" authority, Edison retreated more and more often to the cavernous rehearsal rooms at West Orange, continuing to fill notebook upon notebook with subjective observations. For one recording session alone, on June 14, 1923, Edison consumed thirty pages critiquing every instrument employed for a rendition of *Ave Maria*, specifying the precise distance away from the recording horn each musician should sit: kettle drum, oboe, English horn, bass clarinet, flute, piccolo, viola, snare drum, triangle, and cymbal. Three days later, Edison covered ten more pages comparing the scales of two sopranos, Elizabeth Spencer and Betsy Lane Shephard, at different positions in relation to the recording horn, before deciding which one of them would make the final recording of the aria, although he grudgingly observed the "tendency of *both* to emit a very strong note on high, spoiling the evenness. [T]his is a dramatic note which is out of place in *Ave* [Maria]."

Thomas Edison set forth to ignore the inevitability of a new kind of entertainment, a newfangled "music box" that sent free, indiscriminate tunes and outrageous jazz through the air, unpurified by any governing, rigorous sensibility (such as his own.)

During this same period, Edison shifted away from his much-vaunted intuitive hiring practices—the personal touch, the audience with the "Wizard" exercising his magic powers over wet-behind-the-ears initiates—and *toward* standardized intelligence examination. Like Paul Kasakove only one year before him, A. L. Shands responded to the simple advertisement in the *Times* calling for "young men, college graduates, interested in manufacturing. Salary $40 per week to begin. Will be trained for executive positions." Arriving at West Orange, he was taken up to the third floor of the laboratory. Seated at a long, wooden table with eight other fellows, he was handed a mimeographed sheaf of paper containing one hundred fifty questions, and given from nine-thirty in the morning until noon to complete it. "The knowledge required of one was hardly encyclopedic," Shands noted. "However, you might be well informed and still fail, because the specific data contained in your [university] physics, chemistry, and history text-books had grown a bit vague in your mind." Shands, only a year out of college, "swung into [the exam] like Babe Ruth into a pitch grooved for the plaudits of the customers."

Edison's preemployment screening questionnaire intrigues us today as much for the range and diversity of the topics therein as for their juxtaposition: "Where is the River Volga?" is followed immediately by "What is the finest cotton grown?" "What telescope is the largest in the world?" is followed by "Who was Bessemer and what did he do?" "Who invented printing?" is followed by "How is leather tanned?" "What is the price of twelve grains of gold?" is followed by "What is the difference between anthracite and bituminous coal?" Geography, history, and science all tumble together to form an inventory of factual knowledge, representative tidbits Edison believed should be possessed by every young man applying for a job. Edison had a cogent rationale for instituting this controversial test. It was not merely to prove that a college education did not automatically qualify a person for responsibility in the "great world of struggle" at large, not merely to hammer home Edison's lifelong belief that bookishness could never hold up against a thorough course at the university of life experience, the greatest "institute of learning" of all, a belief he deeply shared with his occasional correspondent John Dewey. To Edison, true scientific knowledge had been and always would be predicated upon a skill in remembering details, and the incessant *cultivation* of that faculty, the literal "exercise" of the brain, in the same way that one stretched and honed other muscles of the human body. To Edison, the single most important quality that an executive had to possess was *"a fine memory."* That was the cornerstone for the ability to make a quick, correct decision, marshaling all the facts at one's disposal.

The only way Thomas Edison felt confident of the viability of a man's memory was to find out the quantity he had remembered—*and* how much

EDISON AT 73

he had forgotten. How well stocked was the prospect's "mental warehouse"? The man with the best memory—the man to end up with a position in the Edison company—would be able to enter that warehouse boldly, go directly to the shelves where the correct facts were inventoried, and pull them out at a moment's notice, to "select and handle" them correctly. In keeping with Frederick Winslow Taylor's ethos, which Edison tried to apply to his shop, he was looking for evidence that the "mind-machine" was efficient, unaffected by any pernicious "atrophy of perception" or shrinkage of useful sinews, so that its energy resources would not be wasted when demanded to play an active part in the "college of business."

Could *anyone* be as stubborn as Thomas Edison when he decided to take a position? "If you listen to an Edison phonograph and a radio side-by-side, with the same tune, singer or instrumentation, you will appreciate the distortion of radio music," he told an interviewer, fearing the destructive influence radio listening would have on the aesthetic aptitude of future generations. The same man who had summarized the keyboard technique of Sergei Rachmaninoff in two words—"He thumps!"—warned that "undistorted music [i.e., in the concert hall or on the phonograph] in time will sound strange to those brought up on radio music and they will not like *the real thing*." But by 1922, no one seemed to be listening to the Wizard, as Americans purchased $60 million worth of radios; four years later, $506 million worth were sold, one receiver for every third home in the country. At any moment, and at will, you could turn a knob and take in more than any single record collection could possibly offer. Ubiquitous antennae sprouted on rooftops across the land; ironically, one of the very first Westinghouse transmitters was perched atop the Commonwealth Edison building in Chicago. Radio rode the crest of "Coolidge prosperity" consumerism, fueled by the advent of credit buying, stock market speculation, and advertising on a grand scale—another vice in which Edison refused to wallow, insisting that his good name and familiar visage on a product had always been sufficient.

In this unhappy context, the old guard of T.A.E., Inc., finally and inevitably clashed with the new. Adversarial strains conspired to split apart this entrepreneurial American family. Watching phonograph cylinder Amberola sales plummet by 90 percent in 1921 alone, Charles was desperate to get out of that vestigial production area altogether. The Old Man insisted they push on; while the disc was more suited to longer-playing musical numbers, Edison believed that as the needle went around in the record groove through smaller and smaller circumferences, a quality of tone was sacrificed, whereas the cylinder, with its unvarying track, provided a truer sound. Despite his belief in the pedagogical value of the phonograph, Edison postponed indefinitely the development of a School Model at the very time when he might have carved out a niche for it. Again in the early twenties as a consequence of radio's accessibility, there was a costly shakeout in the number of available Edison phonograph types; only three of the eleven models were still selling. The Phonograph Committee recommended a new Laboratory Model at an upper-middle price range, but, again, the Old Man vetoed it. In 1925, Victor introduced an Orthophonic line, suited for the latest innovation, electronic recording, and then mounted a campaign for its own Radiola. Thomas Edison would have none of that, of course, seeing his competitor's change of focus as yet another reason for T.A.E., Inc., to hold on to the "straight" phonograph business. From Indianapolis to Chicago to Omaha to Kansas City to Des Moines, throughout middle America, Edison's traditionally strongest sales area, nearly one-third of his longtime regional distribution force deserted him.

Sleepless nights descended upon the once idealistic Charles Edison. At his bleakest moments, he even considered leaving the company. "I would lie awake thinking, 'How can I stop him from making these terrible mistakes? What can I do? He just doesn't understand.'" Once upon a time—it seemed an eternity ago—given the mandate to build a great corporation, Charles Edison now had to preside over retrenchment under his father's unrelenting gaze. In contradiction to the harmonious face they presented to the outside world (and to future corporate history books), the interoffice communications between Charles Edison and the Old Man tell a sad story of micromanagement, note upon penciled note exchanged daily about even the most trivial matters. "I don't want to act without his K-O," Charles confessed to Mina, surrendering to the way Thomas Edison liked to work.

Often, it appears from their exchanges as if Charles is straining to be unqualifiedly certain of his father's approval, even on the subject of whether he can take two days off to attend a jobbers' convention in New Orleans. "Use your own judgment," is the perfunctory, loaded answer. Middle-management discontent, expected under trying times, takes up much of Charles's energy, and the pattern is invariably the son's solicitous caution followed by the father's abrupt decision.

Early in 1922, Charles tried to raise the cautionary flag of potential management defections as parlous economic times loomed: T.A.E., Inc., had worked so hard to "increase the % of good men through the questionnaire and to eliminate dead ones," and now, when there were suddenly far fewer opportunities to move up through the ranks, how could the company hold on to these college graduates who had successfully demonstrated their mental capacities, without "tying a hangman's noose of promises around their necks?" Dismissively came Papa's answer: "Charles—Why worry about these things. Cross the stream when you come to it. I don't know of anybody who has left us that I would want back. I want to keep the system up [i.e., the want ads followed by the questionnaire] of constantly looking out for young educated men who will start and make good. Out of 50, I hope to get six. . . . There is [sic] alot of men here we must lose before we can have a well-managed concern." Better wages were out of the question so long as the personnel flow through the front door continued.

As the twenties wore on, Charles became obsessed with putting out mundane fires—personnel defections, escalating costs for piece work in the Battery Company, delinquent receipts from phonograph demonstration shops, the "rotten practice" of carelessly packaged unsold records returning cracked from stores, shipping delays, dysfunctional trucks, and so on. The Old Man, meanwhile, demanded more and more minute detail from the field, as if there were no support around him. Indeed, telling Charles that the fellow was "hopeless and his men incompetent," Edison fired his chief financial officer, Stephen Mambert, who had worked at the plant for nearly a decade. By mid 1923, the Old Man announced distrustfully that he wanted to "check all jobbers shipping departments from here [i.e., his office] . . . so they can weed out their inefficient men, which they never will do until we prove it."

Cranky, sure of his likes and dislikes, and equally resolute in his bunker mentality in 1924 as he was half a century earlier, Thomas Edison presided over the slow and steady decline of his beloved old-fashioned phonograph with the single-minded mentality he never outgrew. Through Charles, his sounding board, Edison once more chastised the men of Wall Street and blamed "the Bankers" for the "general demoralization of the phono people." Bad times would "continue next year until the money men suck every dollar they can out of the wreck & then they will wind up—in the meantime if we continue our methods of selling & poor organization we too will get in a hole. I am certain next year the bottom will drop out but we can hold on by rigid economy & *better methods of selling*." Would that the culture at large had been willing to buy what Edison persisted in imposing upon them! All he could think about was a longer-playing record; wouldn't it be magnificent to inaugurate such a series with Gibbon's *Decline and Fall of the Roman Empire,* playable by one's bedside on two-hour discs?

When Father in a funk retreated to Fort Myers to celebrate his fortieth wedding anniversary in the winter of 1926, Charles finally admitted to Mother Mina with unusual candor that he was overwhelmed by "a sudden avalanche of things. . . . Perhaps it is because the final responsibility is more up to me when he is away as I don't have him to fall back on for counsel as easily. . . . Perhaps too," Charles added, "we are all keenly anxious to send good news rather than bad." There was precious little good news left.

Like her compulsive husband, Mina Miller Edison was a chronic "doer." Her scrapbooks of letters from Madeleine, Charles, and Theodore are filled to bursting with admonishing messages, warning her to take some time for herself, just lie down on the sofa, rest, and disconnect. On the contrary, just as her "Dearie" was immersed in the thought-work of his imagination and the mechanical work of his factory, Mina plunged wholeheartedly into a staggering range of civic and beautification projects. In the spring of 1920 she founded a Round Table at Fort Myers, inviting twenty of the town's leading lights—including the mayor, several councilmen, and the chairman of the board of commissioners of Lee County—to join her around the polished oak table in the dining room of Seminole Lodge where they discussed "crying needs of civic reform": a refurbished railroad station, a new city park, elimination of unnecessary street noises, conservation of fish in the Caloosahatchie River, and better roadways for increased tourist traffic. In July 1922 Mina gave the keynote address at the first meeting of the Chautauqua Bird and Tree Club, founded in 1913 and chaired since its inception by her sister-in-law, Louise Igoe Miller (brother Robert's wife). Mina "spoke finely about the glory of being one family, one working group, attempting to attain a common end," to protect and preserve the wildlife inhabiting the ten thousand lovely trees of the great upstate retreat, and to establish nature trails and "outdoor museums" for the edification of all visitors.

The same summer, Mina was elected a trustee of Chautauqua Institution, joining her brother Ira on the Board. She dedicated most of her time that year to creating a garden space behind her landmark family cottage on the grounds of Miller Park near the lake. She added tendrils of ivy to the old stone wall out back, and installed a working fountain next to flagstone walks "directing one's footsteps to innumerable corners of cragged picturesqueness." Mina brought her considerable horticultural talents into play, planting forget-me-nots, foxglove, violets, columbine, candytuft, meadow rue, hydrangea, bluebells, petunias, and roses. Today, sitting on a limpid summer evening in the very same green-tinged wrought-iron chairs positioned by Mina on the patio seventy years past, one can listen to her beloved "feathered songsters" in full voice.

In later years, Mina instituted what she called her District Plan at West Orange, Fort Myers, and Chautauqua, challenging neighborhood groups of ten or fifteen families each to raise money through musicals, amateur theatricals, and card parties, and then apply those funds to local-level beautification competitions.

Beyond the primary and most proactive environmental groups closest to her three homes, Mina belonged to the New Jersey Floricultural Society, the Save-the-Redwoods League, the National Association of Gardeners, and the School Garden Association of America. She was also on the board of directors of the National Audubon Societies, the John Burroughs Association, and the National Plant, Flower and Fruit Guild.

Her dedication to community betterment extended to the moral as well as the natural and aesthetic sides of life. Mina Edison was one of the founders of the Red Cross chapter of the Oranges and Maplewood; she supported the Orange branch of the Salvation Army and was on the board of the Women's Exchange of the Oranges. She was a member of the Orange Orphan Society and the Anti-Saloon League of America. For seven years she served as President of the West Orange Improvement League. And for twenty-five years, from 1902 to 1927, Mina was president of the Women's Guild of the First Methodist Episcopal Church of Orange.

During the early twenties, she was also regent of the Essex County chapter of the Daughters of the American Revolution, and eventually chaplain general of the national DAR.

For all her patriotism, civic pride, and abundance of upstanding good will toward those less fortunate than she, Mina Edison was the first to declare primary allegiance to her house and husband, in keeping with the manner in which she had been brought up from childhood in Akron. She was, of course, completely responsible for the social events at Glenmont, a scene in which Thomas Edison felt uncomfortable—he referred privately to dinner guests as "outsiders"—but which at the same time, especially in later life, he recognized as incumbent upon him to keep his image properly burnished. Even as Mina understood her husband's wish to be "protected from unprofitable social activities," he dutifully put on his tuxedo and quietly whiled away evenings with the likes of the King and Queen of Siam, President and Mrs. Herbert Hoover, Orville Wright, Andrew Mellon, Charles and Anne Lindbergh, Helen Keller, Lord and Lady Kelvin, and even the great Hermann von Helmholtz, who, much to Mrs. Edison's dismay, forgot to sign the assiduously maintained guest register.

Mina was scornful of the term "housewife" as "the worst misnomer we have in our language." Rather, she took pride in describing herself as a "home executive." She believed it was an obligation of the true American woman to understand the "science of domestic economy." In her view, it was tantamount to "domestic wreckage" for a wife to stand still while her husband advanced; Thomas Edison had his career, and Mina preserved the dignity of work within *her* rightful sphere. Since the death of her widowed mother, Mary Valinda Alexander Miller, at the age of eighty-two in October 1912; the distancing of her three stepchildren; the marriages of Madeleine and Charles, and Theodore's departure to MIT in his brother's footsteps, Mina spent more and more energy running the daily commerce of the household and ministering to Thomas, a job in itself.

"They need no other rosary whose thread of life is strung with beads of love and thought," ever spiritual Mother Mary Valinda had written the year before she passed away; other-directed to an extreme degree, Mina had always taken these words to heart. As we have seen, her stepchildren as well as her biological children leaned on Mina constantly, afraid to show weakness in front of Papa. As a spouse, he was demanding in many and varied ways. Most familiar, despite his unfulfilled promises to mend his ways, were Edison's chronically unchanging work habits. Muckers at the plant recalled the appearance of the coachman, faithful Mr. Conroy, waiting out in the courtyard for two or three hours in the evening, then being sent back up to the house to bring a sandwich and coffee in a picnic basket. Mrs. Edison would then arrive at some ungodly hour of the night, sometimes toward three in the morning, to stand silently by the laboratory table with Thomas's jacket and overcoat over her arm, hoping her husband would tear himself away and come out to the car.

Mina was proud she "never nagged" this man whose "work was his life," "accomplishing great things for the world. . . . When he started a problem, it had to go out. . . . Compared with what Mr. Edison does in this world, there is exceedingly little I can do," Mina humbly told an interviewer ten months before Thomas Edison died, "and if Fate has placed it within my power to be of some service to him, I feel that I have not striven in vain."

Beyond this unquestioning tolerance (ritually, at least) of her husband's labor at invention, Mina had to remain aware of the endless catalogue of his likes and dislikes, small and large. *The American Magazine* aptly designated her "the custodian of one of the greatest geniuses the world has ever known." After he awoke, most mornings by seven, shed his silk nightshirt (he would not wear pajamas; they were too constricting) and had made his toilet, Edison preferred to have Mina, rather than the chambermaid, lay out his underwear and suit on the bed. When her husband was ready to leave for the laboratory, he expected Mina to be standing ready to open the front door, hand him his coat, and send him off with a hug, a kiss, and a few parting words of affection. He could not bear to depart without this exact succession of events. She was likewise supposed to be the first person he wanted to see when, or if, he arrived home in the evening, entering the foyer and calling Mina's name, even going from room to room in search of her, if by some unfortunate chance the butler were standing there in front of him.

When her husband sat reading his inviolate New York *Times* by the fireplace in his armchair—the chair that was completely off limits to everyone else in the family, even grandchildren—he was not to be disturbed.

READING THE
SUNDAY TIMES
ON THE
SECOND-FLOOR
PORCH WITH
MINA AND
BROTHER-IN-
LAW JOHN
MILLER

When he wanted to hear a tune for relaxation, he expected Mina to place the recording of "I'll Take You Home Again, Kathleen" on the phonograph. Because he could not remember anyone's birthday or anniversary, Mina kept a little notebook in her desk with a list. When in later life he attained global celebrity status, the trusted Mina opened his mail.

She ordered all of her husband's suits, from the same tailor in New York City he had patronized for fifty years; the fabric had to "breathe," always lightweight blue or gray serge in winter and white in summer. Every Christmas, she made a box of India silk handkerchiefs for "Dearie," the only ones he would use, each a yard square. Edison smoked only the mildest Hoffman cigars from Cuba, two for twenty-five cents, four per day, and when the couple were on vacation in Florida, it was Mina's job to make certain his "smokes" were shipped down. At dinner, if Edison could not follow the conversation, Mina at his right hand tap-tapped the chit-chat in Morse Code on his knee, the way he taught her when they were courting.

If his mental preoccupations were cleared away, Thomas Edison could and did find moments to love his wife. The odd balance was such that, while he was self-absorbed by nature, Mina, in turn, was obligated to be absorbed in him. His "Dearest Billy" cherished those times when she and Thomas were alone together, and Mina could draw upon the spirit, as she well put it, of "being quiet with his great soul." She was also subject to private, unavoid-

THE MASTER OF HIS DOMAIN ON THE GLENMONT FRONT LAWN

able episodes of insecurity and self-doubt, revealed to her three children—
"blue," depressed periods when Mina regarded herself with weary depreca-
tion as "a worm in the dust" or of "little value" to them, or anyone. And as
her youngest, Theodore Edison received the strongest expressions of his
mother's deepest need, to feel needed by others.

Theodore, the Rough Rider's namesake, his mother's "golden-headed
ray of sunlight," had arrived in the world even as her adored brother fell in
the hills of Cuba. When his governess was away, "Baba" enjoyed cherished
evenings alone in the big house with Mother reading aloud to him before
bedtime. He stayed close to home for high school in Montclair, then moved
on to MIT from September 1919 to June 1923, specializing in theoretical
physics, and, unlike his older siblings, took to studies with a vengeance,
taking classes all year round the first year: "Have very little time to write,"
he told Mina (although in fact, according to her wishes, he wrote or
telegrammed daily, and she became anxious at the slightest deviation from
this routine). "My day runs about as follows: 5:45 get up, get breakfast.
7:00–8:15 tennis. 8:30–12:30 classes. 12:30–1:15 lunch. 1:15–4:00 classes.
4:00–6:30 errands & study. 6:30–7:00 supper. 7:00–10:00 study. 10:00
bed." Mina pleaded with Theodore not to work too hard and to take a few
moments to write to "neglected" Papa.

Once he had hit high marks in class, Theodore broadened his extracurricular life during sophomore and junior years, still with the same attention
to structure. He was elected treasurer of the Aeronautical Engineering Society. He swam at the YMCA once a week and played seven sets of tennis on
an afternoon when there were no classes. Skating would become a lifelong

passion. The Charles River ran directly in front of his house and when it froze over, Theodore was out early on winter mornings, causing Mina to worry endlessly about his sinus condition and pray that he would not fall through the ice.

Graduating with distinction, the bookish Theodore turned down a position at his father's company, preferring to return to MIT for an additional year of postgraduate studies in engineering, biology, and geology. In the family tradition, he took a summer trip to the Pacific Northwest, where Mina met him for several days of civilized camping. Back at Cambridge in the fall, he registered for eleven courses and audited two more, proudly noting that his class time per week exceeded sixty-five hours.

And then, during a dinner party at the home of a friend, Theodore Edison met Anna Maria Osterhout. Three years his junior and enrolled in a pre-med program, she was the daughter of Dr. Winthrop John Vanleuven Osterhout, distinguished professor of botany at Harvard. Tall, blond, and athletic, Anna shared Theodore's interest in tennis, swimming, and skating. She was effervescent, "happy and full of pep," a wholesome and headstrong girl who drove her own car. She was a good cook, cultured, with a love of music and theater. Her only fault seemed to be an occasional, caustic display of temper.

Within a week of their acquaintance in late January 1924, Theodore took Anna to a Saturday matinee of Oscar Wilde's *The Ideal Husband*, and then the following Sunday to a violin recital by Jascha Heifetz. All these details the good son responsibly reported to his ever more anxious mama, who feared this formidable young lady was drawing him perilously close to being "signed up." Theodore reassured Mina that despite this infatuation—which seemed to be growing serious by the hour—she should not "worry about being left out of my world. You're still far and away my best girl!" However, with the advent of spring came romantic picnic trips to Ipswich in Theodore's flivver, and long walks around countryside ponds. By July the couple were engaged, and the last Edison had left the nest.

CHAPTER

27

FEBRUARY 24, 1926, WAS ONE OF THOSE BALMY WINTER
mornings when you thank the Lord you are in Florida and not the frigid
north. Out at the Fort Myers Fair Grounds, the Philadelphia Athletics were
running through spring training warmup exercises under the vigilant eye of
manager Connie Mack. A stir went among the crowd of onlookers as an old
Ford flivver pulled up alongside the batting cage and out stepped Thomas
Edison, a familiar figure around these parts.

"Think you could hit one?" star right-hander Kid Gleason asked the
Old Man playfully. Smiling, Edison nodded, and shed his Panama hat and
white suit jacket. The warm breeze ruffled his shirtsleeves and a few wisps of
silver hair. "Let's go," said the Kid, grabbing a ball and trotting out to the
mound. "Here, Connie," he said to the boss, "you be catcher." Mack donned
Mickey Cochrane's big glove comfortably indeed; he had been quite a
"backstopper" during his playing days. Edison shouldered slugger Al Sim-
mons's heavy bat and stepped to the plate.

Gleason wound up and delivered a twister that cut across the outside
corner. Edison swung and missed. "Strike one," the call went out from the
crowd. The Old Man, slightly surprised but unfazed, motioned to Gleason
he was ready for another try. Out of the windup shot a smoking low curve.
Edison stepped into it and connected squarely. With a crack, the ball sailed
out of the infield for a Texas leaguer over first base.

"Sign him up, Connie!" came the chorus from the assembled throng, as Edison shook hands all around.

"He must have invented a way to solve my curves after missing that first one," said Gleason with a rueful grin. "I guess he's too smart for me!"

"Pretty good for the first day of training," agreed Connie Mack, tongue in cheek. "Edison is a promising lad, and I predict he will be heard from." Indeed, the octogenarian was already by this time a living legend, not only in the baseball world, but far beyond. Four years earlier, 180,000 bags of Edison Portland Cement at 94 pounds each had poured into the construction of Yankee Stadium; ten million bags of the product were sold annually by the mid 1920s. Perhaps "The House That Ruth Built" was a misnomer. And in that same year, 1922, a New York *Times* poll named Thomas Alva Edison "The Greatest Living American."

Six months after the slugging performance, the Old Man "retired" from his position as president of T.A.E., Inc., declaring he would henceforth spend time exclusively in the lab, handing the chief executive reins over to Charles. However, while publicly relinquishing corporate control, Edison exercised firm authority over the refinement of his ultimate mythic image. The world's greatest inventor, the journalist's dream—good copy since his earliest days—was now far more than "the last great hayseed," or the Promethean, "powerful innocent." Edison as homespun philosopher and visionary prophet increasingly waged a systematic campaign to codify his observations and beliefs on a staggering range of subjects. He deftly played upon the anti-intellectual genius theme of hard work with a blizzard of pronouncements touching upon almost every aspect of American life and thought.

The Edison Pioneers saw to it that their revered mentor's legacy would be preserved for the future. Founded in 1918 with Francis R. ("Culture") Upton as first president, this association saw its mission to bring together "for social and intellectual intercourse" the Menlo Park gang, the fellows who worked with Edison in the days prior to and including 1885, before the move to West Orange, and "to pay tribute to [his] transcendent genius and achievements and to acknowledge the affection and esteem" in which they held him. Edison's secretary and biographer, William H. Meadowcroft, served as historian of the Pioneers, and such noted alumni as Lewis Latimer, Frank A. Wardlaw, John W. Lieb, William J. Hammer, and Samuel Insull held various positions on the executive committee. In 1913 the original Menlo Park lab, that evocative combination of country meeting house and overstocked warehouse, long since abandoned and collapsing, was destroyed by a storm. Four years later, Edison's historic home on Lincoln Highway was leveled by fire. The Pioneers set up a natural boulder on the site with a bronze plaque as a tribute to Mr. Edison. They convened every year on the Old Man's birthday for a black-tie banquet, champagne, cigars, and old-fashioned speechifying, while their esteemed leader, "Wondersmith of the World," made a point of granting his annual group interview with the press, seated at a desk, answering questions handed to him on scraps of paper when his hearing loss became almost complete, while the newsreel cameras whirred.

In the entire decade from 1921 to 1931, while the Edison company submitted only thirty-one patent applications over his signature, Edison the sage was industriously producing pithy opinions for immediate publication in the daily press. On deafness: "In my isolation [insulation would be a better term] I have time to think things out. . . . Most nerve strain of our modern life, I fancy, comes to us through our ears"; on textbooks: "In ten years, [from 1925] as the principal medium of teaching, they will be as obsolete as the horse and carriage are now. Visual education— the imparting of exact information through the motion picture camera—will be a matter of course in all our schools"; on youth: "Somewhere between the ages of eleven and fifteen, the average child begins to suffer from an atrophy, the paralysis of curiosity and the suspension of the power to observe. The trouble I should judge to lie with the schools"; on American music: "Who save a Southerner, can sing effectively *The Swanee River?* And I doubt whether even a Southerner of this day can sing it with all the old-time melody and pathos brought out by the singers of a generation ago. Those who sang this greatly loved ballad during the Civil War were the ones who could get every atom of sweetness out of it"; on disarmament: "If wars are ever done away with, their cessation will not be due to sentimental arguments, but to the fact that science and invention may make war so dangerous to everyone concerned that the sheer patriotism of educated people in all nations, plus their common sense, will be universally against the stupid war idea"; on the inventor: "He is the specialist in high-pressure stimulation of the public imagination. . . . [He understands that] restlessness is discontent—and discontent is the first necessity of progress"; on machines and progress: "We have scarcely seen the start of the mechanical age [1926] and after it is under way we shall discover that it is also a mental age as never has been known before. . . . Man will progress in intellectual things according to his release from the mere motor-tasks"; on wealth: "What does the very rich man get? He is always scheming, always suspicious of the men around him. His money is mostly out, invested. Yes, he lives in a fine house, rides in an automobile, and he eats three meals a day. I defy any one to prove that he gets much out of life. Money doesn't make a man happy and it doesn't make a man a good companion."

As we know, Thomas Edison spent most of his career—from the time he was a small business owner in Boston, New York City and Newark, until he became a virtual media mogul—railing in varying degrees of abstraction against the remote legislators, bankers, and "money men" who controlled the vicissitudes of the American economy and did not understand his self-styled blue-collar plight. In the period leading up to the Great War and into the twenties, as his businesses became more complex and his fortunes swung, Edison became quite outspoken on this topic, lashing out at "the men who made the trust laws" as not capable of distinguishing "pig iron from coffins, so far as the producing business is concerned."

Although he had never been the most articulate economic theorist—with Edison, the indignation was gut-level, a conviction about how the world should operate—in these later years, Edison finally drew together his scattershot ideas in the context of other dicta to focus upon repairing the Federal Reserve system.

In so doing, Edison believed (as did his friend and, in this case, enthusiastic cheerleader Henry Ford) he had alighted upon the true "rascal," as he called it, the institution founded in 1863 with the authority to issue and regulate money. "Gold is a relic of Julius Caesar and interest is an invention of Satan," he announced in the New York *Times.* "Gold is intrinsically of less utility than most metals. The probable reason why it is retained as the basis of money is that it is easy to control. Gold is a trick mechanism."

"What are bills and checks?" he went on, rhetorically. "Mere promises and orders. What are they based on? Principally on two sources—Human energy and the productive earth. Humanity and the soil—these are the only real basis of money."

And so, populist Edison in his waning years proposed adding a *true* foundation to improve upon the Federal Reserve's realms of gold ingots, which he considered a *false* and perilously exclusive basis for the national economy. He looked instead to the fruits of farmers' labor. The farmer would be acknowledged as the authentic backbone of our nation via a network of government-built, sectional concrete warehouses located at strategic shipment points across the land, filled with "base commodities . . . the necessities of life": cotton, wheat, barley, rye, hay, oats, buckwheat, flaxseed, rice, sugar, wool, peanuts, onions, soy and lima beans, Porto Rico coffee, linseed, linen, almonds, cheese, bacon, dried apples, raisins, etc., etc. Bituminous coal could also be stored, in tribute to the miner, and as a way to provide him with continuous work as well. This innovation would, of course, require the implementation of coal storage pits precisely 1,000 feet long, 80 feet wide, and 20 feet deep. Under Edison's revolutionary economic plan, the tables were reversed: The sweat of the brow of the American worker called the shots, not the arbitrary rulings of government bureaucrats. The shift would have a ripple effect over the rest of the world, as other countries were forced to adapt to our system and allow different commodities, beyond gold, to serve as backup to currency, thereby avoiding what Edison called "fictitious prosperity." Paper money thus issued, correlating with goods stored in the national warehouse system, would not fluctuate in worth, as it was predicated upon the present cost of actual, consumable foodstuffs rather than the market value of gold bars.

The farming sector of the American economy, financed by its own commodities, would lift a tremendous strain from the back of the Federal Reserve system and restore a healthy measure of equity to it.

In later life, Thomas Edison also honed his speculative faculties as he ruminated publicly about the world of the future, a place he knew he had helped to shape. But unlike earlier prophets who envisioned fantastical forays into science fiction, Edison assumed a measured, statesmanlike view, seeing his function as assisting the generation of today to plan wisely for tomorrow. In many respects, he was remarkably on target as an urban planner. "Automobiles will change every detail of movement in the cities," he told Edward Marshall in late 1926, as part of a comprehensive series of talks, "and thus is presented one of the greatest problems of the modern city . . . we have only barely begun really to investigate the problem of traffic [which] will be one of the greatest of all those which will trouble us as time advances." To help maintain vehicular flow, Edison advocated establishing "express streets" as opposed to "accommodation streets," and underpasses to segregate traffic surges in the great metropolises. Edison anticipated overnight mail and Federal Express when he spoke of "rapid transport of valuable objects by air" as a way to save money; he predicted helicopter shuttles to airports when he

spoke of "roofs of large buildings in our cities as quick access to people travelling by air"; he anticipated the rush hour on Sixth and Lexington Avenue
subways when he envisioned "all the people in those skyscrapers start[ing]
to flow out into the street at approximately the same moment"; he anticipated deficits in urban budgets when he advocated "eliminat[ing] politics
from the management of American cities," and substituting "big business
management" techniques.

When "The Master Mind of the Machine Age" spoke further to Mr.
Marshall (who certainly was at no loss for words when it came to lauding his
esteemed subject, "the prince of inventors"), he waxed poetic yet cautionary
on the byproducts of technological advancement. To keep up with the spectrum of mechanical innovations that had supposedly begun to emancipate
him, modern man must at the same time develop a new, unfettered intellect: "We want no coolies here," Edison warned. "We shall have none."
Progress in ever more sophisticated machines *demanded* "progress of mind."
Only thus will we come to realize that there can never be such a thing as
overproduction. As long as men want new knowledge, new experiences, and
pleasanter and more comfortable surroundings, the inventor's niche was assured. Edison returned to comfortable, reassuring farming metaphors evok-
ing his Port Huron boyhood: "Are not the productive powers of men as wor-
thy of good fertilizer . . . as our grain fields, fruit orchards, and vegetable
gardens?"

Edwin E. Slosson, Director of Science Service in Washington, D.C.,
"The Institution for the Popularization of Science," visited Edison in West
Orange, interviewed him several times between 1925 and 1931 and published "Q. and A.'s" periodically in his *Science Newsletter*. "Comparing the
electrical industry with the life of a human being," Slosson asked the Wizard on one occasion, "an industry which is your own child and therefore was
an infant forty years ago, at what stage of life do you think it now stands?
Adolescence, manhood, middle age, or past?" Edison replied, "Yelling baby
. . . I am convinced we still do not know one ten-millionth of one per cent
about anything."

We have seen that Thomas Edison was equally at home exploring possibilities in metaphysical "realms beyond" as he was in the absolutely grounded
world of material phenomena. He enjoyed transcribing mind-trips to the
outermost reaches of the cosmos as much as conducting exhaustive invento-

ries in the West Orange supplies room. It naturally followed that as he entered the final decade of life, Edison returned with deep preoccupation to thoughts of existence after death. In so doing, he tapped once again into a long, speculative tradition inaugurated by Gottfried Wilhelm von Leibnitz (1646–1716), the Leipzig-born philosopher, mathematician, and statesman, and the foremost intellectual theologian of the seventeenth century. To Leibnitz, the universe was composed of an infinite number of conscious centers of spiritual force or energy, called *monads* (Greek for "unit"), the harmonious result of a divine plan. This construct was appropriated and elaborated by the physiologist-scientist-engineer Emmanuel Swedenborg (1688–1772), one of the most dramatic mystics of all time, whose writings achieved vogue status among the bourgeois of Edison's era. To Swedenborg, God was the Divine Man; in his major work, *Heaven and Hell*, Swedenborg wrote that every entity in the human (natural and known) universe had a *correspondence* or *harmony* in the larger (spiritual) sphere. It was man's responsibility to seek relationships between the finite and the infinite, between the seen and unseen worlds he inhabited.

Faith in this underlying unity was shared by another of Edison's idols, Michael Faraday, who believed in and wrote often on the subject of the interconnection and correlation of all natural forces—chemical, electrical, and magnetic. Energy in the universe was constant, said Faraday. It was transferred, never created anew or destroyed. "We know matter only by its forces," he declared in his seminal 1857 article, "On the Conservation of Forces."

Ralph Waldo Emerson, Edison's other intellectual paradigm, took up the Swedenborgian banner in his series of lectures published as *Representative Men* (1850), speaking with enthusiasm of the linkage of all things in nature, including human souls: "Every material thing has its celestial side; has its translation, through humanity, into the spiritual and necessary sphere, where it plays a part as *indestructible* as any other." To Emerson, in these essays, the true genius of the scientific age was found in the "inventor" who "began [his] lessons in quarries and forges, in the smelting-pot and crucible, in shipyards and dissecting rooms," and then had the capacity to extrapolate therefrom to the "dim spirit-realm." There, the inquiring man was sure to discover that "nature is always self-similar," reiterating herself in "successive planes" of perception, built always upon aggregate "units" and "particles."

Emerson concurred with Swedenborg that the infinity of God could be understood through the essence of one man. Singling out "Napoleon, or The Man of the World" as his metaphor, Emerson appropriated Swedenborg's "man as type for God" thesis to explain the coming into being of this great statesman, the example of what is best about humanity—because in his deepest self, Napoleon carried the spirit of all the people of France.

By the time he actually delved into "etheric force" and came to know the teachings of Madame Blavatsky, Edison was thus already well versed in the monadic interpretation of the Universe. The "Divine Spark," part of the *Logos* that informed all people and things, was clothed in what the Theosophists called "an atom of spiritual matter." In her seminal work, *The Secret Doctrine*, explicitly invoking Leibnitz's writings, Madame Blavatsky stated with emphasis that monads, to the contemporary occult scientist, embodied *both force (energy) and matter*, and thus were "two sides of the same SUBSTANCE." In her subsequent work, *Isis Unveiled* (a copy of which, we will recall, she gave to Edison), Madame Blavatsky defined "the wave motion of living particles" in spiritual terms—as *atomic energy*.

"People say I have created things," the mystical Edison said in an obscure 1911 essay published in the Gary, Indiana, *Gazette*. "I have never created anything. I get impressions from the Universe at large and work them out, but I am only a plate on a record or a receiving apparatus—what you will. Thoughts are really impressions that we get from outside." Several days later, Edison elaborated in similarly transcendental fashion in *The Columbian Magazine*. While he did not believe in a "Supreme Being," calling such an image of "creedism . . . abhorrent and fallacious," he did on the other hand fervently espouse the existence of a Supreme *Intelligence* ("I do not personify it"), which acted as a kind of "Master Mind" informing all singular intellects on the planet. The crux of individuality, to Edison, was to be found on the most rudimentary and essential *cellular level* of man, as a biological construct, a machine with parts; each and every cell in the human "machine . . . governed by unalterable laws" contained within it the essence of a man's personality: "They, not the men and women, are the individuals." This idea had been and would continue to be the locus of Edison's scientific yet simultaneously religious inquiry into human nature.

Under these (again) monadic terms, Edison proposed that the sum total of all life—the *actual matter of organic life*—was constant and immortal in the Universe (a concept parallel to Faraday's notion of immutable energy), even if there were no such thing as a "soul" in the strict theological sense. He reiterated this theory in 1921, in extensive journal entries toward a planned essay on the quarter-inch-long fold of Broca in the human brain, repository of the faculty of memory. (This was the same year Freud's *Interpretation of Dreams* was published in the United States.) As he often did when following a protracted line of reasoning, Edison numbered his insights as they came to him in a series, trying to make one follow causally from the preceding. To Edison, this particular group of brain memory cells housed human personality. After the person died, and the physical body began to decompose, the Broca cells "abandoned [the] unimportant mechanism they had been inhabiting," and then swarmed "free in space," carrying the entire life history of the deceased person. The cells (he also referred to them as "intelligences") had impressed upon them over a lifetime the aggregate of all experiences assimilated by the person in whom they had resided. They might even carry, Edison wrote, "the subconscious mind so-called."

And where, Edison ruminated, did the cells, far below the perceptual limits of the microscope, come to rest? They could feasibly become affixed to things in nature—to trees, say. And if this were the case, could a mother make contact with the personality that had "fled from the body of her beloved son killed at Chateau Thierry [*sic*]"? In other words, was there a "reasonable and possible" way for loved ones to communicate with the spirit of the dead through the "record" left within the Broca cells, *intelligent, educated entities* that had deserted the body? It seemed to Thomas Edison in his ninth decade that there surely must be, since man had been able to "advance more in the past sixty years than he had for the four thousand previous years."

On into the mid and late 1920s, Edison spoke more and more honestly about his obsession with the existence of fragments of indestructible and undying personality that conformed to the acknowledged scientific standard of enduring constancy of life and matter, a truth in Nature that went beyond—but did not, he was careful to stress, contradict—the Bible. If the great sequoia tree could go on for centuries, why not some essence of human personality?

"The reason why you are you and I am Edison is because we have different swarms or groups or whatever you wish to call them, of *entities*," he wrote in 1920, announcing he was working on a sensitive apparatus with which to detect and record the myriad infinitesimal, immortal monads "prowling through the ether of space."

————

Like electrons tracing odd orbits, the six Edison children maintained inescapable proximity to the unwavering nucleus. Errant Oscar Oeser announced his forthcoming marriage to his longtime mistress, Clara Berger, in the summer of 1924, allowing Marion Oeser to close a painful chapter in her life. "I can never be happy here again," she wrote Mina from her *pension* in Freiburg. "I am coming home." While curious to see her brothers and meet their wives, more than anyone else Dot pined for Father: "It is a great trial that I cannot talk to you," she sighed. "I cannot help but think of old days when I *could* talk to you. I often long to hear you read 'Evangeline' again as you did in the days of my young childhood." Edison bought his eldest child a bungalow and one and a half acres in Norwalk, Connecticut, where she settled down to live for the next forty-one years of her life in quietly reclusive fashion, rising every morning "up and doing" before 5:00 (often as early as 3:00), long anticipating the rooster's crow, taking a brief constitutional around the grounds, then beginning correspondence on her lap board, sewing, sipping tea, and reading *House and Garden* from cover to cover. Marion saw few close friends; she shunned parties and receptions, preferring to indulge her sweet tooth with Schrafft's French Mixture chocolates in solitude, going into town nearby for errands twice a week, and making occasional trips to New York City to meet Mina for the opera—although preferring to catch the last train, rarely staying over night.

After the war Tom, Jr., and the devout Beatrice actually moved to West Orange and effected a mild reconciliation with his parents (more Mina's doing than the Old Man's). Tom was given a job in the Research Engineering Division of T.A.E., Inc., where he explored the feasibility of incoming patents with the potential for product development, and did not seem to mind working for Charles. "Poor boy," he said sympathetically to Mina, "I am afraid he has too much on his shoulders. The strain is gigantic." An uneasy peace reigned, interrupted occasionally by Tom's predictable bouts of vague illness.

With William Leslie and Blanche, the tensions ran deeper. Remaining on their farm in Wilmington, they were permitted infrequent access to Glenmont, again through Mina's efforts, bringing produce and freshly killed chickens as offerings. Unfortunately, Will made another one of his risky forays into technology, investing in a small radio shop. The ancient fear of tarnishing the Edison name rose up, and father erupted. "I thought I received several jolts over in France worth taking notice of," Will wrote, "but they were nothing compared to this. . . . Dad, just what is your reason for trying to throw a wrench in the cog wheel of my honestly built and honestly arrived at radio set?" The unrepentant Will never seemed to learn.

Madeleine and John Sloane, endlessly occupied with their three boys, Tom, John, and Peter, settled in at Tillou Road in South Orange, and still fought to make ends meet; the strong and sometimes feisty Madeleine had trouble asking her papa for financial help. With much agony, copying the letters over and over again and sharing drafts with her mother, she did finally register a plea, and with relief received an open-ended loan to help pay off their bank mortgage more quickly.

Madeleine worked equally hard at keeping a measured distance from her step-siblings, snapping off advice to Mina on how to accommodate Marion, Tom, and Will properly, without risk of igniting Papa's wrath. She also had some telling observations about the newlywed Theodore and Ann (who as of April 1925 had moved to East Orange after a honeymoon on Monhegan Island) finding the couple irritatingly "self-sufficient and egotistical." Madeleine resented their orderly, childless lives; the free time they enjoyed at the latest Charlie Chaplin movie or Anna Case concert; the incessant "busy bee" behavior around their neat apartment; the hours Theodore spent arranging, rearranging and "resorting and sifting" his files, correspondence, and technical magazines, duly assisted by his admiring wife.

Theodore's intellectual prowess seemed to threaten Madeleine, as if he were cut from wholly different cloth than Charles—who as the years passed she grew to admire no end, probably because, in *his* turn, Charles empathized with Madeleine's proud nature, intuiting that "when you go the least fraction of an inch below the surface, she love[d] [Mina] to the utmost." Madeleine could be sharp at times, but she was a faithful, appreciative daughter, a quality that served to draw her closer to Charles.

And what of the Edison rock and foundation, the team player, the good sport counterbalancing his bookworm kid brother? Charles himself had what we might call a "there but for the grace of God" philosophy. As he admitted to his revered mother, "If not for your teaching and letters, I would have followed Tom and William as naturally as going to sleep."

"Do I contradict myself? / Very well, then, I contradict myself"—Walt Whitman's words, but just as easily Thomas Edison's. As the machine age arrived, America began, as Lewis Mumford said, "to make a genuine culture out of industrialization." The *modern era* became more than a philosophical construct—it was a shared state of awareness. "America has again become a new world," enthused the French social scientist André Siegfried in the streamlined, "neotechnic" revolutionary mood of the here and now.

At eighty, partly in belated recognition of commercial realities, and partly because he was still, as ever, starved for something new, Thomas Edison turned his back on the turbine-driven momentum he had worked his entire life to shape. Throughout his long career, had he not made it a habit to drop a "biz" when it no longer captured his imagination? The final transition was upon him, and it would be his most dramatic. He went back to the land, to find, of all things, a natural source for rubber.

True to the character of all his endeavors, Edison's latest venture had a long gestation. At his meeting with Luther Burbank in California in 1915, the men had got to talking about America's dependence upon foreign markets to meet its insatiable demand for rubber—an astonishing 75 percent of total world production. They were joined by Henry Ford, the automobile man, who understood the problem all too well; a shortage was certain to loom as the car became more popular. (Ford fulfilled his own prophecy; with the shift a decade later from high-pressure—resembling bicycle tires—to balloon tires, rubber demands at home escalated even further.) And "The Rubber King of Akron" himself, Harvey Firestone, agreed, a veteran in the business since the turn of the century. During the war, military blockades had been decisive in turning the tide when Germany was prevented from importing rubber and was unable to develop a synthetic source. Rubber was the subject of several campfire discussions between Ford, Firestone, John Burroughs, and Edison during their 1919 camping trip, and Firestone was utterly surprised by the thought and exhaustive botanical study his inventor colleague had been quietly devoting to the matter.

"America should grow its own rubber," Firestone announced, and founded a 1,100-Hevea-tree plantation in Liberia on land abandoned by the British. Not to be outdone, Ford, who had an equally high stake in the commodity, purchased a 17,000-acre plot on the Tapajos River plateau in the Amazon Valley of Brazil and stocked it with imported seedlings from Malaysia and Ceylon. The Intercontinental Rubber Company cultivated the guayule variety in Mexico, imported at a somewhat cheaper level than Hevea. These private, corporate efforts led to small, experimental government-funded plantings in Haiti and Panama that were given added impetus by the British Rubber Restriction Act, which drove prices ever upward and underscored the perilous extent to which America remained too dependent upon outside sources.

Into the fray stepped Edison, with a different angle of vision. First of all, he had been inspired by Luther Burbank to think more broadly about organic alternatives—surely there must be other plants, *on native ground*, that could provide rubber. Secondly, Edison's concern, unlike his friends', was not to become a venture-capitalist in the rubber industry. Those days were gone. Patriot to the end, he wanted to find a plant that could produce "prolific" (his word) supplies in the event of another emergency, as a contingency against the danger of curtailed foreign rubber flow due to crop disease or wartime.

In a limited way, and at first for observational purposes, Edison had been nurturing plots of familiar milkweed, guayule, and Madagascar vine on his Fort Myers estate during the mid 1920s, by which time there was already a documented rubber shortage in America; Lee County's tropical climate, rich soil, and available acreage were identified by the United States Department of Agriculture as ideal for rubber cultivation. Edison tinkered with a vacuum pump for the speedy extraction of latex from the *Ficus elastica* tree, and had lately taken to carrying around a chunk of the resulting crude rubber in his pocket to show to visitors. On Sunday afternoons the Old Man corralled his chauffeur, Sidney Scarth, and the two would meander down back lanes in the big, black Lincoln touring limousine. Every few minutes, Edison tapped on his driver's shoulder, the car pulled over, the Old Man lumbered out, drew his favorite lightweight knife from his pocket, and cut a specimen of leafy stalk from the roadside for analysis back at the lab.

By the time of his eightieth birthday—and birthdays were always opportune moments to make announcements—Edison was ready to act globally. "Bubbling over with enthusiasm" and expressing confidence his experiments with rubber plants would be a success, he took a group of newspaper men on a guided tour of his three-acre cultivation field, and announced with great fanfare the establishment of the *Edison Botanic Research Corporation*, financially underwritten by those champions of progress, Henry Ford and Harvey Firestone. Within the next couple of years, the "Wizard" proclaimed, he would drive out the front entrance of Ford's Detroit plant in the first car ever equipped with tires made of domestic rubber. By early March, a *Tropical News* story had hit the Associated Press wire services, "from New

Orleans to Duluth, from Miami to Vancouver, Portland to San Diego":
Thomas Edison was on the move once more! It was great copy, and, what's
more, it was good for America.

Beyond public relations, Thomas Edison's approach to the problem
was elegant and systematic. He began with the decision to apply himself to
collect and examine as completely as possible "the various species of plants,
bushes, shrubs, etc., which contain a rubber-bearing latex." To accomplish
this, he engaged "field men," as he had done with the search for a bamboo
filament and subsequently in the quest for ore fields. These assistants, in the
early stages, were provided with Ford roadsters and deployed up and down
the east coast, from New Jersey to Fort Myers—that is, from one Edison
laboratory to the other. Promising specimens and cuttings were shipped to
Fort Myers for on-site cultivation. As soon as he returned to West Orange in
late spring, Edison himself set forth on a reading binge. He acquired L. H.
Bailey's *Standard Cyclopedia of Horticulture*, three huge volumes weighing in
at 3,639 pages. Writing a bold note to himself in the flyleaf to "Read for
Rubber," Edison leafed through three times, page by page, marking profusely in pencil the names of plants which held "milky juice" or "sap."

The next stage of the effort was to canvass every botanical garden and
herbarium in America. Through contact with experts in the discipline, Edison intended to educate himself at first hand on every variety of plant with
the potential to provide latex and/or rubber. Naturally, he wrote first to Dr.
Marshall A. Howe, assistant director of the New York Botanical Garden in
the Bronx, enlisting the help of the preeminent repository in the country. In
mid May, Edison sent W. H. Meadowcroft on a preliminary research trip to
the Garden. The reception was enthusiastic, and led to Edison's visit one
month later, specifically to observe the latex-bearing *Ficus* and *Euphorbia*, of
which there were seventy-two varieties growing in the Garden's greenhouses.

Edison settled into a makeshift lab prepared for him in a temporarily
commandeered ladies' lounge converted by obliging Garden administrators.
He spent several days sketching, scraping bark, dissecting pith, grinding
pulp in a coffee mill, measuring residue, mixing chemical solvents, squeez-
ing sap from stems and rubbing the exudation between his thumb and fore-
finger, and taking notes for the compilation of what would eventually
amount to 530 laboratory notebooks devoted exclusively to "Rubber Studies." Over lunch, Dr. Howe introduced Edison to Dr. N. L. Britton, director
of the Garden, and to Dr. John K. Small, curator of the herbarium who, it

turned out, was a widely published specialist on plant life in Florida, devoting two or three exploratory trips a year to the state, finding it to be "the most interesting and prolific plant area in North America." Discovering delightful common ground, the two men became immediate friends and correspondents over the ensuing four years. The following winter, Dr. Small took a leave of absence and enjoyed a month of his colleague's hospitality as "botanist in residence" at Seminole Lodge.

His trip to the Garden—and a subsequent one three weeks later—were a tremendous inspiration to Edison, and caused him to enhance and professionalize his research efforts. On the basis of Dr. Small's willing expertise, and the inspection of Garden specimens, Edison decided to focus his East Coast field men, at least for the time being, on collecting *Euphorbia* (common spurge), with the addition subsequently of *Asclepiadaceae* (milkweed) and *Apocynaceae* (dogbane) in particular. Simultaneously broadening his range in the rest of the country at large, Edison enlisted section foremen of the Union Pacific Railroad to make collections of locally available plants along the right of way. With this dramatic expansion of territory—extending eventually to forty states, as well as Canada and South America—Dr. Small also advised Edison on the all-important proper methods for collecting, packing, shipping, and chemically preserving live cuttings from the field in order to avoid deterioration, rot, and mold in transit.

With an eye to scientific accuracy, Edison decided at this time to establish his own herbarium at West Orange, appropriating the same rigorous labeling, nomenclature, and archival standards as were employed by the New York Botanical Garden. As samples came in, they were immediately identified and tested against the data at Harvard's Arnold Arboretum and the state herbarium at Rutgers. The samples were catalogued in a card-file project paid for by Henry Ford and twice validated after Edison's death by the United States Department of Agriculture and the United States Forest Service. More than 13,000 cuttings were identified as to genus—narrowing down to 2,222 plant species, classified in 977 genera and 186 natural plant families. Of these, thirty families, or 16 percent of all the natural plant families included in Edison's tests, each contained 2 percent rubber or more— Edison's cutoff point for viability. The Edison Herbarium, housed in custom-crafted wooden cabinets in concrete, temperature-controlled, dust-free Vault 32—the former Disc Masters building at the Edison National Historic Site—today still represents the largest group of plant samples ever collected and tested for rubber by any investigator.

The testing procedure was as follows: the specimen was heat-dried, ground up, and treated with acetone to remove the nonrubber ingredients of the plant, precipitated as resin. Then the sample, further extracted with benzol, was left exposed in an open dish until the chemical evaporated. The resulting compound was designated "rubber" and presented to Edison personally for his inspection—as many as fifty samples per day. He recorded the physical characteristics insofar as color, physical nature, and elasticity. If the properties were favorable, Edison instructed field operatives to return to the area where it had been harvested and obtain more quantities for further examination and possible cultivation at Fort Myers. (In later years, the acetone step was abandoned and replaced with more precise bromination for plants already known to contain some amount of rubber, which essentially meant the addition of carbon tetrachloride and alcohol, the resulting solution washed through a filter more rapidly than previously and with a more accurate percentage result.)

The impact of Edison's radical shift in attention was immediately felt throughout the laboratory. Tom, Jr., was right—the Old Man's new cause célèbre did move responsibility for every remaining operation over to Charles's plate, and Theodore's, too, for he had joined the factory team and was "running ragged checking up on cement things . . . taking to it like a duck," his big brother thankfully noted. To Papa, Charles admitted in a rather rueful memo, "I have come to the conclusion that you really do want to concentrate on rubber and not bother with the details of the business." The boys decided for their own good to stop sending reports to the Old Man as frequently, since he had stopped looking at most of them.

A sense of high urgency, at times desperation, was palpable to all who came in contact with Thomas Edison. "Just think," he remarked to Madeleine, "it takes ten years to study one family of plants, and here I am racing with the angel of death all the time." More than science was at stake. "This is the most complicated problem I have ever yet tackled," he told Frank Parker Stockbridge, the editor of *Popular Science Monthly*, saying with a touch of the old insatiable appetite that he had "only started on a task which may yet become the crowning glory of [my] career."

In the new year of 1928, preparing for his longest stay ever at Fort Myers, Edison showed to what lengths he was willing to go in his labors. Along with an entire railroad car full of new equipment, a force of carpenters, electricians, and plumbers descended upon the 42-year-old McGregor Boulevard laboratory in the center of what Edison affectionately called his "jungle," one month before the Old Man's arrival. The sound of pounding hammers and the roar of engines filled the air. The building received new beams, a coat of fresh paint, a new drying room equipped with gas burners, and an electrical stove. A pristine white wooden fence topped with two strands of barbed wire went up around the property, adding to the impression that high-security work was going forth. Fred Ott came down from New Jersey to supervise the proceedings and brought William Benney, the superintendent of the West Orange botanical operation, along with seven additional chemists and aides.

"I'm here to work on my rubber experiments, and nothing else!" Edison told the cheering crowd of fifteen hundred townspeople who gathered to welcome him at the Atlantic Coast Line station in Fort Myers, while a brass band played. Hatless, revealing "the impressive skull which has hatched more ideas than any other of this age," Edison took the hands of his little niece Nancy and nephew Stewart, brother-in-law John Miller's daughter and son, and they smiled and waved to all from the back seat of the sedan.

To begin anew—it was exhilarating, like Menlo Park all over again.

CHAPTER
28

AS IF TO REDRESS THE BALANCE OF YEARS LOST AT THE TURN of the century when he had been sequestered in the gray gloom of North Jersey iron mines, Edison spent more time on the grounds of Seminole Lodge, rising before sunrise to sit in a rocking chair on the broad veranda, sip the first glass of milk (his self-imposed dietary regimen called for one every three hours), puff on a clandestine cigar, and listen to the sound of the Caloosahatchie tide lapping at the wooden pilings of his fishing pier, and the "clack-clack-clack" of the big-veined coconut palms. Daylight expanded, and the climbing bougainvillea tendrils' red blossoms stood out against the deep green of the rubber trees. Shadows of mango branches shrouded the jasmine flowers. Oleander and dwarf palm drooped against each other in the morning sun. Wind off the water shook ripe coconuts, yellow grapefruit, and miniature oranges to the dew-sparkled lawn.

Near seven o'clock, Edison slowly rose and walked to the boulevard over flat stones set into the turf, pausing briefly at the white gate before crossing to the laboratory. He used a cane with self-consciousness; at this early hour, thankfully, no one would see him.

To Emil Ludwig, the visiting German biographer of Napoleon, his stoop-shouldered host was Faust, and Mina his Marguerite. Dressed in dark velvet, gentle, brown-eyed, and smiling, she called Edison "the Patient Impatient" in a low voice as beautiful as that of Eleonora Duse. Ludwig saw Mina Edison's tolerance and tact and heard the clarity and humility with

which she spoke of her husband. On issues of the day about which she felt strong convictions, Mina displayed an unflinching selflessness, throwing her efforts behind Presidential candidate Herbert Hoover and his current prohibition campaign, in memory of her mother, the original temperance crusader; half a century earlier, Mary Valinda had come home drenched by angry buckets of water thrown at her from the saloon doors of Akron.

I took a writer with a European sensibility to compare Edison to an artist. Emil Ludwig cast aside the "laughing philosopher" persona, to find instead a man more truly resembling the aging scholar Goethe, who had once declared, "It was always the same to me whether I made plates or pots." Ludwig saw in Thomas Edison "a man who runs away from his own inventions"; beyond the sheer number of patents was the far more significant thrill of their *making.* And what could be more characteristic of the artist's psyche than the current passion, the suddenness of the rubber quest?

The three went for a drive in the country and stopped for lunch. Mina and Ludwig went alone into the inn while Edison, on an exclusive milk diet, remained alone in the back seat, waiting and withdrawn, thinking quietly, long, scarred hands cupped upon his knees. Within minutes the car was surrounded by a pressing crowd, eager for a glimpse of the Old Man, a few words from the "secret king of America," the workaholic who had no intention of quitting until "the day before [his] funeral."

In the fall of 1928, this "great national asset" (so-called by *The Satur-*

day Evening Post, arbiter of mainstream mores) received the Congressional Medal in a pomp and circumstance ceremony conducted at West Orange, with national radio hookup to the White House, so that President Coolidge could participate: "Noble, kindly servant of the United States and benefactor of mankind, may you long be spared to continue your work and to inspire those who will carry forward your torch," the President said, "Few men have possessed to such a striking degree the blending of the imagination of the dreamer with the practical, driving force of the doer. . . . Although Edison belongs to the world, the United States takes pride in the thought that his rise from humble beginnings and his unceasing struggle to overcome the obstacles on the road to success illustrate the spirit of our country."

"So fast and varied have been his contributions to its use that there are some men who even believe that electricity itself is merely another one of Edison's inventions," added Secretary of the Treasury Andrew W. Mellon in metaphorical remarks. "Edison is set apart as one of the few men who have changed the current of modern life and set it flowing in new channels. They belong to no nation, for their fame, no less than their achievements, transcends national boundaries."

"Won't you let us go into radio?" Charles pleaded with the Old Man at the West Orange train station as his father left for another extended Florida sojourn. Edison turned, and with sudden vehemence, retorted in his high, thin voice, "Well—if you want to be a damn fool, go ahead. You've got my permission, but I'm telling you it's no good!" Charles was relieved, as was Theodore, both eager to take their place on a bandwagon the Edison interests were already far too late in riding. In January 1929, T.A.E., Inc., took over the nearly bankrupt Splitdorf-Bethlehem Electrical Company of Newark, already licensed to manufacture radios and radio-phonographs, as a necessary shortcut to avoid negotiating with RCA for patent rights. Simply assuming Splitdorf's license fees cost the Edison Company more than $100,000 per year. Through streamlining, layoffs, selling the company's languishing inventory (magnetos, spark plugs, generators), and the critical move transferring assembly-line processes out of Newark and into West Orange, the fired-up team of Charles and Theodore managed to retire almost half Splitdorf's debt within one year of the merger. Despite the brothers' alacrity, the new Edison Light-O-Matic radio in its big, ornately carved cabinet was born as a $1,000 dinosaur compared to the versatile "little boxes" that were now invading American homes. The Edison radio was an eventual loser, surviving on the market for only eighteen months. Perhaps a better direction would be Edicraft appliances for the modern kitchen?

Theodore and Tom, Jr., reconvened with West Orange engineers and went back frenetically to the drawing boards, while the Old Man continued revels of fame in Florida. He watched neighbor Henry Ford's outdoor square-dancing and folk-music festivals; and "like a proud father" welcomed President-elect Hoover to Seminole Lodge, entertaining his lofty guest with a boisterous salute by the 116th Field Artillery, a ride through town in an open car followed by the Fort Myers Vocal Chorus rendition of "I Can't Give You Anything But Love, Baby," and a private cruise in search of sailfish. For the first time in all the seasons he had been in Fort Myers, Edison was present for the high school commencement. Mrs. Edison gave the featured address, and the Old Man presented each of the seventy-two graduates with a diploma and a smile. Topping off Edison's joyous spring, a special two-cent stamp commemorating the fiftieth anniversary of the incandescent lamp was issued. And in the spirit of the day, George M. Cohan, himself a native son of Orange, New Jersey, composed a new song, "Thomas A. Edison, Miracle Man," performed at the annual convention of the National Electric Light Association in Atlantic City:

> *What a fame, what a fame, what an aim, what an aim, to live for mankind*
> *To give for mankind*
> *All the joys of living, all he's got he's giving,*
> *Work away, work away, night and day, never play, yet thoughtful and kind*
> *If America needed a king on a throne*
> *Mister Edison sits on a throne of his own.*
> *Oh say can you see,*
> *By the light that he gives you and me*
> *What a man he is*
> *What a grand old "WIZ"*

In Edison's summer 1929 notebooks, as he narrows down the viable prospects for rubber, we detect increasing references to breeding the *solidago rugosa* and *leavenworthii,* two varieties of the goldenrod plant. He sent word up to West Orange ordering two harvesters, George Hart and John Osborn, to cut down as much of the tall weed as possible from two fertile areas in nearby Little Falls and further west, at Chester. The goldenrod was subjected to a further-refined extraction process aimed at efficiency and economy through conservation of materials: drying the leaves at lower temperatures than previously to prevent oxidation; enclosing the resultant powder in sealed cans to prevent reabsorption of moisture; recycling acetone and benzol by slowly pumping their carbon-dioxide driven vapors out of the percolator, and then capturing and condensing them. The remaining coagulant residue gave desired rubber—at least, for the moment. Perhaps yet another "redissolving" step needed to be inserted? "I have not yet gone far enough to find out if this scheme is practicable," Edison wrote, setting up yet another goal just out of reach, another mechanism for survival—all Edison needed to do was redefine his definition of success on a daily basis, and he could keep going, ideally forever. For the cadre of "weed-stalkers," Edison even developed a way to field-test plant leaves out in the forest or swamp, using the tailgate of a truck as a table top, and illustrated the components of the kit in a schematic drawing.

Edison kept scribbling until mid July, when it was time to leave with Mina for the Lewis Miller Centenary at Chautauqua—the one hundredth anniversary of her father's birth on July 24, 1829. A three-decade, uneasy truce had been effected between the Millers and the Vincents since the bishop's moving eulogy at the funeral of young Rough-Rider Theodore, and George Vincent's lovingly published book in tribute to the lad who died so valiantly in Cuba. Every year at Christmas until her passing, John Vincent wrote a cordial letter to Mary Valinda, taking care to pay homage to her "noble husband . . . and the pleasures and friendships of dear old Oak Place. . . . I don't want to drop out as one of your memories and friends." When Bishop Vincent died on May 9, 1920, the obituaries remembered him as founder of Chautauqua, rekindling dormant embers in Mina's breast. In the several months leading up to the centenary, Mina pressured Arthur Bestor, president of Chautauqua, to plan a series of special events for every day of the week commencing July 21. Dr. Bestor complained that George Vincent would not be present, and neither could Harry Emerson Fosdick or Chester Massey, son of Hart A. Massey, one of Lewis Miller's old colleagues, or John Finley, a venerable member of the Chautauqua Press and another Miller

WITH CHAUTAUQUA
PRESIDENT ARTHUR E.
BESTOR AT CHILDREN'S
DAY, SUMMER 1926

ally. Bestor passively asked Mina for any other suggestions of men to invite
who knew her father. She became deeply discouraged, renewing her confidential suspicion to sister-in-law Louise that Bestor "never thinks much of
the idea that Father comes before Vincent."

By the time the great festivities arrived, determined Mina had managed singlehandedly to round up quite a constituency to share the platform
with her, including Mr. and Mrs. Ford, Mr. and Mrs. Firestone, Mr. and
Mrs. Adolph Ochs from New York City, Dr. Elmer Sperry, and Mrs. Anna
Studebaker Carlisle from the old Akron circle. There was a grand reception
hosted by the Bird and Tree Club in honor of the children and grandchildren of Lewis Miller, where "Auntie Ga," as Mina was called by the little
ones, felt securely at home. Mr. Alvin I. Findley, editor-in-chief of *The Iron
Age Magazine*, delivered a rousing speech in tribute to Edison the industrialist. Mary Miller Nichols, Mina's sister, spoke at a meeting of the Chautauqua Women's Club. Albert Stoessel conducted the symphony orchestra
in a broad repertoire of Brahms, Debussy, Berlioz, Wagner, Gluck, and
Handel. The huge amphitheater echoed with a sacred song service attended
by six thousand worshipers.

Chautauqua had come a long way since its early era as the inspiration of Lewis Miller, son of German-Bavarian farmers, father of eleven children, inventor, churchman, educator, "constructive radical and Christian idealist." During that celebratory summer of 1929, more than twenty satellite Chautauquas thrived across the land, conducting fifty-one circuits, each in turn representing scores of spinoff tent meetings, for a few weeks "changing parish to planet." Thousands of cities, towns, and villages were touched by Lewis Miller's informing vision.

The round of photo sessions, receptions, and hand-shaking took its toll on the Old Man. He returned to West Orange in August suffering from what he assumed (or pretended) was a cold, and tried to work through it in the lab. Pneumonia developed, and Edison was confined to bed for three weeks, recovering his strength with barely enough time to spare before a trip via Port Huron, where a riverside park was dedicated in his name, and then onward to Detroit, where he was expected to preside as the centerpiece of "Light's Golden Jubilee," an elaborate historic milestone concocted by General Electric and Henry Ford.

Henry Ford was responsible for the creation of one of the most oft-quoted clichés of the twentieth century: "History is more or less bunk." This unfortunate sentence was uttered in the course of courtroom testimony in the summer of 1919 during a million-dollar libel suit brought by Ford against the Chicago *Tribune* in retaliation for that newspaper's branding him an "ignorant idealist" and an "anarchist," because he was outspokenly opposed to United States involvement in border disputes with Mexico. Fourteen weeks of detailed questioning by the *Tribune*'s adept attorney, Elliott G. Stevenson, sought to establish Ford as pitiably naive, ill-educated, and unpatriotic. It became clear that the Flivver King was ignorant of most basic textbook facts: the fundamental principles of government, the causes of the Revolutionary War, the identity of Benedict Arnold, and so forth.

Twenty years later, Ford was still trying to clarify what he meant by that infamous statement. He was really talking about—had *always* been referring to—the history *books* that "weren't true. They wrote what they wanted us to believe, glorifying some conqueror or leader or something like that." In the intervening decades, Ford proved his point to the degree of overkill. He sought the artifacts of history, the things themselves that really mattered, documentary evidence that made American heritage come alive,

EDISON AND
HENRY FORD ON
THE RADIO AT
CHAUTAUQUA,
SUMMER 1929

the plows and furniture and milk pails and butter churns and china sets and flintlock rifles—not reductive dates and places on a page, kings and queens, diplomatic initiatives—and by God, he had the wherewithal to drive the message home!

Ford started with his birthplace in Dearborn, threatened with demolition by a road construction project in 1920. He moved the entire spread two hundred feet to the east and restored the farmhouse and barns, furnishing them with period pieces obtained by ransacking the inventories of every antiques shop he could find in the state. He went on an antique-collecting binge, ordering all Ford dealers to do likewise, and rapidly accumulated what Ford scholar Reynold Wik aptly calls "a mountain of miscellany," tokens of Americana large and small.

By the mid-twenties, warehouses and old tractor barns overflowing, Henry Ford defined a more ambitious and egocentric concept. He would establish at Dearborn a tripartite institution to tell the three-dimensional story of America: there would be an eight-acre Industrial Museum (thus originally named, but changed to The Edison Institute by the fall of 1929) of inventions and artifacts on display, every kind of article used in America since the Puritan era; an actual town, at first called The Early American Village, dramatizing our nation's most memorable structures, based upon the village green layout, with restored landmark buildings, workplaces, and homes of great men; and a Scotch Settlement School where children would learn according to Henry Ford's philosophy, by direct experience and perception of the places and things before them.

And what better centerpiece for this gargantuan Institute of Technology than a reverently reconstructed shrine, a replica of the entire Menlo Park complex in tribute to his friend and mentor, Thomas Alva Edison? Ford tracked down a blacksmith on Hendry Street in Fort Myers named

Wilbur Ross, who had purchased old, abandoned tools from Edison six years back—drill presses, lathes, engines and other machinery, much of it used for incandescent light and phonograph experiments in the 1880s. Ford paid Ross an attractive price and had the stuff crated and shipped to Dearborn, to join his most recent acquisitions, two New England plows. Mr. Edison expressed no nostalgic interest in keeping rusted machines and chemical bottles, and these materials left town to be included at Dearborn without so much as a whisper. One summer later, Henry was back with a vengeance. This time, he had his eye on the original Edison laboratory workshop at Fort Myers, scarcely more than a garage outbuilding almost obscured by vines and shrubbery. The Fort Myers chamber of commerce and the city fathers protested this was going too far, and Mrs. Edison was *said* to be quite peeved at the idea (although she did not make a public protest)—all to no avail. Two more years passed, and Edison built his new, larger lab for rubber work. Henry took possession of the old, vestigial structure, and had it disassembled, board by board, and loaded on flat cars. In October 1928, at Ford's insistence, Edison and son Charles accompanied him out to Menlo Park to supervise salvaging original bricks still remaining among ruined debris and charred timbers. These shards, too, were sent to Dearborn, along with seven railroad carloads of red Jersey clay to provide an authentic foundation. From the New York Edison Company came Jumbo dynamo Number 6, the last remaining of the eight units once operating in the old Pearl Street Station.

Henry Ford accelerated the Dearborn construction pace after he announced that the opening and dedication of his village and museum would coincide with the fiftieth anniversary of Edison's breakthrough on the incandescent lamp, the Golden Jubilee, scheduled for October 21, 1929. A procession of buildings, whole and in parts, arrived with great fanfare: the 1854 General Store from Waterford, Michigan; the splendidly colonnaded but dilapidated Eagle Tavern from Clinton, Michigan, an original stagecoach stop on the Detroit-Chicago road; the 1832 Loranger Grist Mill, from Stony Creek, near Monroe, Michigan; the 1840 Logan County Court House, site of Abraham Lincoln's earliest law practice; and the 1858 Smith's Creek Grand Trunk Railroad Depot, where, once upon a time, "Little Al" Edison and his chemistry paraphernalia had been thrown from the train.

Ford brought in a replica of the Detroit workshop where he built his first car in 1896; the 1855 Trip Saw Mill from Lenawee County, Michigan; the New Haven, Connecticut home of Noah Webster; Wilbur and Orville Wright's bicycle repair shop from Dayton, Ohio; replicas of the homes of

Edgar Allan Poe, Walt Whitman, and Patrick Henry; an Ackley Covered Bridge from southwestern Pennsylvania—and finally, in 1944, three years before his death, Ford moved his own birthplace to Greenfield Village, cutting the house into two neat halves and hauling it in by truck.

Construction on the vast museum building to house the two million dollars' worth of artifacts Ford had by now amassed did not begin until April 1929. The 700-foot-long facade went up as a reproduction of Independence Hall in Philadelphia, complete with bell tower, and two extended wings representing Congress Hall and Old City Hall. The completely enclosed museum structure, supported by 180 columns, boasted the largest inlaid teak floor in the world—350,000 square feet in all—and had no balconies and no basement, following Ford's insistence that he be able to "see everything" in one sweep. It was not completed for nearly another decade, by which time the entire saga of farming, manufacturing, and transportation in America was told through the relics and icons that had aided man's manual labor. Here were farm steam engines, threshers, all manner of conveyance from stagecoach to buckboard to Model T, as well as every conceivable manifestation of domestic life: beds, grandfather clocks, quilts, and rocking chairs.

The Museum and Greenfield Village served as Ford's refuge from the pressures of his business and the noise of automobile production at the River Rouge plant. Rejecting the past as defined by scholars, Ford demonstrated he had no tolerance for highbrow culture or social change, instead forging a personal model of what the British historian Eric Hobsbawm has called "invented tradition": a fixed, catalogued codification of older, safer, *purer* values in defense against the "shifting landscape" of modernity. He went about the task—for him, a delight—of capturing and then reconstructing rural America, on his own terms, through a stubborn faith that was his most important point of consonance with Edison.

And so, Light's Golden Jubilee and the new, clean, detailed version of Menlo Park, from Mrs. Jordan's boarding house to the machine shop to the laboratory and office buildings and even the little shed where carbon for filaments was scraped together out of lampblack, all went forward with Thomas Edison's benign approval. But within the Edison family, the elaborate effort was privately met with skepticism. "It became just a publicity thing, and we didn't care for it," said Madeleine in later years, feeling that

Papa just passively let himself be "taken in on these things." Mina had a quiet mistrust of Mr. Ford, his "advertising" instincts, and his ostentatious displays of money and material trappings, going back to the much touted camping trips with Harvey Firestone, which she described confidentially to Theodore as "two rather shrewd men using Father for the newspapers." She enjoyed sharing the Fords' comfortable and luxurious way of life, but it was not for her, and she tired of the endless litany of their worldly possessions, constantly on display. To Mina, money could *not* buy everything, and she hungered for some spiritual side to Mr. Ford to help offset her suspicion that he was exploiting his relationship with "Dearie."

On the Saturday night before the Jubilee was set to begin, Mina set "mixed feelings" to paper in the quiet of her bedroom at Fair Lane, the Ford mansion. She did not approve of "disturbing" historical sites, felt that it had been wrong to move the little laboratory from the grounds of Seminole Lodge, and now, saw keenly the irony that corporate cosponsor General Electric was making such a tremendous fuss over Edison, coopting his presence as a symbol, especially considering the inauspicious circumstances under which her husband had long since parted with the company. Menlo Park, after all, represented a part of Edison's life that predated his marriage to her. "Dearie is so much more than the electric light," she wrote. To Mina, Llewellyn Park should be glorified as "the real shrine," in the here and now.

October 21 dawned in cold, driving rain and fog. Joining Mr. and Mrs. Edison on a special, wood-burning train from Detroit to the Smith Street Station replica were Mr. and Mrs. Ford, and President and Mrs. Hoover on their first trip out of Washington since Hoover's inauguration. Edison picked up a basket of fruit, magazines, and newspapers and walked down the center aisle of the train as he had done as a news "butch" sixty-seven years earlier. More than five hundred invited dignitaries then toured the Greenfield Village buildings, shuttled from one to the other through the mud in covered, horsedrawn carriages.

That evening, in the still unfinished foyer of Independence Hall, a formal, candlelight dinner was held. The three children of Mina and Thomas Edison were there with their spouses, but Marion, Tom, Jr., and William Leslie were conspicuous by their absence. Marie Curie, John D. Rockefeller, Jr., Charles Schwab, Adolph Ochs, Walter Chrysler, George Eastman, Will Rogers, and Julius Rosenwald were joined with a national and international radio audience listening to a special hookup with Albert Einstein, who was measured in his praise, avoiding the hyperbole of the day in deference to the scientific long view: "The great creators of technics, among whom you are one of the most successful, have put mankind into a perfectly new situation, to which it has as yet not at all adapted itself."

Owen D. Young, chairman of the board of General Electric, remarked to Edison that "the same unconquerable will and unquenchable fire which took you out of [the Smith Street] depot, and set you on your career, brings you back to it today," and read toasts in tribute to Edison from the Prince of Wales and German president Paul von Hindenburg.

"When you honor me," Thomas Edison responded with his usual brevity, in words drafted for him by Charles, "you are also honoring the vast army of workers but for whom my work would have gone for nothing."

The apotheosis came after dinner, when Edison, Ford, and Hoover took carriages over to the dim Menlo Park laboratory building and ascended to the second floor where, among rows of glass jars and ancient instruments, Edison and his old assistant, Francis Jehl—a mere lad of twenty when he first met the Wizard, and now on Ford's payroll—reenacted the lighting of the sacred lamp. Announcers Graham McNamee and Phillips Carlin of NBC radio set the mood for listeners, asking them to sit by candlelight or gas flame while attending the ritual. "But here is Mr. Edison, again," said McNamee. "While he was at the power house, Mr. Jehl sealed up the old lamp, and it is now ready. Will it light? Will it burn? Or will it flicker and die, as so many previous lamps had died? Oh, you could hear a pin drop in this long room." Becoming impatient, leaning back in a hard wooden chair, a replica of the original from the old laboratory, Edison, hard of hearing, asked what all the fuss was about. "Light 'er up, Francis, light 'er up!" he urged, as the small group gathered about the vacuum pump. "Mr. Edison has two wires in his hand," breathed McNamee, "Now he is reaching up to the old lamp. Now he is making the connection. *It lights!*" The entire laboratory building was bathed in searchlights. Applause rippled through the crowd at Independence Hall, and a replica of the Liberty Bell pealed in the darkness, as sirens and whistles blew through the city of Detroit, and planes flew overhead.

Meanwhile, back in West Orange, T.A.E., Inc., was on that same day in the process of shutting down its entire domestic phonograph recording operation, a decision that had been in the works since the summer. Three days later, on October 24, the stock market crashed. On October 28, all phonograph manufacturing ceased, and sixty-two employees at the plant were let go. On November 1, the last Diamond Disc Record Release List was published, including Stephen Foster's classic, "Oh! Susanna" and a fresh hit by Frankie Marvin, "She's Old and Bent (But She Just Keeps Hoofin' Along)." Finally, the news made headlines with a jolt: RADIO FORCES EDISON TO QUIT RECORD FIELD, blared the New York *Telegram* of November 7, 1929. "In the give and take of modern business, there is little room for sentiment," philosophized staff writer Edward O'Toole, who broke the story.

Thomas Edison always had a knack for using new discoveries to overshadow his failures. As an ancient invention was feted, a favorite one faded—and a new idea sprang to the fore. Five freight cars loaded with goldenrod arrived in Fort Myers as the snow swirled in New Jersey, and with them the "silver-haired creative genius," who asked William H. Meadowcroft to make the official announcement that yes, "goldenrod is the weed in which Mr. Edison has now found a satisfactory rubber content . . . ending his long search" in the winter of 1929—or did it? "I may say that the patience of Job has been considerably overrated," Edison said soon thereafter. "He did not know what patience was." He was coaxing one hundred pounds of rubber out of one acre of the weed, and he hoped to increase this to one hundred fifty pounds, developing a *leavenworthii* variety that contained a very good 12.5 percent rubber in its upper leaves. In the aftermath of this discovery, Edison pushed for a new Plant Patent Bill, to be passed by Congress, as protection for botanical breeders and experimental farmers. "Luther Burbank would have been a rich man if he had been protected by such legislation. The plant breeder needs a grubstake and this will help him out," said Edison, the farmer's friend to the end.

Despite Meadowcroft's forthright declaration to the press, a survey of Edison's last surviving notebooks reveals that he continued his quest for other varieties of rubber-bearing weeds well on into the summer. A month after the "discovery" statement, Edison "started [his] Campaign," in familiar rhetoric, to find another "practical acetone-benzol mixer." The handwriting is so tremulous as to be nearly illegible. With springtime, the notebooks shift to accounts and checklists, seventy-four varieties of "herbs collected by Mr. Edison" along various roads, down McGregor Boulevard, and farther afield, into Venice, Labelle, Rio Vesta, and Naples.

In a further testament to his growing legend, Edison achieved another milestone of recognition, a *New Yorker* profile—in three installments, no less—by Alva Johnston, dubbing him "the old Prometheus, for no article on

Edison is constitutional without an allusion to the theft of fire from heaven." With old age came a denial of the Wizard sobriquet. "I have never liked that word applied to me," the inventor told Remsen Crawford of *The Saturday Evening Post* in a long, rambling conversation in August. "It bears an ironic inference that my creations came by magic, or sleight of hand, or some incomprehsible legerdemain, or jugglery with natural laws and physical forces. Whatever I may have accomplished in my life has been the result of hard, persistent labor." The old man who as a boy never even came close to completing grade school received an honorary Doctor of Science degree from Rollins College in Winter Park.

But Thomas Edison was napping more, had taken to wearing wire-rimmed spectacles while reading. It was rumored that he fainted after his latest ordeal with the press—this he strenuously denied—and had not Mrs. Edison given him a yellow wheelchair for his eighty-third birthday? Tourists with cameras lined McGregor Boulevard every morning and waited for the Old Man to be trundled across the road by an attendant. Wrapped in a linen duster, with a floppy old straw hat pulled down over his brow, Edison, seated, slouched and covered his face with his arms so as not to be photographed.

CHAPTER

29

TOWARD CHRISTMAS, EDISON, WITH HIS WIFE, TOOK THE TRAIN
down to Pittsburgh for Grace Miller Hitchcock's sixtieth birthday weekend. Mina's sister Mary and sister-in-law Louise joined the party, as did
Henry and Clara Ford. The main course at dinner was an elaborate baked
squab, but Edison, sticking to his milk diet and ever mindful of stomach
troubles, wandered around the dining room inspecting Grace's porcelain
collection. "I came in with milk, and I guess I'll go out with it," he murmured to no one in particular.

After cake and simple presents—a poem from Louise, a handillumined book from Mary, a corsage of gardenias and lilies of the valley
from Mina—the family group adjourned to the living room, where Edison
settled into a corner of the sofa. The soft glow of a bridge lamp accentuated the lines on his face, and fell upon his white hair, thinning in front,
but still plentiful, and his dark, bushy eyebrows shrouding deepset eyes.
In the past year, he had lost more than ten pounds. There was a perceptible gap between his neck and the starched white collar, tie as usual loosely
knotted.

Mina sat to her husband's left. As the conversation washed over him,
she tapped its outlines onto his hand, resting on the cushions. Edison responded with slight nods and dreamy smiles, and then his mind, like a
slow-starting engine, engaged with fragments of a past life. He began to

speak, and for the next two and one-half hours, the stories and free-associated vignettes came forth.

He remembered when the big Jumbo dynamo had to be shipped from New York City to the Paris Exposition in 1889. Edison's men were working on it until the last minute and tied up all traffic in lower Manhattan when they rushed the machine from Pearl Street to the steamer across town—"We don't want those Frenchmen to beat us," Edison told the mayor. Jump-cut to little Al as a boy in Milan, feeding Seidlitz powder to his friend to see if he would fly—"I wanted to make a dirigible of him, but it didn't work!" Remember those great days at 65 Fifth Avenue, when Sammy Insull arrived in town from London? Someone came in to the first-floor reception area looking for Edison's secretary; there was a man at a desk stamping envelopes at a furious pace, who looked up briefly to direct the visitor to the fourth floor. Tramping up there, the fellow found Insull, who brought him back down four flights and introduced him to the man stamping letters, none other than Thomas Edison.

He drifted again, back to the traveling days as a Knight of the Key. Edison was in Fort Wayne, Indiana, on the Wabash Railroad: "The boys were all talking on the wire, and one fellow broke in and called up a station where a woman was an operator, and he said, 'Would you cut out for a little while? What I am going to say is not very proper for a lady to hear.' When those fellows got through talking, she said, 'Well, I didn't hear anything very awful that a woman shouldn't hear.' " He chuckled and went on, "Once I was working in Cincinnati. The fellow working with me was always fighting with a fellow out at Morrow, Ohio, six miles away. One night they were struggling to get the [telegraph] key, and had been calling each other names. The one at Morrow said, 'What's the difference between you and a damned fool?' And the Cincinnati fellow signaled back, 'Six miles.' " He paused, but only for a moment: "During the Civil War, the operator at Louisville where I worked heard that one of the operators at another station had the smallpox, and the operator said, 'Aren't you afraid you will catch it?' And the other fellow said, 'Not on *this* rotten wire.' I worked at Louisville for about a year," he continued, "and the office was up a very rickety flight of stairs on the second story. There was no efficiency at all. No discipline. The manager was a man by the name of Ellsworth, and he had a great crony named Billy Lewis. Billy was a fine operator but he got drunk every three months. One night we heard tramp, tramp, tramp, coming up the stairs and the door was flung violently open, and there was Billy

Lewis—no sleeve on his coat and the end of it all torn off. His eyes were all bloodshot and he was wild. We were paralyzed. We didn't know what he was going to do. He went and kicked the stove over and the pipe came down and the soot went over everything. He pulled out the wires from the switchboard and threw them in the middle of the floor. Then he went into the next room and there was a bottle of nitric acid. He threw that down and it leaked through the floor and into the room below, eating its way through the floor. Then he went away. The next morning at about seven-thirty the manager came in, and he was a bosom friend of this fellow. He looked around the whole place. It was a sad wreck. He said, 'Who did this?' and we said, 'Billy Lewis.' 'Well . . . Billy Lewis,' he said, 'if he ever does such a thing again, I will discharge him!' " Laughter, which Edison could not hear. He was already on to the next town.

"I was treasurer for the telegraph crowd in Memphis and they would ask me for money when they wanted to go out for a drink, and I doled it out to them, but one man demanded all the money, so he knocked me down, but the other fellows jumped on him and they made a wreck out of him. I resigned from treasurer then." Next scene: Boston. "Next to the Western Union telegraph office there was an old restaurant with millions of cockroaches. In the night they would run up the walls. You could see them by the thousands. I had been used to them—they didn't faze me at all. The boys wanted to know if I could figure out some way of getting rid of them, so I fixed up two strips of tinfoil with the small apparatus, and connected them with two cells {batteries} we had in the cellar. When the cockroaches started in, their front feet would strike the one strip of tinfoil and their back legs the other, and there was a flash, and they disappeared into gas. The police made us stop it."

The new year 1931 was ushered in with the birth of Edison's fourth grandson. Madeleine was still in the hospital when a telegram arrived from her father with dozens of suggestions for names—anything, anything at all but plain "Michael," which Madeleine and John Sloane were seriously considering. How about Victor, or Claude, or Percy, the Old Man wrote, something unusual? Madeleine compromised with Michael Nicholas *Samuel* Edison, after her father's father. When the baby was six months old, she brought him to Glenmont, and Edison held the boy for the first and last time. The baby had his eye on a pencil near the telephone, attached to the table by a long chain. Michael reached over and grabbed the chain, and reeled it in, link by link, until he held the pencil in his hand. Edison was delighted at the ingenuity. "Bright boy!" he said with pride.

Waiting until the baby had arrived safely in the world, Edison and Mina made their habitual move to Florida. For the first time in the memory of any members of his staff, Edison did not visit the rubber laboratory immediately after his arrival. Mrs. Edison sent word that her husband would be unavailable for several days, and it was not until his birthday—which had by now achieved the status of a national holiday—that he appeared in public. Sleeping late, Edison awoke for one of his rare shaves, and then was taken by limousine to dedicate a new bridge in his name spanning the Caloosahatchie River. That night, he sat down before the fire in his private office for a brief colloquy with Harvey Firestone, captured on newsreel film: "What do you think of the Einstein theory [of relativity], Mr. Edison?" "I don't think of anything of Einstein's theory, because I can't understand it," came the husky, nasal response. "What do you think of the sound pictures of today?" "You ask what I think of the talking pictures?" "Yes." "Ah, well, I don't know, I never heard one." And finally, "How does it feel to be eighty-four years old?" "Well, it feels very fine to be eighty-four years of age, if you don't have anything the matter with you. But I have a little trouble now and then, but that's because I'm getting old. But I've got a lot of ginger yet!"

Ginger, indeed, for he was talking about building a factory near Savannah, Georgia, on land owned by Henry Ford, where the Edison Botanical Research Corporation would commence the full-scale processing of goldenrod rubber. A prototype model had already been pieced together in Edison's green frame workshop, down to the smallest detail.

Even as Papa had designs to become a self-made captain of industry in the middle of his ninth decade of life, son Theodore was on the verge of departing. For several years, since he had dutifully joined the company as a natural family obligation, Theodore had acted to the best of his ability as a loyal sidekick to his big brother, at a time when Papa was pulling out, and the Edison interests needed all the brainpower they could muster. An obliging but poor administrator, Theodore felt like a third wheel when in the presence of his father and brother. His heart and intellect were not really with the routine side of the show. "I cannot help bringing a great deal of humanity into my business decisions," he told Charles. "Possibly this makes me unfit to have an executive position. . . . I have some rather radical ideas of my own which I want to try out."

The pressure of Theodore's weekend hours at the factory, required by the radio launch, led to a repertoire of health problems, ranging from rheumatism to the grippe, impetigo, insomnia, and, most disturbingly for someone so fragile to begin with, a primitive form of bulimia. "I have done so much writing and figuring in the past few days that I'm almost cross-eyed! . . . I go to the lab fully determined to set aside a whole morning for the work, come what may, and then one thing after another turns up which I have to attend to, and by the end of the day I haven't even looked at my original task!" There was more scrambling and consolidation once the decision had been reached to exit the radio business: reorganizing the cement company, trying to promote a new calculating machine, pushing work on electrifying the Ediphone dictating machine, finding new applications for an Edison heater.

Poor Theodore had already been taking furtive leaves of absence when he decided in February 1931 to rent a small office place to use as a workshop, within walking distance of his apartment in East Orange, where he could have his own drafting board and woodworking bench—a place apart, where he often just sat and wrote down private thoughts and tried to unwind the intolerable springs of tension coiled in his brain.

Madeleine, ever attentive, already had typecast Theodore as the maverick of the family, "the queer child . . . the square peg in the round hole." Her brother would never be happy until he found his own path. In early March, he did just that, filing articles of incorporation at the county clerk's office in Newark for *Calibron Products, Inc.*, a little company chartered to do general experimental work and research, while providing engineering consulting expertise. Its first product was a special graph paper for making perspective drawings. BREAK IN RANKS OF EDISON CLAN, blared the headlines across the entire front page of the Orange *Courier* a week later. This was powerful grist for the rumor mill—had the sacred dynasty finally cracked? "This is not the first time that a member of the Edison family has helped to organize an outside venture. My interest in the new company does not affect my connection with Edison Industries," Theodore said in a prepared statement, advising Charles that he was resigning as a director of the Edison companies and also dropping off the executive board, continuing, only for the present—until Charles found a qualified successor—in his research and development job as technical director of the laboratory. "As far as general business goes," Theodore said to his brother, "you will probably find that you have greater freedom of action if I am not around." Charles saw the move for the best and was not surprised. He wished Theodore well. Mina, Madeleine, and Ann had also been expecting the transition.

Theodore had put off telling his father for months, fearful of disappointing him, afraid the Old Man would respond wrathfully as he had long years ago when Tom, Jr., and Will had tried to set up their own shops, diluting the pure name of Edison. Theodore implored Mina to intercede, to show Papa the various memoranda and prospectuses he had developed, and to convince him that if Calibron proved a flat failure, the stigma would be on Theodore's head alone. Ann joined the lobbying effort, reminding Mother Edison that "Children with as forceful and distinct personalities and abilities as your three must fulfill their ambitions each in their own way." In his wife's opinion, Theodore's new trail was blazed in the pioneering spirit and faith in the soundness of his own ideas inherited from Edison senior: "Isn't that just what appealed to Father Edison in his youth, new fields to conquer and as *he* saw fit, in *his own* way? Certainly! . . . Fortunately you have a son of strong character and not one who clings like a parasite, which is so often true of young men whose fathers have been successful."

It was late May before Mina was able to accumulate enough courage to break the news to "Dearie." Theodore's trepidation was unwarranted. The mellow patriarch dismissed his youngest son's move lightly with a casual wave of the hand, too much lost in thought preparing for his now traditional radio remarks to the National Electric Light Association by direct radio wire to Atlantic City. It was his last public address, and it had the tinge of valedictory, delivered in thin, reedy tones—a dramatic contrast to Edison's humorous birthday greetings. With Mina beside him, he leaned toward the microphone and said: "My message to you is to be courageous. I have lived a long time. I have seen history repeat itself again and again. I have seen many depressions in business. Always America has come out stronger and more prosperous. Be as brave as your fathers before you. Have faith. Go forward."

On June 15 Edison departed for West Orange, instructing the laboratory and field personnel to ship all his rubber equipment north. Back at Glenmont he took to bed for a few days, then made haphazard visits down to the lab, where he began to doze at his rolltop desk. On July 30 he composed a codicil to his will. Two days later Edison collapsed in his chair in the parlor over the porte-cochere. His longtime personal physician, Dr. Hubert S. Howe, on holiday at Sands Point, Long Island, chartered an amphibious plane and flew straight to Newark, where a police escort sped him to the Edison home.

Dr. Howe had a challenging patient in Thomas Edison—a totally deaf, eighty-four-year-old man who did not bathe more than once a week, did not believe in exercise, still (by his own account) "chewed tobacco continuously" and smoked several cigars a day, and whose only foods were milk and the oc-

THE LAST
PORTRAIT, JULY
1931

casional glass of orange juice. As to Edison's digestive ailments, including constipation, Dr. Howe could only conjecture—because despite the doctor's reassurances, the Old Man refused to be x-rayed—that he had an ulcer of the duodenum, with contraction of the pylorus. In later years, Edison also suffered from diabetes and Bright's disease.

After the weekend of his sudden decline, Edison began to pick up again, rallied by the attentions of his family, Mina's constant encouragement by his bedside, eight hours of sleep a night, a few sessions with the daily papers, and, lo and behold, a cup of tapioca. He walked unassisted to an open touring car and climbed into the back seat for a half-hour ride through the countryside. A freshly tarred road was encountered along the way. The driver wanted to turn aside, and assumed that a MEN WORKING sign partially blocking the road would be enough of an obstacle for the Old Man. "The road is closed," Mina told Edison as the car backed up. "It's not *closed*!" he retorted. "The sign only said there were men working." That night, during his examination before retiring, Edison said to Dr. Howe, "My friend Rockefeller seems to have passed these dangerous stages successfully. I feel if he can do it, I can."

Despite these encouraging flashes of familiar spunk, Dr. Howe told Charles privately that he did not think Edison would "ever be out of danger. . . . The wear and tear of time" had finally made its mark on the Wizard.

By early September uremia had set in, and Edison took a turn for the worse. An oxygen tank was installed in his bedroom for relief during dizzy spells; of course, he refused to use it. The concrete garage a few hundred yards from the house had been converted into a press bivouac, as no newspapermen were allowed into the house. Each morning and afternoon, Charles and Dr. Howe released bulletins on Edison's condition, which were immediately transmitted by wire and print to an anxious world. A large box of flowers arrived from President and Mrs. Hoover, and a message of hope from the Vatican. Daily automobile rides over the comforting landscape continued, rain or shine, until the end of the month, when the leaves turned brittle and the Orange mountains reddened.

The winds grew sharp, and Edison remained indoors. On October 3 the good citizens of Fort Myers declared a citywide prayer day for their old friend, "the foremost winter resident of the City of Palms." Day and night nurses were in constant attendance as his sleep became fitful and his disposition lethargic. Edison no longer asked to be taken to a chair for his meals of milk toast and a few teaspoonsful of stewed fruit. Harvey Firestone stopped in to pay tribute to his old crony, knelt by the bedside, and was greeted with a wan smile, but little else.

Aside from pensive walks around the Glenmont grounds with Madeleine and her grandsons, and Charles or Theodore, Mina remained with her husband, composing herself for the inevitable. Tom, Jr., William Leslie, and Marion were summoned to Glenmont on October 17 when Edison's pulse accelerated and his respiration grew shallow.

Toward the end, he emerged from a coma, opened his eyes, looked upward, and said to Mina, "It is very beautiful over there."

He died at 3:24 a.m. on Sunday, October 18, 1931.

———

A death mask and casts of Edison's hands were made that morning by James Earle Fraser, a friend of the family and the designer of the American Buffalo nickel and Robert Todd Lincoln's sarcophagus at Arlington National Cemetery. "He had a marvelous, powerful face," the artist said, emerging from the bedroom. "The beautiful, full forehead, the nose, the mouth, the chin— withal show great power." In accordance with a plan worked out a month previously by the children in consultation with Mina, the body lay in state in the vaulted library of the West Orange laboratory for two days and nights. The tiered shelves filled with bound volumes formed a final image of Edison's crowded life: Patent Office *Gazettes*, *The Chemical News*, the archives of the National Electric Light Association, geological surveys, a complete run of the *Illustrated London News* and *The English Mechanic*, histories of the Oranges, *National Geographic*, Theosophical tomes, *Proceedings* of the Newcomen and Linnaean societies. At center stage stood the rolltop desk with its five pigeonholes labeled simply "Money," "Cement," "Financial," "Amberola," and, naturally, "New Things." In the lower right bottom drawer was an open carton of Muriel Perfecto cigars and a box of wooden Staylit matches.

Admission for the first hour was limited to Edison employees and their families—the Pioneers, the muckers, the chemists, the lab boys. Then, the factory gates were thrown open to the general public. More than fifty thousand people silently filed by. Henry Ford could not bring himself to look into the bier. "I want to remember him the way he was," he said.

"Farewell, Thomas Edison," intoned the narrator of the Paramount newsreel, *Edison Dead!* "The nation places you among the immortals of America—and of all time." The tributes came in a torrent. A somber, chiaroscuro cartoon in the New York *World-Telegram* showed an angel of death about to extinguish a solitary light bulb hanging over a bed in which the Old Man lay, eyes closed. "Prospero is dead," said the New York *Times*. "The light-bearer has gone into darkness. . . . Ours was wondersmith of the world." The Kansas City *Times* called Edison "one of the chief architects of the modern world." "He helped liberate man from the shackles of time," said the Omaha *World-Herald*. "It can be said of Edison, as it was said of Lincoln, and can be said of very few others: 'Now he belongs to the ages,' " wrote the New York *Evening Post*. "Yes, the world pauses for the passing of Edison," wrote the *Morning Oregonian*, "as it might not pause for any other, of whatever station—the world he so zealously served." "Years must elapse before the true measure of Edison can be taken. . . . In the nature of the case, there can be no successor," the Cleveland *Plain Dealer* declared. "By his work we know him," said the St. Louis *Post*. "If we had a mythology, Mr. Edison would be placed in that gallery of gods which includes Prometheus." "Edison was in nothing more profoundly American than in the honor he paid in word and deed to the gospel of work," the Chicago *Tribune* announced.

He was "The Inspirer," "One of Our Immortals," the "Conqueror of the Unknown," "The Genius of Light," "The Revolutionizer," "Foremost Among Creators."

"It is impossible to measure the importance of Edison by adding up the specific inventions with which his name is associated," Walter Lippman wrote, in one of his first "Today and Tomorrow" columns for the New York *Herald Tribune*. To Lippmann, Edison's career was an example of supreme synergy, four decades before that word was even created. Edison was the personification of the power of human intelligence; through his "propagandist" spirit, he showed that "anything could be changed and everything could be controlled." And by defining the potential energy of technological change, Edison broke down the limits of blind optimism in popular expectations about science.

Walter Lippmann was the only commentator who saw—beyond the immediate rhapsodies—that Edison had achieved far more than bring light to industrial-age darkness. He had also lifted the lid of Pandora's box and released a permanent legacy of questions about the art of invention. Lippmann correctly predicted that the "wisdom" modern Americans would require to fully comprehend the rapid changes around them would grow far more slowly than the inventions themselves.

On Wednesday morning, October 21, Edison's body was carried back to Glenmont for a 3:00 p.m. private funeral ceremony. Mina went down to the conservatory an hour early, to be alone with her "Dearie." Charles had arranged the seating, placing Mina in the first row of chairs, Theodore, Madeleine, and himself to her right; and then Tom, Jr., William, and Marion in the second row. However, the jockeying for position continued even after the Old Man's death, as Mina changed the order, placing Madeleine directly to her right, then Theodore, then Charles, and adding Tom, Jr., to the first row. In the second row of seats were Marion, Will, Blanche, Carolyn (Charles's wife), and Beatrice.

The service began with Alexander Russell, organist of Princeton University, accompanied by Arthur Walsh, violinist and manager of the Phonograph Works, playing their chief's two all-time favorite tunes, "I'll Take You Home Again, Kathleen" ("Across the ocean wild and wide, / To where your heart has ever been, / Since first you were my bonny bride") and "Little Grey Home in the West" ("When the golden sun sinks in the hills, / And the toil of a long day is o'er, / Though the road may be long, / in the lilt of a song I forget I was weary before.")

A musical medley reflecting Edison's popular classical taste followed, including the first movement of Beethoven's "Moonlight" Sonata and Wagner's "To the Evening Star."

Dr. Stephen J. Herben, former pastor of the Orange Methodist Episcopal Church, where Mina was such an active member, read the Twenty-Third Psalm, and then added a personal note, invoking "one who has lived long and has toiled faithfully. . . . His fertile brain has ceased to do its accustomed work. His gleaming eye is dimmed. His quick hand is folded in rest."

Dr. Lewis Perry, headmaster of Phillips Exeter Academy and another family friend, read a rousing eulogy composed by Arthur J. Palmer, an employee of the Edison Industries, concluding, "Of this man, this super-being who defies classification, what more can be said, what greater tribute can be paid than this: *He is humanity's friend.*"

To the strains of "Now the Day Is Over," four hundred guests left the house and grounds. Only the minister, Mina and the Edison family, Mr. and Mrs. Firestone, Mr. and Mrs. Ford, and Mrs. Hoover formed the cortege leaving through Eagle Rock Gate, into Montclair, through the wooded residential sections of Orange Road and into Rosedale Cemetery for the interment as twilight gathered.

The coffin slowly descended. Dr. Herben recited a brief poem, "Earth to earth, and dust to dust,/ Calmly now the words we say,/ Left behind, we wait in trust/ For the resurrection day./ Father, in Thy gracious keeping/ Leave we now Thy servant sleeping."

In the gathering moonlight, each person stepped forward and cast a single white rosebud into Thomas Edison's grave.

CHAPTER

30

WORK ON EDISON'S RUBBER EXPERIMENTS CONTINUED UNDER the supervision of trusted brother-in-law John Miller. Botanical cultivation was focused in Fort Myers, with testing at the West Orange lab. Yields continued to rise, exceeding three hundred pounds of rubber per acre of goldenrod by 1934, at which time the Edison family decided to close out the project.

All in all, fifty-six different species of *Solidago* had been sampled. Henry Ford continued financing rubber cultivation in tribute to Edison's memory on his own land near Richmond Hill, Virginia, in conjunction with the United States Department of Agriculture, which issued the final reports in the late 1930s, upon which much of the information in this book has been based. When the war came, the government's research funding shifted to synthetic rubber, Ford's attention waned, and Edison's momentum rolled to a halt.

Under the terms of his will, 80 percent of Thomas Edison's $12 million estate was divided outright between Charles and Theodore, the two favored sons; they were named co-executors of the remaining assets, with authority to distribute them among the other children. There was no bequest to Mina, because she had "already [been] adequately provided for" through transfer of corporate shares over the years as well as real estate.

The only other direct cash beneficiaries named were John Ott, the oldest living employee of the Edison industries, meant to receive $10,000, but who died in his Glen Ridge home within hours of the news of Edison's passing; his brother, Fred, who received $8,000; and William H. Meadowcroft, the Old Man's secretary, who received $10,000.

Despite the guarantee of a trust fund of $50,000 a year that would come through the distribution of shares of the Edison Portland Cement Company (stipulated in the July 30 codicil Edison had composed soon after his final return from Fort Myers), angry William Leslie took the lead in threatening to contest the structure of the estate outlined in his father's will as prejudicial against him and the other three children. But Will was unable to get Tom, Jr., Marion, and Madeleine—resentful as they were—wholeheartedly behind him. Tom even went so far as to issue a public statement putting distance between himself and the others: "I loved my father as the dearest and most god-like parent a man could have; as a genius, with the world sharing my pride."

Three months after Edison's death, Charles was able to prevent William from filing an imminent suit, about which Charles remained taciturn, except to say emphatically to the New York *Times* that it was an "understanding, *not* a settlement," and once his brother was shown the actual figures of the will, "[he] acted like a gentleman and admitted that he was satisfied with his share."

Tom, Jr., died incognito in a Springfield, Massachusetts, hotel room on August 25, 1935, after registering under the fictitious name of "J. J. Byrne" (or "J. J. Griffin," depending upon which newspaper account one read). Tom had been suffering from high blood pressure for several years, and, indeed, the cause of death was listed as heart disease by the Hampden County medical examiner. Edison biographer Matthew Josephson writing in 1959 stated that Tom "committed suicide in 1936 [*sic*]," and Robert Conot took up this refrain two decades later in his *A Streak of Luck*, likewise noting that Tom "killed himself."

William Leslie Edison died two years later, after a long struggle with cancer.

Mina carried on alone for four years after her husband's passing, then married Edward Everett Hughes, a retired lawyer and steel manufacturer who had been her childhood playmate in Akron. In her seventies Mina took up piano lessons again in Fort Myers, insisting the servants leave the house when she was practicing so they would not hear her awkward efforts. She continued to entertain friends, but on a more intimate level—the Hitchcocks, the Fords, the Lindberghs, the Mercks. During the war years, Mina invited enlisted men stationed at Fort Myers over for Sunday buffet dinner, after which she gathered the G.I.s in the parlor and asked them to talk about their families, wives, and girlfriends back home, then sent them off to town with theater and movie tickets while she conscientiously wrote to each of their mothers, reassuring them the "boys" were all right.

Edward—Mina called him "Mr. Hughes"—was a companion, nothing more or less. He gave her a respectful peck on the cheek before retiring to his separate bedroom upstairs. "Dearie was the only husband I ever had," Mina told her personal secretary, Jeannette Perry, as they sifted through the morning mail together.

After Mr. Hughes's death in 1940, Mina kept up with her educational, religious, and social causes. She taught men's Bible study classes at the Fort Myers Community Congregational Church. Standing side by side on the receiving line with Spencer Tracy, she gave a huge reception at Glenmont for the world premiere of *Edison the Man*. She was made honorary president of the Community Concerts Course Committee for the Oranges. She was a sponsor for the United States destroyer *Edison*, christened in Kearny, New Jersey. She dedicated a set of chimes at the Orange Methodist Church as a memorial to her brother John. In June 1944 she was awarded an honorary Doctor of Humane Letters by Mount Union College in Alliance, Ohio, where the Miller family had long been active.

On February 11, 1947, the one hundredth anniversary of Thomas Edison's birth, Mina officiated at a ceremony opening her husband's rolltop desk, locked since his death by Charles.

She died on August 24, 1947, the very day the Chautauqua season was closing, as seven thousand people, gathered in the outdoor amphitheater, sang "Abide with Me" in tribute. Mina Miller Edison was buried adjacent to her husband at Rosedale Cemetery. Sixteen years later, Edison and Mina were reinterred in a shady grove within sight of their Glenmont home.

Quaintly emotional to the last, Marion Edison Oeser died in 1956, leaving instructions that "I would like my ashes put in a pot of pink roses so as to make them more beautiful, as potash would be good for them."

After the radio debacle, Charles kept the storage battery business going, as well as a line of medical and industrial gases, the Edison Voicewriter (a variety of dictaphone), and a successful line of juvenile furniture. Inspired by the New Deal, Charles stepped aside from direct management of the Edison Industries. He changed political parties, served in various positions in the Roosevelt administration, and was named assistant secretary and then secretary of the navy from 1937 to 1940. In 1934 his name had been bandied about as a likely candidate for Democratic senator from New Jersey; instead, Charles became regional director for the Federal Housing Authority and New Jersey regional head of the National Emergency Council. From 1941 to 1944 he was governor of New Jersey. After the war and retirement from politics, Charles returned to T.A.E., Inc., and took it public. He and Carolyn also became active in historic restoration efforts at Sag Harbor, on Long Island's East End, subsidizing the preservation of the Old Custom House and the adjacent Hannibal French House.

In 1956, the Edison laboratory was given to the Federal government by T.A.E., Inc., to become part of the National Park Service; Glenmont followed three years later. Together, they now constitute the Edison National Historic Site.

In 1957, T.A.E., Inc., was sold to the McGraw Electric Company, and the new company became known as McGraw-Edison, with Charles assuming the title of chairman of the board. Like his civic-minded mother, Charles Edison enjoyed many outside activities. He was a trustee of the China Institute, and served on the National Commission on Organization of the Executive Branch of the Federal Government, and the National Municipal League. He was one of the founders of Young Americans for Freedom. In Fort Myers, he crowned the "King and Queen of Edisonia" at the annual Pageant of Light parade.

"Looking back over my own life," Charles said in a speech at a testimonial dinner in his honor at the Plaza Hotel in May 1963, "I think I have, on the whole, been faithful to the family tradition: I don't go looking for battles, but I always seem to be in one."

Charles Edison died of a heart attack on July 31, 1969. Following his instructions, funeral services were supervised by sister Madeleine, with whom he had effected a reconciliation; she would outlive her younger brother by another decade. "Is it a help or a hindrance being the son of a great man?" the New York *Daily News* asked, in one of his last interviews. "The question is one that cannot be answered with any measure of certainty," Charles replied, "Being the son of a great man is, in a way, like being the nonprofessional husband of a motion picture actress. Such a man shares a reflected glory but also is relegated to being known as 'the husband of Greta Greatstar.' So it is with many sons of famous fathers. They are never able to step out of the shadow of their fathers' fame."

And what of Theodore, the last Edison? Money had never been of great import to him. He resigned as co-executor of the troubled estate, leaving all disputes and controversies to Charles. A year later he obtained his first independent patent completely separate from his father's auspices, for a device used in eliminating vibrations in machinery. Theodore and Ann established a quiet home and exquisite English garden down a steep hill off Glen Avenue in a thickly wooded section of Llewellyn Park. A secluded trail led to the back entrance of the Calibron building. Ann earned a pharmacist's degree from Rutgers in 1935 and went to work for the Merck Institute for Therapeutic Research in Rahway.

In the late forties, Theodore took more than one million dollars inherited from Mina to establish the Edison Industries Mutual Association, to distribute shares of the company and divide the profits among the labor force of T.A.E., Inc., caring little for his material success.

Long before it had become a popular cause, Theodore became devoted to environmental issues, picking up his mother's banner. He founded a preservation society in Florida to save a three-square-mile area of bald cypress trees near Fort Myers that included large rookeries of wood ibises and American egrets. In 1954 he founded Monhegan Associates on the beloved island off the coast of Maine he had visited first as a child. The group was dedicated to preserving for posterity the natural, wild beauty, "biotic communities," and desirable historic features of the island's rugged headlands portion facing the sea, "as well as the simple, friendly way of life that existed" on Monhegan as a whole. The Associates, under Theodore's leadership, further committed themselves to establishing records and archives documenting the flora and fauna of the island. Over the decades Theodore systematically purchased parcels of land there, donating the property to the Associates in a steady crusade to maintain the spirit of the place.

Again, before it became part of a countercultural revolution, Theodore Edison was an early and outspoken opponent of America's involvement in Vietnam. "Let's search for better approaches to the problem," he declared in a full-page New York *Times* advertisement on October 16, 1966. "I cannot believe that killing, maiming, and starving our opponent, at great cost in men and money to ourselves, is the best way to convince him that we are not the monsters he has been told we are."

In later years, Theodore the scientist delighted in revisiting his father's reputation, giving speeches and publishing papers intended to enhance and solidify Thomas Edison's story, rather than see it become obscured in the misty realm of American fairy tale. "Mr. Edison's *generalist* approach to problems helped him to make inventions," Theodore told the MIT Club of Northern New Jersey. "He pioneered in establishing organized research as a business." Striving to refocus his audiences on the forgotten importance of his father's mechanistic approach to technology, Theodore said, "Perhaps I am simply not attuned to the efficiency of modern progress, but I can't help feeling that the present generation is suffering real loss by not paying more attention to what was good in the past."

On Friday, June 19, 1992, I drive up to the security guard's kiosk at Llewellyn Park. There is a sign of recognition in his eyes; by this time, there should be, after my years of working on the Edison saga. I tell him I am expected at Theodore Edison's house. "Fourth driveway on the right on Glen Avenue after you cross Honeysuckle," he says. "Is it marked?" I ask. It is not. I glide up the rain-soaked green roads past set-back houses, not a soul in sight. Rounding the curve past Glenmont, which hovers red through the fine mist, I cross Honeysuckle and begin to count, losing track, having to back up as a couple of luxury cars pass.

Finally, at what seems like a hundred yards down an unmarked driveway, I see the brick house, an English cottage, slate-roofed and glowing. I knock, and Nancy Miller Arnn, Edison's niece, answers. Wearing a blue, flower-print dress and looking apprehensive, she lets me in.

Stacks and stacks of newspapers and magazines. I have never seen so many back issues of *Scientific American* in one place. Daily mail, bills, letters, cards, flyers, piled on the dining room table. Boxes filled to overflowing with papers and file folders. In the dusty, book-lined living room a grand piano with a scattering of games, brain-twisters, interlocking rings, optical illusions and other complex diversions Theodore collected. There is a familiar portrait of T.A.E. nearby, and a photo of "Ann and Ted" from 1975 or so. He is wearing a corduroy jacket, rumpled brown trousers, hands thrust in jacket pockets, smiling and tweedy. The intellectual light is there.

A card table is set up at the other end of the room, and Ann Osterhout Edison, gray hair boyishly cut, is eating toast and tea and watching television. She is in a wheelchair. To maneuver, she does not push the wheels in the conventional manner, but propels herself with her feet firmly planted on the floor. A strong handshake. "Take a look at the garden out back," she says immediately, before even saying hello. A sloping, manicured lawn edges up against neat hedges. "And how is Mr. Edison today?" I say. There is a silence. "Well . . . he's fair," says day nurse Patty Brune. "Would you like to meet him?"

Nancy and Patty lead me down the length of the living room to a small, dark bedroom, shades drawn, made even darker with the gloom outside. Two smallish beds, parallel to each other. More walls of books. Between the beds is a tiny night table with an impossibly antique black cradle phone, its wire frayed.

In the bed on the right, a hospital version with railings along the sides, is a slender, still figure. The reading light on the night table illuminates Theodore Edison, aged 94, his frail body covered with a thin quilt. He is propped up on several pillows, and the bed has been cranked so that his large, oblong head, crowned with a bolt of whitish, faded-yellow hair, faces me. His eyes are half-closed. There is stubble on his chin. His mouth is barely open, and he breathes laboriously.

"Mr. Edison, there is someone here to see you!" Patty calls to him. A pause. His eyes open a fraction wider. His hands are arthritically cupped across his chest, long, thin fingers welded together, motionless. Parkinson's disease has him in its thrall. Suddenly, Theodore turns his head slightly to look at me, and begins talking, a monologue about "sound" and "frequency." He speaks as if breathing the words, rather than saying them, with the exhaling a rush of language broken by ellipses of silence.

"Father . . . R.C.A. . . . importance of frequency in middle registers . . . not as simple as turning up the volume . . . Father . . . realized . . . hearing . . . was difficult . . . " I assume he is talking about the production of hearing aids, and fall into a digression with Nancy and Patty—one stationed opposite me across the bed, the other at the foot of the bed. No, says Patty, who heard this story about a week ago, this is the technological shift at the phonograph company, the difficulty Thomas Edison had in making the transition in the studio from one primitive cornucopia horn to electronic recording and the positioning of many microphones in a session, one for each instrument. I think, how interesting that of all the places to start, Theodore would launch into this particular chapter without prompting.

"Father . . . knew the problems . . . " comes the refrain. The tone is of respect and admiration, and a quiet knowledge that T. A. E. was ultimately smarter than all the rest of them. But I cannot get at the intervening words. "Tell him about your father's favorite holiday," Patty says to Theodore, while looking at me.

"The Fourth of July," he breathes. "We made our own fireworks . . . on the driveway . . . at Glenmont"—he pronounces it *Glehmah*—"Father . . . made them with me . . . Mother sat on the porch and watched . . . Chasers . . . " I interrupt, ask what they are. "Explode . . . go on . . . explode again."

He is back there, eight decades and more, the prancing, golden-haired boy. "We almost blew ourselves up . . . We were using TNT . . . We found out later . . . Mother didn't approve . . . " I ask Patty to tell Theodore that I am the person who is writing the book about his father, the author who sent him a copy of my biography of William Carlos Williams last year. Theodore seems to have heard me. He is concerned: "So much mail . . . I can't keep up with it . . . I have fallen behind . . . "

Suddenly, Ann wheels in. "You're all standing *around* him and I can't hear a *word* he's saying!" "We were talking about the fireworks," I say. "Oh yes, of course, the fireworks!" she replies with a trace of annoyance, as if this were the most obvious subject in the world right now. She strikes me as very different temperamentally from her husband. Even with his debilitating illness, I sense in Theodore a clarity of mind and an inner activity. In that shell of a still body, there is a sensibility, a mind at work. Ann, protective, rolls noisily, impatiently this way and that over the parquet. She will not leave Theodore. Now that he is bedridden, she refuses to go out even for one minute, even to be taken around the garden. And no outsiders are allowed in, except caregivers and family.

"I sent you a book, last year," I say to Theodore.

"I haven't read it yet . . . I'm sorry," he says, tiring, "I wanted—to write my *anecdotes* . . . ," he looks right at me with stress on the final word. "Now, I *can't* . . . and I had—I *have*—so many *anecdotes*." His palpable frustration drives me away. I turn to leave, say good-bye and walk toward the bedroom door.

"Thank you for the book!" Theodore Edison calls out with utter clarity, smiling.

I never saw him again. He died on November 24, and Ann, as had been her custom for seventy years, followed her husband, succumbing to heart failure on January 28, 1993.

& SELECTED BIBLIOGRAPHY

CHAPTER 1

PAGE

3   Ted Brush and Mead Stapler, "The Edisons, One Family, Twice Exiled," *The North Jersey Highlander*, Summer 1976 (Newfoundland, N.J.), 3–16, North Jersey Highlands Historical Society.

3   J. L. Dillard, *American Talk: Where Our Words Come From* (New York, 1977).

3   *Correspondence*, TAE:Holland Society of New York, June 7, 1889. Wm. Ogden Wheeler:TAE, October 26, 1893. G. W. Beardsley:TAE, September 23, 1925. A. R. Ogden:Arthur E. Bestor, July 23, 1929. J. M. McCaughey:Mrs. TAE, October 26, 1929. ENHS.

4   Frank Lewis Dyer and Thomas Commerford Martin, *Edison, His Life and Inventions* (New York, 1910). Two volume, authorized biography. The original narrative was laid out by Edison himself in a series of vignettes, some fragmentary, some well worked, but all from his memory and with his official imprimatur.

4   *Checklist, The Ward Harris Collection*, Accession #377, Edison National Historic Site, West Orange, New Jersey. Biographies and other first edition titles about TAE. Important repository for the mythic subtext of Edison's life.

4   Arthur J. Palmer, *Edison, Inspiration to Youth*, Thomas A. Edison, Inc. (West Orange, N.J., 1928). The comic-strip, vignette version of TAE's life in pictures. An adaptation of the serial entitled "The Life of Thomas A. Edison," which appeared in the daily newspapers of the United States in the months of August, September, October, and November 1927.

4   Madeleine Edison Sloane, notes from oral history interviews with her

father, c. 1920. Collection Edison National Historic Site, West Orange, N.J. Hereafter referred to as ENHS.

4 Correspondence, Margaret Nagy to the author, re Edison genealogy. Including notes by Marietta Wadsworth, curator of the Edison home in Milan, Ohio, for thirty years.

6 The Edison Sites, Digby County, Nova Scotia, covering letter and 15-page typescript documentary account from G. D. Anderson, Commercial Officer, Nova Scotia Light and Power, to Edwin Vennard, Edison Electric Institute, July 5, 1961. Collection Edison Electric Institute, Washington, D.C.

7 Lyal Tait, *The Edisons of Vienna, being the history of the branch of the Edison family which resided in and around Vienna, Ontario,* privately published (1973). From the personal library of Theodore M. Edison. Loaned to the author by Nancy Miller Arnn.

9 Correspondence, Harry Douglas Elliott:Madeleine Edison Sloane, March 17, 1976, re Elliott family lineage pertaining to Edison family. There is still some ambiguity about Nancy Mathews Elliott's age at the time of her marriage to Samuel Edison. Correspondence dated April 8, 1927, in the Archives of the Henry Ford Museum and Greenfield Village, from Winfield Scott, Commissioner of the Bureau of Pensions, to Mrs. E. D. Wheeler, a descendant of Thomas Edison through his oldest sister, Marion, assiduously traces Nancy Edison's father's career in the Continental Army during the Revolutionary War, as follows: "[Ebenezer Mathews Elliott] enlisted at Claremont, New Hampshire, in 1775 [at which time he would have been seventeen years old] and served five months in Captain Wetherbee's Company, Colonel Wyman's New Hampshire Regiment; and enlisted in the spring of 1777 and served three years in Captains William Scott's and Isaac Farwell's Company, Colonel Joseph Cilley's First New Hampshire Regiment; and from the spring of 1780 was sergeant under Lieutenant Jonathan Cilley in Colonels Cilley's and Scammell's Regiment. He was at the surrender of Burgoyne, Battle of Monmouth, wounded at the Battle of Horse Neck, Battle of Chemung, and surrender of Cornwallis, and was discharged June 7, 1783.

"He was allowed pension on his application dated May 1, 1818, while a resident of German, Chenango County, aged sixty years. In 1820, he referred to his wife (name and age not given) [Mercy Peckham] and to *two daughters, aged sixteen and twelve years, their names not stated.*"

Assuming that Nancy was the younger of Ebenezer's two daughters, she would have been born in 1808, not 1810, which would have made her nineteen, not seventeen years old at the time of her wedding.

10 Donald Creighton, *A History of Canada, Dominion of the North* (Cambridge, 1944).

10 Kenneth McNaught, *The History of Canada* (New York, 1970). The above two titles provide background on the Mackenzie-Papineau Rebellion of 1837.

11 George S. Bryan, *Edison, The Man and His Work* (New York, 1926). Source of the anecdote about Samuel Edison's overland run to the United States border.

CHAPTER 2

12 David T. Glick, *Thomas Edison, The Port Huron Years*. 4 pp. typescript summary and chronology, 1982.

12 Helen Endlich, *A Story of Port Huron*. A local history published at Port Huron, by Endlich, 1981. Anecdotal and somewhat episodic, but a rich source of local lore.

13 James A, Ryan, *The Town of Milan*, The Ohio Historical Society (Columbus, 1974). Reprint of original 1928 publication.

13 Charles E. Frohman, *Milan and the Milan Canal*, published by Frohman (Sandusky, Ohio, 1976).

14 *Harvest: Gleanings from Ohio's Farm Village Heritage. Number VII, "Milan."* Part of a series of monographs "exploring the beginnings of Ohio's farm villages which importantly steered the development and character of the Buckeye State . . . and still do." Richly illustrated. J. M. Smucker Company (Orrville, Ohio, n.d.).

15 Linn H. Westcott, "Structure with Character: Edison's Birthplace,"

PAGE

Model Railroader Magazine, October 1963. Provided to the author by Robert Crim.

15   "The Birthplace of Thomas Alva Edison." Brochure published by the Edison Birthplace Association.

15   Lawrence J. Russell, curator, the Edison Birthplace, et al., Guide Sheet, 1990. 34-page typescript provided to the author. An exhaustive description of the home, room by room, item by item, with historic context for each item and art work.

     The following two books are conscientious if somewhat labored local histories substantiated by extensive research and documentation from nineteenth-century newspapers and journals.

15   The Charles Edison Fund, *Edison Genealogy* (n.d.). A family tree beginning with John Edison in 1727 and ending with Thomas and Mina Edison's great-great-grandchildren.

15   *Edison Family Birth and Death Records.* Photocopied from family Bibles at the Fort Myers and Milan homestead collections.

16   Bruce Hawkins and Richard Stamps, *Report of the Preliminary Excavations at Fort Gratiot (1814–1879) in Port Huron, Michigan,* Odyssey Research Monographs, II.1, 1989, Oakland University, Rochester, Michigan.

17   Letter from William A. Galpin to Thomas Edison, January 11, 1923, with photograph of Dr. Leman Galpin, the physician who delivered Edison. ENHS.

18   Arthur J. Palmer, *So Rich A Life.* 40-page typescript with pencil and ink corrections by Palmer, loaned to the author by Charles Hummel (c. 1929). Anecdotes of Edison's early years, told in embellished, worshipful style.

19   W. E. Wise, *Young Edison: The True Story of Edison's Boyhood.* (Chicago, 1933). From the personal collection of Theodore Edison. Of the many "boy's lives" of Edison, this one is distinguished by its facsimile copy of one of the two editions of the Grand Trunk *Herald* known to be extant. Many of the stories herein are based upon Wise's personal interviews in Port Huron "with men who had been boys with Edison." The author also notes that "Mrs. Thomas A. Edison graciously checked certain points upon which previous biographies of the great inventor differed."

20   Richard Lingeman, *Small Town America: A Narrative History, 1620–Present* (New York, 1980).

21   George W. Knepper, *Ohio and Its People* (Kent, Ohio, 1989).

The following three books must be referenced by any author doing work on Edison. The first two, the most recent biographical studies, have their flaws. Josephson is too worshipful and pays scant attention to the late years; Conot is too skeptical, and his title alone misrepresents Edison's processes. Nevertheless, these are useful studies:

Matthew Josephson, *Edison, A Biography* (New York, 1959).

Robert Conot, *Thomas A. Edison: A Streak of Luck* (New York, 1979).

Finally, an ambitious, comprehensive scholarly undertaking, the Edison Papers Project, under the direction of Professor Reese V. Jenkins of Rutgers University, aims to codify on microfilm and in book form the vast holdings of the Edison National Historic Site in West Orange, New Jersey, as well as to survey other Edison repositories around the United States and the world. As of this writing, Volumes I and II have been published, covering the years 1847–76. Begun in the late 1970s, this monumental endeavor is expected to continue well into the next century.

Reese V. Jenkins et al. eds., *The Papers of Thomas A. Edison: Volume I, The Making of an Inventor* (Baltimore and London, 1989).

CHAPTER 3

24   Daniel Calhoun, *The Intelligence of a People* (Princeton, 1973).

24   Carl F. Kaestle, *Pillars of the Republic: Common Schools and American Society, 1780–1860* (New York, 1983, 1991).

25   William M. Van der Weyde, ed, *The Life and Works of Thomas Paine* (New York, 1925). "Introduction" by TAE. And correspondence, Van der Weyde: TAE, June 19, 1924. ENHS.

The following three books provide an overview of the prevailing context for Thomas Edison's earliest education in pre–Civil War midwest America.

PAGE

25 Michael B. Katz, *Reconstructing American Education* (Cambridge, 1987).

27 Richard B. Stamps and Nancy Wright, *Thomas Edison's Boyhood Years: A Puzzle*, Michigan History Magazine, May/June 1986.

27 Port Huron Museum of Arts and History, *Edison's Boyhood Home, 1854–1863* (n.d.).

27 Personal communications in Port Huron with Professor Richard B. Stamps of Oakland University, Rochester, Michigan; and Steven Williams, Director of the Port Huron Museum of Arts and History, including Edison sketches provided by Dr. Stamps, were essential to the writing of this chapter.

28 Albro Martin, *Railroads Triumphant: The Growth, Rejection and Rebirth of a Vital American Force* (New York, 1992). A more conventional, anecdotal account of the railroad saga from a determinedly American point of view. A fine, general overview.

28 John A. Lent, "News Butches, Presses on Wheels, and Reading Travelers," Antiques and Collecting Magazine, May 1988. The only traceable contemporary article on Edison's first real profession. Again, provides the tradition and context for his efforts.

28 Charles Batchelor, excerpt from personal diary, October 7, 1886. Provides anecdotal information about the episode on the Milan train platform in 1859, as told by "Dad" Edison during reminiscences over dinner in Batchelor's home. ENHS.

29 Horace Porter, *Railway Passenger Travel, 1825–1880*. Reprinted from Scribner's magazine, September 1888, by Chandler Press (Maynard, Mass., 1987).

29 Wolfgang Schivelbusch. *The Railway Journey: The Industrialization of Time and Space in the 19th Century* (Munich, 1977; Berkeley, 1986). This seminal work relies heavily upon archetypes and the mental and emotional consequences of technological change to state its bold thesis from the European tradition of critical sociology.

30 Thomas A. Edison, Grand Trunk *Herald*. Facsimile copies of February 3 and "June" issues, 1862. ENHS.

30 Elliott West and Paula Petrik, eds., *Small Worlds: Children and Adolescents in America, 1850–1950*. Specific reference was made to Dr. Petrik's article, "The Youngest Fourth Estate: The Novelty Toy Printing Press and Adolescence, 1870–1886," which provided the

historical precedents for TAE's early efforts in publishing his own newspaper.

CHAPTER 4

33 Victor S. Clark, *History of Manufactures in the United States, Volume II (1860–1893)* (New York, 1949). Reprint by the Carnegie Institution of Washington of original 1929 edition. Takes an industry-by-industry, sector-oriented look at the fundamental trades and producers in the American economy, predicated upon the jingoistic theory that "by the middle of the nineteenth century the political consequences of the industrial revolution were manifest . . . manufacturing had become in a new sense a basis of world power."

34 Alexis de Tocqueville, *Democracy in America*, ed. J. P. Mayer, trans. George Lawrence (New York, 1966). Provides a background of American values as noted by a sentient visitor in 1831, setting the stage for Edison's entrance.

34 Marshall McLuhan, *Understanding Media: The Extensions of Man* (New York, 1964), and, with Bruce R. Powers, *The Global Village: Transformations in World Life and Media in the Twenty-First Century* (New York, 1989).

McLuhan has much of relevance to say—directly and by implication—about the role of Edison's inventions as signposts in the span of technological innovation, before and after he arrived on the scene. He is especially helpful in placing Edison on the chain of communication progress.

35 Joel Mokyr, *The Lever of Riches: Technological Creativity and Economic Progress* (New York, 1990). The author aptly insists that in the age of invention, insofar as progress represents true forward movement, "Ideas themselves are not good enough." Edison compared inventors to poets in their capacity for imagination, and Mokyr supports that analogy.

35 Substantiating TAE's remembrances of his busy adolescence is a faded, handwritten letter in the Archival Collections of the Henry

Ford Museum and Greenfield Village in Dearborn, Michigan, dated August 10, 1862, to Edison's friend, William Brewster, which reads in part, "My time is all taken up with my business on the cars. . . . You see I am on the Grand Trunk road. I don't get home until ten in the evening [an hour and a half earlier than Edison noted in 1910, but still a late hour for one so young] and have no time to write except Sundays. . . . I have no more time to write you [The letter was finished at 11:00 p.m.]."

36   James M. McPherson, *Battle Cry of Freedom: The Civil War Era* (New York, 1988). "The United States at Midcentury" (chapter 1) provides a concise analysis of growth, progress, industrialism, population shifts, transportation advances, expansion of market economies, the gospel of success, the rise of the banks, and what the author neatly identifies as "the democracy of consumption," all as a prelude to the great conflict.

37   Edwin Gabler, *The American Telegrapher: A Social History, 1860–1900,* Class and Culture Series, Rutgers University Press (New Brunswick and London, 1988). A vastly entertaining study of this representative and heretofore overlooked sector of Gilded Age working society. Gabler takes the itinerant class of telegraphers, a wandering band of young, single men constantly in motion, "drifting, swimming in the period's great streams of migration"—Thomas Edison among them—and demonstrates how their lives and experiences were symptomatic of the era; how they evolved from a specialized, entrepreneurial, and decentralized profession into well-organized wage earners with their fingers on the pulse of American commerce: "Our Fathers gave us liberty, but little did they dream, / The grand results that flow along this mighty age of steam; / For our mountains, lakes and rivers, are all a blaze of fire, / And we send our news by lightning, on the telegraphic wire." [Popular song, "Uncle Sam's Farm," c. 1860, cited in Gabler, p.37.]

37   Emil Ludwig, "Ludwig Calls Edison Secret King of U. S.: 'Everything About Him Reveals His Originality' " [Newspaper article, source unknown], March 1928. ENHS.

37   Correspondence, James H. Fullwood: TAE, November 1930. Reminiscences of their early days in Port Huron, when "from the spring of

PAGE

1865 to 1867, [Fullwood] was a wheelsman on the old *W. J. Spicer*, the Grand Trunk Passenger Ferry between Point Edward and Fort Gratiot across the St. Clair Rapids." Fullwood, repatriated to England, evocatively recalls "the sound of the caulking mallets at the mouth of the Black River when they were ship building." ENHS.

39 Alva Johnston, "Thomas Alva Edison; The Wizard," Three-part profile, *The New Yorker*, December 28, 1929; January 4, 1930; January 11, 1930.

42 John Meurig Thomas, *Michael Faraday and the Royal Institution: The Genius of Man and Place* (Bristol, 1991). A popular, schematic biography of the great English scientist, drawing heavily upon his journal writings and setting forth his most prominent chemical and electrical discoveries. With relation to my archival research in London on Faraday's role in the evolution of modern science, I am grateful for personal assistance from Mrs. I. M. McCabe, Librarian/Information Officer, Royal Institution of Great Britain, and Susan Devlin and Susan Moak of her staff; and to Mrs. Lenore Symons, Archivist, The Institution of Electrical Engineeers, and her assistant, Oliver Wooller.

42 TAE's 1867 pocket notebooks, containing book lists to be referenced, lists of tools to be purchased for his makeshift workshop in Cincinnati, and various drawings for "polarized repeaters" and other improvised telegraphic machines. ENHS.

42 Interview with Ezra Gilliland, "Thomas Alvey Edison [*sic*]," Cincinnati *Commercial*, March 18, 1878. In Charles Batchelor's scrapbooks, ENHS.

43 Correspondence, John V. Miller (TAE's nephew): Dr. Elliott P. Joslin, August 5, 1931. A summary medical history of the Edison family, with reference to Nancy Edison's "nervous trouble" and sister Harriet Ann's death in "childbirth." ENHS.

CHAPTER 5

44 Frederick Jackson Turner, *The Frontier in American History* (New York, 1962), reprint of original 1920 edition; and *Frontier and Section:*

Selected Essays, ed. Ray Allen Billington (New York, 1961). These are the classic and definitive dicta on the theory of the frontier as it relates to the psyche and identity of America.

45 Horatio Alger, *Ragged Dick and Mark, The Match Boy* (New York, 1968). Edited and with an introduction by Rychard Fink. These two early works are set in context by Professor Fink's richly anecdotal and immensely helpful introduction, illustrating that there are as many myths *about* Alger as he himself created.

45 Richard Weiss, *The American Myth of Success: From Horatio Alger to Norman Vincent Peale* (Urbana and Chicago, 1988). It is impossible to come to terms with Edison's rise to prominence without understanding the same myths that he grew up with and in many ways personified. Weiss establishes a tradition for the myth and in so doing helps us place Edison that much more specifically.

46 TAE pocket notebooks, 1870, n.d.; July 28, August 5, 1871. ENHS. Edison here makes the first of innumerable exhortatory and cautionary notes to himself on the virtues of keeping a "full record" of all his ideas and inventions.

47 Byron M. Vanderbilt, *Thomas Edison, Chemist* (Washington, 1971). This is still the best single-volume approach to Edison's involvement in chemistry, his first love, from boyhood through to his experiments with synthetic rubber in the 1920s. While at times technical and aimed at the specialist, the book is important because it throws into relief another of Edison's facilities and reminds us, as the author points out, of his "place in the chemical fraternity."

48 John F. Kasson, *Civilizing the Machine: Technology and Republican Values in America, 1776–1900* (New York, 1976). Kasson is expert at drawing together the pastoral and technological strains in nineteenth century American history as well as pointing out Emerson's role in delineating the parallels between Industry and Nature.

48 Alfred North Whitehead, *Science in the Modern World,* The Lowell Lectures (New York, 1925). For the purposes of this study, Chapter VI, "The Nineteenth Century," is most useful. However, Whitehead's views on the changing definition of, again, "progress," are illuminating insofar as he illustrates the tie-in of this concept to the birth of Darwinism in Edison's time.

50   Correspondence in ENHS archives, Daniel H. Craig:TAE, August 12, 1870. TAE to Craig and Harrington, various dates throughout 1870. Some of these letters are also cited and footnoted very helpfully in *The Papers of Thomas A. Edison,* Vol. I, The Edison Papers Project.

51   John T. Cunningham, *Newark,* New Jersey Historical Society (1988) and *New Jersey: America's Main Road.* Good general overview of the city, state, and their histories.

51   Samuel H. Popper, *Newark, New Jersey, 1870–1910: Chapters in the Evolution of an American Metropolis,* unpublished dissertation, New York University, submitted February 1952. Exhaustive and detailed, industry by industry account of the city.

51   Herbert Gutman, *Work, Culture and Society in Industrializing America: Essays in American Working-Class and Social History* (New York, 1976). The title essay, as well as "Protestantism and the American Labor Movement," "The Reality of the Rags to Riches Myth," and "Class, Status and Community Power in Nineteenth-Century American Industrial Cities," all discuss previously unexamined aspects of American working-class life in the years immediately following the Civil War. Anecdotal information on the special situation in Paterson, New Jersey, as a metaphor helps us understand what was happening in neighboring Newark at the same time.

52   Correspondence, Samuel Edison:Mary Sharlow Edison, June 30, 1887 and March 19, 1893. Refer to her as his "dear wife." Archives of Henry Ford Museum and Greenfield Village, Dearborn, Michigan.

53   Benjamin Marshall Stilwell, *Early Memoirs of the Stilwell Family* (New York, 1878). This volume was the property of Alice Stilwell Holzer, Mary Stilwell Edison's older sister, and was donated to the ENHS archives by the Stilwell family. Family trees for the Stilwells compiled by the Charles Edison Fund are also in the ENHS collections.

53   John E. Stilwell, M.D., *The History of Captain Nicholas Stilwell and His Descendants* (New York, 1930). A privately printed compendium and genealogy in the collection of the New Jersey Historical Society, as is the Newark City Directory for 1869–72, wherein may be found the occupations and street addresses of all Newark residents.

53   The Newark *Daily Journal* for December 26, 1871, may be found on microfilm in the Newark Public Library.

54 Walter L. Welch, *Charles Batchelor, Edison's Chief Partner* (Syracuse, 1972). The papers of Charles Batchelor (journals, notebooks, patents, unbound documents, letterbooks, accounts, and scrapbooks), preserved by the Charles and Rosanna Batchelor Memorial, were deposited into the custody of the Edison National Historic Site in West Orange, New Jersey. This brief monograph covers the years of Batchelor's association with Edison into the 1890s and relies heavily upon Batchelor's voluminous, compulsive record-keeping. The citations from Batchelor's correspondence and journals as well as from newspaper articles pertaining to Edison's inventions are all derived from the Batchelor collection at ENHS.

55 Robert A. Rosenberg, Paul B. Israel, et al, *The Papers of Thomas A. Edison,* Vol. 2, *From Workshop to Laboratory, June 1873–March 1876* (Baltimore and London, 1991). Takes Edison from his early quadruplex work through the Newark years and up until the verge of his move to Menlo Park. In scope and scholarship on the nature of Edison's inventive progress, the study is unrivaled, comprehensive, and exhaustively detailed. The scholarly emphasis is decidedly toward the history of technology rather than the psychology of invention.

61 Robert V. Bruce, *Alexander Graham Bell and the Conquest of Solitude* (Ithaca and London, 1973). A meticulous disaggregation of the intertwining developments leading to the invention of the telephone in the early and mid 1870s. The author's prejudice is unabashedly toward Bell rather than Edison, whom Bruce portrays as an ingenious interloper infringing upon Bell's staked-out territory.

63 Charles L. Dana, M.D., *Dr. George M. Beard, A Sketch of His Life and Character* (Chicago, 1923). Very little has been written on the elusive Dr. Beard. This pamphlet sheds light on some of his lesser-known idiosyncrasies and makes it clear from the point of view of one who knew Beard that "Nervousness, nervous exhaustion and the symptomatology of what we now call psychoneuroses was his longest and deepest love."

64 A. D. Rockwell, M.D., *The Late Dr. George M. Beard, A Sketch* (New York, 1883). Tipped in to the presentation copy of this little pam-

phlet, New York Academy of Medicine Special Collections Library, is
a revealing note to Dr. Dana (above) from Dr. Rockwell dated November 3, 1905, stating, in part, "Beard's diary reminds me very
much of that kept by George Eliot in her early days. Both of them
had been brought up according to the strict letter of the law, both
were decidedly morbid and both gradually grew into a saner atmosphere. Beard worried about his soul a good deal, and scored himself
for loving the vanities of life."

64 Handwritten correspondence from Dr. George M. Beard at the
Archives of Electrology and Neurology offices on West 22nd Street
in Manhattan to TAE in Newark, December 1, December 5, December 10, December 14, 1875; ENHS.

65 Articles on "Etheric Force," December 1875–February 1876, from
The New York *Sun*, the New York *Daily Tribune*, the New York
World, and the *Telegraphic Journal* are excerpted from Charles Batchelor's scrapbooks in ENHS. The illustrated pamphlet, "Experiments
with the Alleged New Force" (November 1875) is located in the
George Beard Papers, Manuscripts and Archives, Yale University Library. The articles, "Etheric Force" and "Mr. Edison's Curious Experiments" appeared in *Scientific American* for December 25, 1875; January 15, 1876; and January 22, 1876, and were provided to the author
by Mr. Neil Maken from his collection.

The following six works cited together provide an overview of the
prevailing themes of the Gilded Age from a stimulating variety of
perspectives ranging from social history to in-depth biography.
Peter N. Carroll and David W. Noble, *The Free and the Unfree, A New
History of the United States* (New York, 1977). Especially helpful with
respect to the roots of the Panic of 1873 and the role of the big capitalists therein.
Harold G. Vatter, *The Drive to Industrial Maturity, The U. S. Economy,
1860–1914* (Westport, Conn., 1976). Analysis of the Panic from a
mainstream economist's stance, illustrating convincingly that there
were many shades to the so called "depression" which ensued, some of
them deeper and more pervasive than others. This book helps explain
the survivalist character of Edison's manufacturing.
Robert H. Wiebe, *The Search for Order, 1877–1920* (New York,

1967). Vividly portrays the dramatic shift in American life from "the values of the village" to industrialism, urbanization and technology. Discusses the pros and cons of the syndrome of progress which took such an influential role in the American psyche during this crucible time.

Ron Chernow, *The House of Morgan, An American Banking Dynasty and the Rise of Modern Finance* (New York, 1990). A view of the period through the coming to power of the quintessentially capitalist family, and an instructive illustration thereby of the dangers inherent in ceding financial control to outside investors. This was a vicious circle Edison entered the hard way: how to maintain precious creative autonomy while at the same time allowing others to provide necessary infusions of support.

Sean Dennis Cashman, *America in the Gilded Age, From the Death of Lincoln to the Rise of Theodore Roosevelt*, 2nd ed. (New York, 1988). Pulls together the strands of industrial America, the paradoxes of reconstruction, and the increasingly dark tones of dissent leading to the Populist revolt.

Herbert G. Gutman, *Power & Culture, Essays on the American Working Class* (New York, 1987). The last work by the labor and social historian, focusing upon the true ways of working life during the Gilded Age of American corporate growth. These essays shed light on the role of the disenfranchised laborers especially in the northeast, and their search for a voice in lifting American industry to new heights.

CHAPTER 7

66 The quotation from Ralph Waldo Emerson is on page 19 of a booklet privately printed at the Golden Hind Press in Madison, New Jersey by Charles Edison in 1944, entitled, *Thoughts: Favorite Quotations of Mr. and Mrs. Thomas A. Edison*, dedicated "To Mother and my Two Sisters." Charles Edison notes that the quotations particularly favored by Mr. Edison are indicated with an asterisk (*).

66 I have taken the well-worked concept of "the American experiment"

from Arthur Schlesinger, Jr., *The Cycles of American History* (Boston, 1986). Schlesinger views the past as prologue. With respect to Edison, we must put him in a social continuum and begin to treat him as part of a much wider tradition of the kind defined by Schlesinger.

67 Thomas J. Schlereth, *Victorian America: Transformations in Everyday Life* (New York, 1991) examines the four decades between the Philadelphia Centennial in 1876 and the Panama-Pacific Exposition of 1915.

68 Juxtapositions of industrial and rural landscapes and their tradition of interrelationship in American literature and culture are discussed in the classic *The Machine in the Garden: Technology and the Pastoral Ideal in America,* by Leo Marx (Oxford, 1964).

68 My descriptions of Menlo Park and environs have been enhanced immeasurably by a visit to the Henry Ford Museum and Greenfield Village in Dearborn, Michigan, where the entire Edison complex has been replicated. William S. Pretzer, curator of programs, and John Bowditch, curator of industry, as well as Luke Swetland, Robert Casey, and the staff of the archives provided guided tours and research assistance. Mr. Pretzer has edited the single most definitive collection of essays on Menlo Park, *Working at Inventing: Thomas A. Edison and the Menlo Park Experience* (Dearborn, 1989), which contains work by scholars Bernard S. Finn, Andre Millard, Paul Israel, W. Bernard Carlson and Michael E. Gorman, Edward Pershey, and David A. Hounshell.

68 The Henry Ford Museum and Greenfield Village has also published an illustrated guidebook to the site, *Thomas A. Edison's Menlo Park Laboratory including the Sarah Jordan Boarding House* (Dearborn, Mich., 1990).

68 The letters from Thomas Edison at Menlo Park to his father in Port Huron, Michigan, are in the Archives of the Henry Ford Museum and Greenfield Village. Responding letters from Samuel Ogden Edison to his son are more difficult to find, but one example in the ENHS collections, complete with misspellings, expresses the father's strong ambivalence about coming east to live permanently with the young family and the grandchildren he adores: "I know that Dot is very smart and I think Tommy will be equaly so. I would be very

glad to com and see you all but fate has ordered it other wise and I must submit. . . . After I came home from Menlo I bilt a store and used up all the redy money I had. . . . God only knows how I shall pay my taxes for money hear is very scarce. I have plenty owing to me but the people canot pay until spring opens and business revives. I have 3 cows and fifty hens and one pig—I would like to go to Menlo and spend my days for I like that part of New Jersey very much but I have too much property hear to leave and be of no value for Fort Gratiot is bound to be a big place and out gro Port Huron."

69   The importance of Menlo Park is well conceived in Jonathan Hughes, *The Vital Few: American Economic Progress and Its Protagonists* (Oxford, 1973). Hughes argues that "progress" is not an abstract concept by any means, but rather the result of concrete accomplishments by specific individuals. He includes a discussion of Edison alongside the lives of William Penn, Brigham Young, Eli Whitney, Andrew Carnegie, Henry Ford, E. H. Harriman, and J. P. Morgan.

70   Documentation of Charles Batchelor's experiments with the electric pen, including schematic sketches for the implement, test drawings, advertisement circulars, autographic notes to Edison, cartoons, menus, and letters to his father are found in the Charles Batchelor papers at ENHS.

74   The citation of Walt Whitman's "Song of Myself" is from Whitman's *Complete Poetry and Selected Prose*, ed. James E. Miller, Jr. (Cambridge, Mass., 1959).

75   The best theorist on the American transcendentalists is Perry Miller, whose work provides here the context for Edison's position with relation to the ideas of Emerson; and, prior to that, the logic for Edison's cultural debt to Benjamin Franklin. I have used Miller's *The American Transcendentalists: Their Prose and Poetry* (New York, 1957) as the source for Emerson's writing and for an overall vision of transcendentalist themes; and his groundbreaking work, *Nature's Nation* (Cambridge, 1967), the story of America's thrust into the wilderness from Puritan beginnings.

Two excellent scholars on Edison are drawn into play in this chapter and will reappear throughout the story. They take radically different approaches. The more traditional, history of technology, route may be found in a fine pamphlet by Thomas P. Hughes originally published in London by the Royal Science Museum: "Thomas Edison: Professional Inventor" (1976). Hughes's analysis of the telephone is quite accessible. A more theoretical, archetypal slant has been taken in David E. Nye, *The Invented Self: An Anti-Biography of Thomas A. Edison* (Odense [Norway] University Press, 1983). Nye examines the mythic components of Edison's career.

A pioneering work by Norbert Wiener with special homage to Edison, *Invention: The Care and Feeding of an Idea*, has just been brought out (1993) by MIT Press based upon an unpublished manuscript written in the 1950s. The emphasis is on the historical circumstances which foster innovation. Wiener speculates on "what history would have been like without Edison" and underscores the importance of the craftsman tradition in a consideration of Edison's work. Paying homage to Menlo Park, Wiener declares that "Edison's greatest invention was not scientific but economic. It was the invention of the industrial scientific laboratory in which a moderately large trained crew of technicians was directed by a central mind toward the making of inventions as an everyday business."

CHAPTER 8

77   Francis Jehl, *Menlo Park Reminiscences*, Vol. I. (New York, 1990). This Dover edition is an unabridged republication of the original 1937 edition of the work published by the Edison Institute, Dearborn, Michigan. It contains an illuminating and cautionary introduction by William S. Pretzer, curator of the Henry Ford Museum and Greenfield Village, who advises the reader to take Jehl's "filiopietistic adulation" with a few grains of salt. While there are errors of fact and deeply romanticized stretches in the *Reminiscences*, their value resides in the way Jehl captures the flavor of the time and place.

77   The on-site chronology of Edison's intensive work on the phonograph

and its interplay with Bell's progress and publicity around the telephone is drawn from the following primary sources: Charles Batchelor's *Diary* in ENHS, each day (including day of the week) documented; autographic pen copies of Batchelor's letters home and to other correspondents in England, in ENHS; clippings from Batchelor's Scrapbooks, ENHS, from the New York *Sun,* New York *Herald,* Newark *Daily Advertiser,* New York *Daily Graphic,* Albany *Daily Evening Times,* and *Scientific American.*

77   E. C. Baker, *Sir William Preece, F. R. S.: Victorian Engineer Extraordinary* (London, 1976), in the collections of the Institution of Electrical Engineers, London, especially the chapter, "The New World," containing excerpts from Preece's journal of his first trip to America in 1877. Further information on Preece and his relationship with Edison is held in the Royal Post Office Archives, London: Private Collections, Post 106. Preece was one of the early pioneers of wireless telegraphy and was consulted by Marconi when conducting experiments in England. His pervasive interest in the development of electrical science and telegraph engineering in particular led him to introduce many new ideas and developments in fields akin to Edison's, including duplex telegraphy, railway signaling, and telephony.

79   The formal articulation of the phonograph's ("speaking machine") mechanical function is drawn from *Thomas A. Edison, the Complete U. S. Patents,* seven volumes, ENHS. Also on microfilm.

80   Edward Jay Pershey, "Drawing as a Means to Inventing: Edison and the Invention of the Phonograph," in Pretzer, et al. This is a painstaking reconstruction of Edison's visual thought processes throughout the crucial year 1877 when he was developing the first phonograph.

82   The author would like to thank Charles Hummel of Wayne, New Jersey, Edisonia collector and museum coordinator for the Charles Edison Fund, for personally demonstrating the original foil phonograph, complete with "Mary Had a Little Lamb" dictation.

83   Joseph Brent, *Charles Sanders Peirce: A Life* (Bloomington, Ind., 1993). A fresh look at the heretofore-neglected and tragic life of this American philosopher. The circumstances surrounding the birth of pragmatism dovetail with the evolution of the phonograph. Perry

Miller, *American Thought: The Civil War to World War I* (New York, 1954), touches upon Peirce's seminal 1877 essay, "The Fixation of Belief," espousing the virtues of "tenacity" as symptomatic of "competitive America."

85 Thomas A. Edison, "The Phonograph and Its Future," *North American Review*, May–June 1878, 527–536. "Of all the writer's inventions," the essay begins, "none has commanded such profound and earnest attention throughout the civilized world as has the phonograph. This fact he attributes largely to that peculiarity of the invention which brings its possibilities within range of the speculative imaginations of all thinking people."

The essay was also published in Detroit and Toronto by the Rose-Belford Publishing Company that same year as part of their International Religio-Science Series, as the lead essay in a pamphlet also containing an anonymous article on "The Auriphone and Its Future" (lauding Edison's development of a primitive intercom system leading from the first floor to the second floor of the laboratory building); and a lengthy piece by T. H. Huxley, "On the Hypothesis that Animals are Automata, and its History."

86 "A Night with Edison" (anon.), *Scribner's Monthly*, November 1878, makes the telling analogy with Nathaniel Hawthorne in its concluding paragraphs: "Edison has great heaps of notebooks which, technical and abstruse as they are, send one's thought for a moment to Hawthorne. . . . It is the study of both alike to place their characters in unusual circumstances and watch the result."

"The Hall of Fantasy" and "The Celestial Railroad" follow each other in Hawthorne's *Complete Short Stories*. In the first tale, the narrator finds himself in a noble edifice dedicated to the memory of men of genius who throughout the ages have dominated the realm of the imagination yet who today wish to be considered on the same general social level as their peers—"the incongruous throng"—not placed on a pedestal. In the second, a Bunyanesque journey through the "Valley of the Shadow of Death" is made less gloomy by the illumination of "fiercely gleaming" gas lamps artificially but nonetheless reassuringly transforming night into day.

87   This chapter draws its anecdotal flavor from numerous reports by
contemporary journalists clipped and filed by Charles Batchelor and
preserved in his *Scrapbooks* housed in ENHS. These include especially,
but not exclusively: "Edison's Phonograph," *English Mechanic*, January 25, 1878; "Edison's Phonograph," *Evening Bulletin*, Providence,
Rhode Island, January 30, 1878; "Thomas A. Edison, The Electrician
and Inventor," *Phrenological Journal and Life Illustrated*, February 1878
("Among the many ingenious men which electric art has brought to
public knowledge."); "That Wonderful Edison, Perfecting the
Phonograph Beyond Even the Dreams of the Caricaturists," New
York *World*, March 24, 1878; "Bottled Talk," Philadelphia *Times*,
March 9, 1878 ("Showing how to stereotype your vocal utterance,
making a plate from which a perfect reproduction of your speech can
be had whenever you are in the grave."); "Marvellous Mechanism,"
Philadelphia *Press*, March 9, 1878; "The Speaking Phonograph,"
New York *Daily Graphic*, March 12, 1878; "A Wonderful Machine
That Will Be Of Immense Value For Various Purposes," New York
World, March 20, 1878; "The Aerophone," New York *Times*, March
25, 1878; "Awful Possibilities of the New Speaking Phonograph,"
Daily Graphic, New York, March 21, 1878; *Punch's Almanack for
1878*; "The Napoleon of Science, Early Days of the Marvellous Man
at Menlo Park," New York *Sun*, March 10, 1878; "Professor Edison
Exhibits the Phonograph to Visitors," *Frank Leslie's Illustrated News-
paper*, New York, March 30, 1878; "The Papa of the Phonograph, an
Afternoon with Edison," New York *Daily Graphic*, April 2, 1878;
"The Wonderful Edison," Cincinnati *Courier*, April 8, 1878; "The In-
ventor of the Age," New York *Sun*, April 29, 1878; "Invasions by the
Telephone," New York *Daily Graphic*, April 30, 1878; "A Visit to
Edison," Philadelphia *Weekly*, April 22, 1878; "A Wonderful Ge-
nius, a Life of Eccentricity, Study, and Usefulness," Boston *Herald*,
April 14, 1878; "An Interesting Session Yesterday at the National
Academy of Sciences," Washington *Star*, April 19, 1878; "The Man
Who Invents," Washington *Post & Star*, April 19, 1878; "The Amer-
ican Mechanic," New York *Telegram*, April 25, 1878; "Edison's Ma-

chine, How a Piece of Mechanical Ingenuity Reproduces the Human Voice," Newark *Morning Register*, May 3, 1878; "Thomas A. Edison," Chicago *Tribune*, May 4, 1878; "Uncle Sam Astonishing the Crowned Heads of Europe," New York *Daily Graphic*, May 28, 1878; "A Visit to Edison, How the Great Inventor Lives and Works," Boston *Daily Globe*, May 28, 1878; "The Phonograph, Shaking Hands with Inventor Edison," St. Joseph (Missouri) *Herald*, May 23, 1878; "Edison at Home," Boston *Post*, May 24, 1878; "All About the Phonograph," *Christian at Work*, New York, May 23, 1878 ("Mr. Edison, who had been advised beforehand of the proposed visit, and extended an invitation to 'come on any day,' [*don't go without it, reader*] sat down at once at his Phonograph."); "The Phonograph," New York *Public*, May 2, 1878; "Edison in a Restaurant, His Theory of Eating," Cincinnati *Commercial*, May 3, 1878; "Is the Brain a Phonograph?" *English Mechanic*, June 14, 1878; "Ears for the Deaf," New York *Daily Graphic*, June 5, 1878; "A Voice of Thunder," Rochester *Daily Union and Advertiser*, March 20, 1878; "Edison's Ear Telescope, Conversation Carried On at Ease Between Persons Two Miles Apart," New York *Sun*, June 8, 1878; "A Visit to Edison," *Daily Fredonian*, June 6, 1878; *Scientific American*: "Mr. Thomas A. Edison," and "Edison's Telephonic Research," July 6, 1878; "An Hour with Edison," July 13, 1878; "Edison's Phonometer," July 27, 1878; "Edison's Megaphone," August 24, 1878; "A Night with Edison," *Scribner's Monthly*, November 1878.

88   The "Thoreauvian" letter in which Edison threatens to "take to the woods" was a handwritten note to "Friend [J. W.] Arnold" dated March 20, 1878, in ENHS Archives. Arnold was one of the cadre of phonograph demonstrators.

92   For extensive background pamphlets and research assistance with relation to the life and work of Mme. Helena P. Blavatsky, I am grateful to Nancy Cotter, circulation librarian at the Olcott Library and Research Center of the Theosophical Society in America in Wheaton, Illinois.

93   H. P. Blavatsky, *The Secret Doctrine*, ed. Elizabeth Preston and Christmas Humphreys (Wheaton, Illinois, 1992). 6th ptg.

93   Lewis Spence, *The Encyclopedia of the Occult, A Compendium of Informa-*

tion on the Occult Sciences, Occult Personalities, Psychic Science, Magic, Spiritism, and Mysticism (London 1988). Contains historic overview of Theosophy and its major tenets.

93 Sylvia Cranston, *HPB: The Extraordinary Life and Influence of Helena Blavatsky, Founder of the Modern Theosophical Movement* (New York, 1993). An exhaustive and compelling, if overly reverent account which attempts to redress some of the misperceptions about Mme. Blavatsky and the authenticity of her theories.

93 The correspondence precipitating the dispute between Edison and Olcott regarding Edison's membership in the Theosophical Society is in ENHS: Elliott Cowes:TAE, October 16, 1889, with newspaper clipping citing TAE as Theosophist. Cowes was a member of the American Academy of Sciences and had met Edison at his phonograph presentation in the Smithsonian Institution in April 1878. Alexander Hamilton Church:TAE, November 14, 1889. Writing from Birmingham, England, Church had attended a recent lecture by Colonel Olcott in which the Colonel stated in reply to a question that Edison was a member of the Theosophical Society, "which, if true, goes far to dispel any doubts which I had respecting the genuine character of the movement." H. S. Olcott:TAE, December 20, 1889, "A Manchester paper prints a letter from your Secretary in reply to one addressed to you from Birmingham, to the effect that I was mistaken in stating that you are a member of the Theosophical Society. . . . This puts me in a false position." To set the record straight, Olcott encloses a copy of Edison's covering letter to Olcott of April 4, 1878, thanking Mme. Blavatsky for the copy of *Isis Unveiled*, and enclosing the membership forms for the Theosophical Society.

94 Henry Steel Olcott, *Old Diary Leaves, The History of the Theosophical Society As Written by the President-Founder Himself, First Series, America, 1874–1878*, Theosophical Publishing House (Adyar, Madras, India, 1974).

95 H. P. Blavatsky, *Collected Writings, 1874–1878,* vol. I, Theosophical Publishing House (Adyar, Madras, India, 1966). Mme. Blavatsky kept a daily diary of her time in New York City, including her thoughts on Edison and the phonograph.

PAGE

95 H. P. Blavatsky, "Magic" (essay), *The Dekkan Star*, Poona, India, March 30, 1879. *Collected Writings*, vol 2.

95 H. P. Blavatsky, "The Cycle Moveth" (essay), *Lucifer*, vol. I, no. 31, March 1890. *Collected Writings*, vol. 3.

95 H. P. Blavatsky, "Kosmic Mind" (essay), *Lucifer*, vol. 6, no. 32, April 1890. *Collected Writings*, vol. 3. This essay in defense of Edison's "flights of fancy" is headed with an epigraph from Cicero's *Tusculan Disputations*: "Whatever that be which thinks, which understands, which wills, which acts, it is something celestial and divine, and upon that account must necessarily be eternal."

97 Thomas S. Kuhn, *The Copernican Revolution: Planetary Astronomy in the Development of Western Thought* (Cambridge, Mass., 1957, 1985). Stimulus for the theory of Galileo and the telescope as a model for Edison and the phonograph.

98 Norman R. Speiden, curator, laboratory of Thomas A. Edison, West Orange, New Jersey, "Thomas A. Edison, Sketch of Activities, 1874–1881." *Science*, Edison Centennial edition, February 1947. Details about Edison's western trip.

98 Correspondence, Rev. John Heyl Vincent:TAE, February 25, February 27, April 12, April 25, and August 12, 1878, regarding commitment from Edison to be at Chautauqua during August, ENHS. On the second page of Vincent's April 12 letter, Edison has written a draft response for his secretary: "Friend Vincent: I will come and furnish a man to do the talking." The April 25 letter is a confirmation of the date of the lecture, set for August 20. However, S. L. Griffin telegraphs Vincent on August 13 that "Mr. Edison is in Ogden, Utah. Doubtful about his being with you twentieth." Edison had become deeply involved in his post-eclipse vacation jaunt and was unable to honor the arrangement with Vincent. The vicissitudes of circumstance would soon connect Edison far more intimately with Chautauqua than he could ever have imagined.

98 Nathan Reingold, *Science in Nineteenth-Century America, A Documentary History* (Chicago, 1964). Correspondence from Edison to Henry Draper one year before the eclipse expedition regarding spectroscopic research.

PAGE

99   Thomas A. Edison, "On the Use of the Tasimeter for Measuring the Heat of the Stars and of the Sun's Corona, being a report to Dr. Henry Draper, and presented by his permission." Offprint from the *Proceedings* of the American Association for the Advancement of Science, St. Louis Meeting, August 1878. Contains lithographic representation of the tasimeter and cutaway diagram.

99   Thomas A. Edison, letter to the editor of the New York *Daily Tribune*, June 8, 1878, is a lengthy defense of his innovative use of the carbon button in the tasimeter. Batchelor Scrapbooks, ENHS.

100   Correspondence, J. M. Davis:TAE, November 24, 1926; and TAE: J. M. Davis, November 26, 1926, ENHS, provide reciprocal accounts of the rifle shooting episodes in Rawlins. "Those were glorious days!" Edison concludes.

101   John A. Eddy, High Altitude Observatory, National Center for Atmospheric Research, Boulder, Colorado, "Thomas A. Edison and Infra-Red Astronomy," reprinted from *Journal for the History of Astronomy*, vol. 3 (Cambridge, England, 1972). Definitive essay on Edison's tasimeter, bringing to bear modern technological knowledge on astronomy to validate Edison's claims and making use of ENHS correspondence as well as primary and contemporary scholarship on infrared research in general and the tasimeter in particular.

101   The New York *Daily Graphic*, July 13, 1878; *Scientific News*, n.d. (summer), 1878; New York *World*, August 27, 1878 ("Tom Edison Back Again, The Marvellous Man Returns to His Workshop Full of Prairie Breezes"); and New York *Sun*, August 28, 1878, contain lengthy stories about the eclipse trip and Edison's descriptions of subsequent travels to San Francisco.

CHAPTER 10

102   The epigraph from Karl Friedrich Gauss may be found in W. I. B. Beveridge, *The Art of Scientific Investigation* (New York, 1950). The book itself begins with an appropriate quotation from William

Henry George: "Scientific research is not itself a science; it is still an art or a craft."

103   Charles Batchelor's *Scrapbooks* in ENHS once again chronicle rich background detail derived from stories in the daily press pertaining to Edison's work on the electric light: "The New Electric Light, Lamps That Outshine Canopus Piercing the Autumnal Fogs of London," London *World,* August 31, 1878; "A Great Triumph, an Invention That Will Revolutionize the Motive Power of the World," New York *Daily Mail,* September 10, 1878; "Edison's Newest Marvel," New York *Sun,* September 16, 1878; "Edison's Electric Light," New York *Herald*, October 12, 1878; "Fiat Lux! How the Gas Companies Feel Concerning Edison's Reported Invention," ibid, October 14, 1878; "Gas vs. Electric Lighting," *Telegraphic Journal*, London, October 15, 1878; "Edison's Electric Light," New York *Sun*, October 20, 1878; "Edison's New Light," New York *Daily Graphic,* October 21, 1878; "The Man Who Moves the World," ibid., October 26, 1878; "The Electric Light, An Authoritative Explanation of Edison's Great Invention," New York *Herald*, December 11, 1878.

106   Correspondence, Charles Batchelor:James Batchelor, September 16, 1878, ENHS, refers to "solving the problem of the subdivision of the light, but it will not be shown until all the patents are secure." Clearly, Batchelor was seized by Edison's premature fervor that a solution was near.

107   An exchange of telegraphs between Drs. Ward and Daily in Newark and Thomas Edison in Menlo Park clarifies the course of events around Edison's "neuralgia" and the birth of William Leslie Edison, October 24–26, 1878, ending with Edison's statement that he has "got a twelve pound boy," ENHS.

108   Francis Upton's appearance on the scene at Menlo Park toward the end of 1878 is well documented in a variety of sources: In the collections of the New Jersey Historical Society in Newark, there is one box of manuscript material and other ephemera donated by Upton's daughter, Eleanor, including a typescript reminiscence, "Memorandum Regarding Thomas A. Edison," dated December 7, 1908, which traces the roots of their relationship back to its beginnings: "It was a

wonderful experience to have problems given me out of the intuitions of a great mind, based on enormous experience in practical work, and applying it to new lines of progress. . . . I have often felt that Mr. Edison purposely got himself into troubles by premature publications and otherwise, so that he would have a full incentive to get himself out of the trouble."

108 The Francis R. Upton Collection, 1878–1918, housed in ENHS, was donated in 1963 by Paul Kruesi, the son of John Kruesi, who had also received some materials from Eleanor Upton. These include a valuable series of letters from Upton to his father, including the following, excerpted for this study: November 7, 1878; March 23, 1879; June 15, 1879; July 6, 1879; August 24, 1879; October 19, 1879; November 2, 1879; November 16, 1879; December 21, 1879; and December 28, 1879.

108 The Francis R. Upton Notebooks, November-December 1878, housed in ENHS, provide details on Upton's literature search for Edison. The first notebook is inscribed, "Mch. 1905. This was my first work for Mr. Edison. FRU." The notebooks were used in both directions.

109 Upton also left a handwritten and typescript manuscript of observations on the character and methodology of Edison, as well as on the first theories formulated while the search for a viable electric light was being conducted, which from internal evidence dates from the time of the founding of the Edison Pioneers in 1918. Upton was the first president of the group: "When I first went with Mr. Edison," Upton recalls forty years later, "It was common to make comparisons of electro motive [*sic*] forces by stating that the tension was equal to so many batteries and to speak of resistance in terms of so many miles of wire. The unit for measurements of electric current could be found only in the thoughts of comparatively few mathematicians. . . . At Menlo Park I worked hard to sharpen the definitions of electrical units and to fix reliable standards for physical measurements.

"I always prefer to think of electricity as a fluid," he continues, "though I know it is not, yet such a conception seems to render it more clear in my mind. . . . Electricity as we know it generally is energy in a very agile form. There are many reasons to think that it is in the form in which latent energy combines with matter in its ordinary states. When once the bond between matter and this energy is broken, we find electricity distributing the energy over the surrounding medium and allowing it to be given off as heat, light, motion, etc."

110   "The Edison Light," *Journal of Gas, Lighting, Water Supply, & Sanitary Improvement*, London, February 18, 1879, a harsh and disdainful critique of Edison's threat to the gas industry; "The Edison Electric Light," *Scientific American*, March 22, 1879; "Edison Going to Try the Light in Metuchen," New York *World*, March 23, 1879; "Edison Still Hard at Work," New York *Daily Graphic*, April 11, 1879; "Edison's Electric Light, The Great Inventor Frankly Declares It Entirely Successful," New York *Sun*, April 12, 1879; "Edison's Long Delay," and "Wanted, A Platinum Mine," ibid., July 7, 1879; "The Wizard's Search," New York *Daily Graphic*, July 9, 1879; "Edison in His Workshop," *Harper's Weekly*, New York, August 2, 1879; "Edison's Light, The Great Inventor's Triumph in Electric Illumination," New York *Herald*, December 21, 1879; "A Lucky Horseshoe," New York *Herald*, December 22, 1879; "Edison's Lamp Yet Burning," New York *Sun,* December 24, 1879; "Edison's Success," New York *American*, December 27, 1879; "Menlo Park Illuminated Last Night with Forty Electric Lamps," New York *Herald*, December 28, 1879; "A Night with Edison, Some Flashes from A Laboratory Symposium," ibid., December 31, 1879.

112   Charles L. Edgar, W. W. Freeman, et al., Lamp Committee Report, Association of Edison Illuminating Companies, *The Development of the Incandescent Electric Lamp Up to 1879* (New York, 1929). A pamphlet digest of inventive progress from Sir Humphry Davy to Edison, with affidavits attesting to Edison's consummate role in the history of the development of the lamp. "At the end of 1878," the authors conclude, "electric lighting of general applicability did not exist. At the end of 1879 there was a practical lighting system which was the prototype of that now in use. That system was Edison's!"

112   Robert Friedel and Paul Israel, with Bernard S. Finn, *Edison's Electric Light, Biography of an Invention* (New Brunswick, N.J., 1987). An exhaustive, day-by-day chronicle of the development of the light drawn from laboratory notebooks and contemporary accounts. Probes

painstakingly beneath the popularly accepted scenarios to discover the actual, detailed procedures followed by Edison and his colleagues.

113   In addition to Edison's prescient discussion of "systems" in the *North American Review* of May-June 1878, we refer to his reply to H. G. Colton, dated February 3, 1912, citing "electric systems" as his greatest invention. In ENHS there is also an undated pencil manuscript in Edison's handwriting on the letterhead of "Laboratory of Thomas A. Edison, Orange, N. J.," (meaning that it postdates 1886), meant as a one-paragraph biographical summary apparently for inclusion in a scientific reference work, which begins: "Inventor—Remembered for his energy in the application of science to Industry. *His greatest success was the invention of the incandescent light system.*" Finally, an undated typescript in the Subject Files of the Edison Electric Institute Library, Washington, D.C., "Short History of the Companies Organized Under Mr. Edison's Direction For Distribution of Electric Current for Light and Power," provides background on the founding of the Edison Electric Light Company and its original incorporators.

113   Thomas A. Edison, *The Beginning of the Incandescent Lamp and Lighting System* (Dearborn, Mich., 1926). Republished in 1976 by the Edison Institute, with a foreword by Theodore M. Edison and an introduction by Robert G. Koolakian. Written at the request of Henry Ford, this remains the only autobiographical account by Edison of the early process leading up to the bulb in the late 1870s.

113   Wolfgang Schivelbusch, *Disenchanted Night: The Industrialization of Light in the Nineteenth Century* (Berkeley, 1988). Translated from the German, *Lichtblicke. Zur Geschichte der kunstlichen Helligkeit im 19. Jahrhundert* (1983). As with *The Railway Journey*, cited earlier, there is no one as apt and incisive as Schivelbusch when focusing upon the archetypal resonances of cultural symbols, and the interpenetration of public and private lives influenced by technological development.

114   Thomas P. Hughes, *Networks of Power, Electrification in Western Society, 1880–1930* (Baltimore, 1983). The magisterial study of the dissemination of the commodity of electricity throughout the western world and of Edison's role in empire building toward an international syndicate based upon selling the light.

114   Thomas P. Hughes, *American Genesis: A Century of Invention and Tech-*

nological Enthusiasm, 1870–1970 (New York, 1989). While recapitulating many of the arguments posed in a more theoretical manner in his earlier work, Hughes here popularizes the story of "practical ingenuity" in America by opening it up to other makers of our modern world, including Edison, Elmer Sperry, the Wright brothers, Lee De Forest, and Henry Ford.

CHAPTER 11

116 The Charles Batchelor Scrapbooks at ENHS continue their day-by-day tracking in the newspapers of Edison's progress: "Depredations of Progress," New York *Herald*, January 3, 1880; "Mr. Edison," *Whitehall Review*, London, February 5, 1880; "The Electric Light as Produced at Menlo Park," *Times and Expositor*, March 8, 1880; "Edison at Home, He Talks Freely of His Progress and Plans" ("The Wizard defends the electric light and says, 'Just you wait.'"), Denver *Tribune*, April 25, 1880; "Jerome Pays Another Visit to the Wizard of Menlo Park," New York *Journal*, April 16, 1880; "Mr. Edison at Work," *Hand and Heart* magazine, July 20, 1880; "An Inventor's Workshop," New York *Times*, August 9, 1880.

118 For early Edison sketches of ore milling, Edison Laboratory Notebooks, ENHS, dated April 20, 1880, and signed by Charles P. Mott, Edison, and Batchelor. Edison's patent No. 228,329, for Magnetic Ore-Separator, dated June 1, 1880, based upon application filed April 7, 1880. Contemporary articles on the ore-separator in *The Engineer*, July 2, 1880; and *Scientific American*, July 17, 1880. On the electric railroad, lead article in *Manufacturer and Builder*, June 1880, vol. 12, no.6; The New York *Daily Graphic*, July 27, 1880, includes Edison's earliest pencil sketch of the 100-horsepower engine "to run between Perth Amboy and Rahway," front and side views of the locomotive and car; "An Inventor's Workshop," New York *Times*, August 9, 1880; "Everyone Hard at Work, Edison's Electric Railroad in Working Order," New York *Herald*, August 10, 1880; "Edison's Electric Locomotive," *The Railroader*, Toledo, Ohio, November 1880,

PAGE

vol. 3, no.11 ("To train-boy Thomas Edison is the world indebted for many useful and wonderful inventions and scientific discoveries"). Batchelor Scrapbooks, ENHS.

120   The story of Thomas Edison and *Science* is summarized in an unsigned essay, "Thomas Edison and the Founding of *Science*, 1880" ("Documents recently uncovered in the Thomas A. Edison Library at West Orange, New Jersey, show clearly for the first time the relationship between a journal, *Science*, published in New York in 1880 and 1881, and the present *Science*"), *Science* Centennial Issue, New York, 1980. A complete run of the first volume of *Science*, beginning with Vol. I, no. 1, July 3, 1880, is held at ENHS. Also the following articles in *Science*: "The Edison Light," July 10, 1880; "The Photophone," September 11, 1880; "An article in the *North American Review* . . . ," September 25, 1880; "Two or three weeks ago, we complained of the coldness of British writers . . . ," October 9, 1880; "Since we last referred to Mr. Edison and his incandescent lamp . . . ," April 30, 1881.

122   Thomas A. Edison, "The Success of the Electric Light," *North American Review*, October 1880, 295–300, includes Upton's defense of bamboo as the newest and most viable filament. The most authentic account of the origins of the bamboo search is found in a bilingual pamphlet by Sabura Tatemoto, "Thomas A. Edison and Japanese Bamboo," published by the Otokoyama Edison Historical Club (Kyoto, Japan, 1989). Shortly after Edison's death in 1931, the Japan Electric Association, the Japan Communications Association, and the Japan Telephone Association held a memorial service at Hibiya Hall in Tokyo. At that time, according to Mr. Tatemoto, they decided to erect a memorial to Edison on sanctified grounds at the peak of Otokoyama Mountain in Yawata, original source for Edison bamboo. Dedication ceremonies were held May 1934. Fifty years later, the memorial was replaced and rededicated. Grateful acknowledgement is made to Paul J. Christiansen, president, the Charles Edison Fund, for making the author aware of this narrative account.

123   The most complete description of the Edison–Swan rivalry is in the pamphlet by a professor in the department of electrical engineering, University of Newcastle, England, Diane Clouth, "Joseph Swan, 1828–1914: A Pictorial Account of a Northeastern Scientist's Life

and Work," published by the Gateshead Metropolitan Borough Council Department of Education, Local Studies Series (1979). This booklet includes the letter from Swan to Edison dated September 24, 1880, which was never sent, as well as the descriptions of Swan's speeches in Newcastle. Grateful acknowledgment to George Frow, president, City of London Phonograph and Gramophone Society, for sending the publication to the author. The comments in *La lumière électrique* were in response to reports of Swan's lectures on the bulb. "T.A.E.'s Lament" is in Batchelor's Scrapbooks. Coverage of the Edison–Swan litigation actions is in *The Electrician*, London, January 21, May 27, August 5, and September 25, 1882, Institution of Electrical Engineers archives, London. Transcript of Swan's lecture, "Electric Lighting by Incandescence," given at the weekly evening meeting of the Royal Institution of Great Britain, March 10, 1882, is in the library of the Royal Institution, London. Contemporary account of Swan's lecture, "The Engineer," London, March 24, 1882, is in Batchelor Scrapbooks, ENHS.

126   George Bernard Shaw, *The Irrational Knot: Being the Second Novel of His Nonage* (New York, 1931), contains a preface with Shaw's amusing anecdotes about his brief period working for the Edison telephone interests in London. The interview with E. H. Johnson at the Edison telephone headquarters in London is excerpted from "The Telephone," *Gravesend Argus, Kentish Chronicle and Port of London Journal*, no.1, May 15, 1880. The dispute between the Edison Telephone Company of London, Ltd., and the Attorney General in the High Court of Justice, Exchequer Division, Westminster Hall, is detailed in the judgment of that Court, December 20, 1880. These items are in the collections of the British Telephone Archives and Historical Information Centre, London. The author is grateful to archives manager David Hay, and to archives assistants Norman Harris and Yvonne Smith for their kind assistance.

127   Gertrude Himmelfarb, *Darwin and the Darwinian Revolution* (New York, 1959); Richard Hofstadter, *Social Darwinism and American Thought* (Philadelphia, 1944); and Sigmund Diamond, ed., *The Nation Transformed: The Creation of an Industrial Society* (New York, 1963), which includes a slightly abridged version of William James's

October 1880, *Atlantic Monthly* essay, were useful in formulating Edison's position relative to the doctrines of Darwinism and the ethos of progress.

128 Marion Edison Oeser, *Early Recollections*, voicewritten in Wilton, Connecticut, March 1956, ENHS, is a poignant description reaching back to her childhood days growing up in Menlo Park and then New York City as the first child of Thomas and Mary Edison. Sarah Bernhardt's visit to Menlo Park in November 1880 is reconstructed from Marion's *Recollections* and from the thorough description provided in Arthur Gold and Robert Fizdale, *The Divine Sarah: A Life of Sarah Bernhardt* (New York, 1991). A contemporary newspaper article dated November 29, 1880, housed in the archives of the Henry Ford Museum in Dearborn, Michigan, describes Bernhardt's recitation of excerpts from *Phèdre* as well as *Hernani*.

129 Details on Edison's move to New York City in the new year of 1881: "Edison's Light," *Engineering News*, December 25, 1880; "Lighting over Snow, Midwinter Wonders of Edison's Light at Menlo Park, Why Do Our Aldermen Withhold a Permit?" New York *Herald*, January 20, 1881; "Invention as an Art," *The Indicator*, Indianapolis, Indiana, January 1881 ("Imagination and industry are requisite for finding or inventing; acuteness and penetration for discovering"); "The Inventor Moves His Establishment from Menlo Park to New York City," New York *Herald*, March 1, 1881.

129 An unsigned, typewritten manuscript in the library files of the Edison Electric Institute, Washington, D. C., "Short History of the Companies Organized Under Mr. Edison's Direction for Distribution of Electric Current for Light and Power," provides a chronological account of the corporate entities set up by Edison beginning with the Edison Electric Light Company in 1878 and concluding with the Edison General Electric Company in 1889. This document also breaks down the distribution of all the shops in New York City organized to handle the components of the Edison lighting system, coordinated by the 65 Fifth Avenue "home office." The "New York Letter" in the *Union and Advertiser*, May 23, 1882, takes the reader into Edison's new laboratory in Sigmund Bergmann's building on Avenue

B, where, upon first meeting the man, "One is inclined to say, 'Here is an actor, or a musician, an artist of some sort.' "

129 Samuel Insull, *Memoirs*, ed. with additional information by Larry Plachno (Polo, Ill., 1992). This is a transcript of Insull's previously unpublished 285-page typewritten manuscript, housed for half a century in the archives of Loyola University of Chicago, repository for the Insull Papers. Dictated by Samuel Insull from memory, the book is flavorful and evocative, especially in its recreation of the interactions between the young Insull and the sanctified Edison.

130 The definitive overview of the machinations of Edison's representatives in Paris before and during the International Electricity Exposition is Robert Fox, "Edison et la presse française a l'exposition internationale d'électricité de 1881," in Fabienne Cardot, ed., *Un siècle d'électricité dans le monde, 1880–1980* (Paris, 1987). Through the good offices of Lenore Symons, archivist at the Institution of Electrical Engineers, London, the author was put in touch with Professor Fox of the modern history faculty, University of Oxford, who sent an offprint of his article, as of this writing unpublished in English. See also: "Une revolution dans l'éclairage," *Le Figaro*, July 14, 1881; "The Electric Exposition," *Evening Telegram*, London, August 11, 1881; "L'exposition d'électricité," *L'Illustration*, August 13, 1881; "The Rival Electricians," New York *Evening Post*, August 24, 1881; "The Electrical Exhibition at Paris," New York *Herald*, August 28, 1881; "Les lampes électriques à incandescence," *La lumière électrique*, October 1, 1881 (Theodore Du Moncel's article); "The Large Edison Dynamo," *Engineering*, London, October 21, 1881; "Edison, ou La Lampe Merveilleuse," *Le Figaro*, October 24, 1881; "The Paris Electrical Exhibition," *Metropolitan*, London, October 27, 1881; "L'éclairage électrique du foyer de l'Opéra," *L'illustration*, November 10, 1881; "Le systeme Edison," *L'électricien*, January 1882; "Electric Light in the Opera House, Paris," *The Electrician*, London, May 13, 1882. All Batchelor Scrapbooks, ENHS.

132 Charles Batchelor's correspondence to Edison from September 1881 through December 1882, in ENHS, provides a chronicle of his many activities in Paris and elsewhere. ENHS also houses the first pub-

lished bulletins from the Compagnie Continentale Edison and the Société Électrique Edison, from February 1882.

132 George Frow, "The Edison Electric Light Station at Holborn Viaduct, 1882–1884," *Hillandale News* ("The Official Journal of the City of London Phonograph and Gramophone Society"), Issue Number 193, August 1993. Provided to the author by Mr. Frow. *First Bulletin*, Edison Electric Light Company, 65 Fifth Avenue, New York, January 26, 1882; and *Seventh Bulletin*, April 17, 1882, contain narrative progress reports on the Holborn Viaduct and Pearl Street projects. ENHS. "The Edison Light in New York," *The Engineer*, January 6, 1882; "Crystal Palace Electrical Exhibition," and "The Installation of the Edison Light at Holborn," *The Electrician*, February 11, 1882; "The Edison Electric Light," London *Times*, April 13, 1882; "Edison's System of Incandescent Lighting at Holborn," *The Electrician*, April 15, 1882; "The Edison Light at Holborn," ibid., April 22, 1882, includes summary of number of lights used per building; "The Edison Light at the General Post Office," *The Metropolitan*, London, August 26, 1882.

134 George Beard, *A Practical Treatise on Nervous Exhaustion (Neurasthenia): Its Symptoms, Nature, Sequences, Treatment* (New York, 1880); and *American Nervousness, Its Causes and Consequences (A Supplement to Nervous Exhaustion—Neurasthenia)* (New York, 1881), in New York Academy of Medicine Library, New York City. Also, "American Nervousness, A Lecture Before the Philosophical Society by Dr. George M. Beard of New York, Some of the Symptoms of the American Branch of the Neurological Tree," New York *Herald*, November 8, 1880, Manuscripts and Archives, Yale University Library, New Haven, Connecticut.

134 Correspondence pertaining to schooling for the Edison children, October 21, 1881; September 20, October 2, 1882. Correspondence, Dr. Leslie Ward:TAE:Dr. Ward, January 18, 1882. As was his custom, Edison's reply is written on Dr. Ward's letter for Insull to draft a response. This letter is followed by two telegrams, from Mrs. Edison:Samuel Insull, February 23 (Aiken, South Carolina) and March 13, 1882 (Green Cove, Florida), reporting her whereabouts. ENHS.

CHAPTER 12

135 The path to the Pearl Street station proceeds directly out of the Paris
Electrical Exhibition. "Edison's New Steam Dynamo," *Scientific
American*, December 10, 1881, describes in detail the configuration
of mutually shared driveshaft between steam engine and dynamo.
"The Installation of the Edison Light in New York," *The Electrician*,
London, January 7, 1882, includes a map of the square-mile Pearl
Street district bounded by Wall, Nassau, Spruce, and Ferry Streets,
Peck Slip, and South Street.

137 "The Dangers of Electric Lighting," *The Electrician*, January 28, 1882;
"A Correspondent Visits Edison," Cleveland *Herald*, February 13,
1882; "The Doom of Gas," St. Louis *Post-Dispatch*, May 1, 1882
("Millions of Sewing Machines to Be Run for Five Cents a Day per
Machine. . . . How the Great Inventor Has Worked Twenty Hours
out of Twenty-Four to Master Every Detail and Overcome Every Dif-
ficulty"); "Edison Light at Railway Stations," *American Gas Light
Journal*, June 2, 1882; "Laying the Electrical Tubes," *Harper's Weekly*,
June 24, 1882 (graphic illustration of Edison's men digging
trenches, splicing connections, heating the insulation mixture, and
pouring it into the conduits); "The Edison System of Electric Lighting," *American Contract Journal*, September 16, 1882; "The Edison
System," *American Gas Light Journal*, October 16, 1882; "The Electric
Light in Cotton Mills," *The Engineering and Mining Journal*, November 11, 1882.

137 "The Edison System of Incandescent Electric Lighting," a pamphlet
issued September 7, 1882, by the Edison Company for Isolated
Lighting, organized November 1881 as a licensee of the Edison Electric Light Company, entitled to do business under the Edison patents
for electric lighting. This document contains the sales pitch for isolated lighting and includes illustrations of all the components, the
lamp, detached from and screwed into its socket, and the "Z" Dynamo with its specifications, as well as a summary of all isolated
plants installed in the United States. The Edison Electric Light Company also published excerpts from media coverage of the inauguration of the Pearl Street Station for use in selling its systems: "Edison's

Illuminators, The Isolated System Is in Successful Operation," New York *Herald*, September 5, 1882; "Electricity Instead of Gas," New York *Tribune*, September 5, 1882; "Edison's Electric Light," New York *Times*, September 5, 1882. ENHS.

Also in ENHS is a 4-page undated handwritten summary, "Actual Cost of Installation" pertaining to the Pearl Street station, including all real estate and construction costs, operating expenses, and payroll for the Central Station for one year, as well as a summary "Report for 24 hours ending at 12 noon December 5, 1882," which provides data on amount of coal received, ashes removed, cubic feet of water used, and a breakdown of engines running and their hours of operation. Appended to this report is a short memorandum to Edison from S. B. Eaton, dated December 21, 1882, which says in part, "[Subscribers in the First District] were told they would have to pay for their light on and after this date. He finds almost unanimous satisfaction with the light and thinks that all the subscribers will become customers." Samuel Insull's business correspondence from this period in ENHS is voluminous, and two letters of October 30, 1882, to William H. Rideing and Charles Batchelor, provide detailed summaries of all Edison activities at the time. Insull was equal to Batchelor in his painstaking chronicle of Edison's movements.

138 Thomas J. Schlereth, *Victorian America, Transformations in Everyday Life, 1876–1915* (New York, 1991), contains the anecdotal description of the first Labor Day on September 5, 1882.

139 Reyner Banham, *The Architecture of the Well-Tempered Environment* (Chicago and London, 1974). Banham (1922–1988) was professor of art history at the University of California, Santa Cruz. He insisted that technology, human needs, and environmental concerns be considered integral to architecture. His portrayal of Edison in this light is imaginative, taking the inventor from the "dark, satanic century" into his "kit of parts," the systematic approach to selling light as a commodity and inspiring a domestic revolution.

139 Edward H. Johnson, "Personal Recollections of Mr. Morgan's Contribution to the Modern Electrical Era," typewritten manuscript dated November 1914, collections of the Pierpont Morgan Library, New York City: "The knowledge that Mr. Morgan had given his support

[to Edison] brought in its train the capitalists of all the leading cities of the country," Johnson wrote. Two books in the Pierpont Morgan collection were also helpful: Herbert L. Satterlee, *J. Pierpont Morgan, An Intimate Portrait* (New York, 1940); and Vincent P. Carosso, *The Morgans, Private International Bankers, 1854–1913* (Cambridge, Mass., 1987). I am grateful to David Wright, archivist and registrar, the Pierpont Morgan Library, for his kind assistance.

The definitive work on the Morgans is Ron Chernow, *The House of Morgan, An American Banking Dynasty and the Rise of Modern Finance* (New York, 1990). In an interview with the author on December 5, 1991, Chernow shed light on J. P. Morgan's personal affinities with Edison leading to the bank's financial involvement, which eventually extended to holding companies and electrical utility underwriting in thirteen states.

139 Covering letter from Marjorie C. Wilkins, librarian, Edison Electric Institute, to Thomas Ewing Dabney, New Orleans Public Service, Inc., August 27, 1952, conveys a summary of electric lighting company incorporations in the United States up to the end of 1881. Collection of Edison Electric Institute.

140 The John W. Lieb Collection of Leonardo da Vinci is in the Samuel C. Williams Library at Stevens Institute of Technology, Hoboken, New Jersey. The author is grateful to Mrs. Jane Hartye, curator of special collections, for her kind assistance in access to the Lieb Collection; and for providing back issues of the Stevens Institute newspaper, *The Stute*, October 30 and November 6, 1929, containing extensive biographical information and appreciations of John Lieb. Correspondence, Lieb (in Milan):TAE, January 13 and July 1, 1883 is in ENHS. With reference to his frustration in completing construction of the Milan central station, Lieb writes, "I have experienced great difficulty in getting men to assist me in my work as those possessing intelligence are above working themselves and not well enough posted practically to see that the work is properly done." Also, "Edison Central Station at Milan," *Engineering*, August 31, 1883.

Thomas Edison's resonances with Leonardo da Vinci would make a fascinating subject for another book, unfortunately outside the purview of the study at hand. Sigmund Freud, in his classic

Leonardo da Vinci and a Memory of His Childhood (New York, 1910), has some provocative things to say about da Vinci's compulsion to alter, conceal, and even mythologize facts about his upbringing, including "the instinct to look and the instinct to know"; his lifelong inclination to surround himself with acolyte-like and adoring assistants; his desire to "probe the secrets of nature while relying solely on observation and his own judgement"; and his enduring affection for play and frivolity, games and pranks. See also Kenneth Clark, *Leonardo da Vinci* (New York, 1939), on "Leonardo as a scientist rather than a mathematician"; as obsessive sketcher, exhausting the visual possibilities of his ideas; as pack-rat hoarder of every scrap of writing, keeper of the "full record" of his activities; as supreme egotist; as student of human movement and the flight of birds; as engineer and botanist; as motivator of craftsmen; "obsessed by vital force . . . he pursues science as a means by which these forces can be harnessed for human advantage."

141   Edison's place in the development of American management is clarified in Alfred D. Chandler, Jr., *The Visible Hand, The Managerial Revolution in American Business* (Cambridge, Mass., 1977). The edition of William Dean Howells's *The Rise of Silas Lapham* cited herewith is The Library of America (New York, 1982), edited and with notes and chronology by Edwin H. Cady.

142   The tragic chronology of the final months of Mary Stilwell Edison's life was reconstructed from material in ENHS as follows: "Inventory of Furniture at 25 Gramercy Park," 1882. Correspondence, TAE: James W. Pryor, Esq. (his landlord), September 20, 1883; TAE: R. Tobin, Esq. (his attorney), October 14, 1883; Mary S. Edison: Samuel Insull, February 27, 1884, postmarked Palatka, Florida; Mary Edison: Insull, April 30, 1884; TAE: Rupert Schmid, August 1, 1884 ("My family is away from the City just now"). Contents of ENHS document file D-84-014, "Edison—Family." Telegram from Edison to Insull, April 7, 1884 ("person out of mind"). Edison laboratory notebooks, July 28–30, 1884. Announcement of Mrs. Edison's death, August 9, 1884 ("congestion of the brain"). Personal reminiscences of Marion Edison Oeser, op. cit.

PAGE

Edison's statement to Marion that his wife died of "typhoid fever" may be viewed as a rationalization. It is highly unlikely that unhygienic circumstances were the cause of death, as no one else in the house was ill. As noted, Mary Edison was a chronically depressed, nervous, alcohol-dependent, overweight, lonely young woman. The possibilities exist, judging from the symptoms she herself discussed three months before her death, that she may have died of an acute infectious disease, or even by her own hand.

143 Funeral and interment of Mary Edison: "Mount Pleasant Cemetery," Newark *Daily Advertiser*, August 20, 1873; "Oldest Newark Cemetery," Newark *Evening News*, October 12, 1959; "Mrs. Edison's Funeral," Newark *Daily Advertiser*, Wednesday, August 13, 1884; Newark *Evening News*, August 13, 1884, all on microfilm at Newark Public Library; cemetery records, Fairmount and Mount Pleasant cemeteries; correspondence, Marion Edison Page (older sister): Thomas Edison, August 11, 1884. ENHS.

144 Edison's relocation to 39 East 18th Street, noted in lease agreement, August 26, 1884, for "private dwelling apartment by his family, consisting of himself and three children"; correspondence to various addressees from Edison, and from Samuel Insull on Edison's behalf, August 28, September 15, and September 17, 1884; Marion's handwritten comments in her father's laboratory notebooks, October 8, 1884; comprehensive report on "The Edison Exhibit at the Philadelphia Electrical Exhibition," Scientific American, October 18, 1884. ENHS.

CHAPTER 13

147 Elwood Hendrick, *Lewis Miller, A Biographical Essay* (New York, 1925), published twenty-five years after Miller's death, and long out of print, is the only book-length biography of the man extant. It features an introduction by Thomas Edison, who identifies strongly with his father-in-law: "He grew up in the 'School of Hard Knocks,'"

Edison writes. "He had an extraordinary amount of energy and was one of the most industrious of men. . . . He did not believe that recreation consisted in wasting valuable time in doing nothing or useless things." Lewis Miller left a brief, untitled autobiographical memoir ("My earliest recollections are . . . ") which is housed in typescript form in the Edison–Miller Collection (hereafter referred to as E-MC) at the Chautauqua Institution Library, in Chautauqua, New York. During my visit to Akron and environs, Mrs. Dorothy Stevens was kind enough to guide me through the original Miller house on Cleveland Avenue in Greentown, Ohio, where Lewis Miller spent his boyhood; Ruth and Howard Miller welcomed me to the homestead and farm purchased by Lewis's brother, Abraham, from their father's estate in 1844, close by in North Canton. Oats, wheat, corn, hay, and pumpkins are still raised there.

149   Aultman, Miller & Co., Buckeye Mower and Reaper Trade Catalogue, and Buckeye Binders, Table Rakes, Droppers and Mowers Trade Catalogue, (Akron, Ohio, 1878; 1887), in the collections of the Research Center, Henry Ford Museum and Greenfield Village, Dearborn, Michigan. These rare, profusely illustrated examples of nineteenth century marketing provide firsthand information on the workings of Miller's great inventions. I am also grateful to Mr. George Ball, of Akron, Ohio, for providing me with his assiduous compilation by number, name, and date of Lewis Miller's ninety-two patents relating to the Buckeye Mower and Reaper.

151   Karl H. Grismer, *The Story of Fort Myers, The History of the Land of the Caloosahatchie and Southwest Florida* (Fort Myers, 1982). Facsimile of 1949 edition. The Southwest Florida Historical Society reprinted this book as an important document of local history. It contains good anecdotal information on Edison's first visit. However, the most thorough contemporary research on Fort Myers and Edison's relationship to the place has been conducted by Dr. Leslie H. Marietta, official historian of the Edison winter home. Dr. Marietta's compilation from newspaper accounts and other archival material of Edison's many visits to the area provides the basis for the narrative here.

153   Correspondence, Ezra Gilliland:TAE, April 28 and June 18, 1885,

regarding Edison's trips to Boston; TAE: Samuel Insull, June 27, 1885, ENHS. Edison's journal from this courtship phase has been published as *The Diary and Sundry Observations of Thomas Alva Edison*, ed. Dagobert D. Runes (Westport, Conn., 1968; reprint of 1948 Philosophical Library edition). There is also a photocopy of the manuscript in ENHS.

154   The early history of Chautauqua presented here is derived from a mixture of texts and archival material. I am grateful to Alfreda L. Irwin, historian of Chautauqua, and her staff at the Chautauqua Institution Library for their kind assistance during my research residence: Barbara Haug, archivist; Nathalie Leonard, and Margaret Wade, curator of photographs.

     Edwin P. Booth, "The Biography of Lewis Miller." A lecture delivered at Chautauqua on July 17, 1964, typescript, E-MC.

     Theodore Morrison, *Chautauqua, a Center for Education, Religion, and the Arts in America* (Chicago and London, 1974).

     "Chautauqua," article in *Harper's New Monthly Magazine*, no. CCCLI, August 1879 (n.a.). This contemporary account provides immediate flavor of Chautauqua in its formative period: "During the past few years the word which appears at the head of this article has been growing gradually familiar to the abstract personages known as the general public." E-MC

     John H. Vincent, *Autobiography*, published weekly in 1910 in twenty-four installments. Paragraph excerpts, typescript, E-MC.

     John H. Vincent, *The Chautauqua Movement* (Plainfield, New Jersey, 1886).

     Leon H. Vincent, *John Heyl Vincent* (New York, 1925).

     Alfreda L. Irwin, *Three Taps of the Gavel: Pledge to the Future, The Chautauqua Story* (revised and expanded edition, Chautauqua, N.Y., 1987).

     Cathleen Schurr, "Chautauqua: Yesterday and Today," in American History Illustrated Magazine, Summer, 1992.

     Fannia Weingartner, editor, special issue, *Henry Ford Museum and Greenfield Village Herald*, vol. 13, no.2, 1984, "Chautauqua." A superb collection of essays on the history of the institution and the ensuing movement. Of particular use for this chapter were the essays by Alan Trachtenberg and Thomas J. Schlereth.

155   The account of Miller and Vincent's convergence in 1868 and afterwards, and of their at-times stormy discussions regarding the establishment and mission of Chautauqua may be found in the following documents in E-MC: Mina Miller Edison, "Chautauqua Address," typescript (n.d., but from internal evidence c. 1922): "Fifty years ago, Bishop Vincent came to my father's home in Akron to ask his cooperation in acquiring the privilege of holding a two weeks' Sunday School convention in our church. "; typescript remarks by Lewis Miller (n.d.): "My daughter fully confirms the statement I made about the talk we [Vincent and Miller] had in her hearing."; Nancy Miller Arnn, Mina Miller Edison's niece, in personal communication with the author on August 24, 1992, spoke of her aunt's remembering years later these "vehement arguments well into the night at Oak Place"; Lewis Miller's prepublication typescript emendations to his introduction for John Vincent, *The Chautauqua Movement*, dated January 27, 1886; "The Early Days of Chautauqua," typescript, by Kate (Patterson) Bruch, August 1, 1897, with marginal notes by Mina Miller Edison.

159   The unofficial courtship of Thomas Edison and Mina Miller in the summer and fall of 1885 has been reconstructed from the following sources in E-MC: "Thomas A. Edison," article in the Chautauqua *Assembly Herald*, Saturday, August 15, 1885; "Mr. Edison at Chautauqua," *The Chautauquan Weekly*, October 22, 1931; "The Maid of Chautauqua," Miller family history, typescript (n.d.): "A man not yet 40 and a maid not yet 19 could not be alone."; typescript excerpts from *The Diary of George Vincent*. Also, correspondence, Ira M. Miller:Cora Wise, August 14, 1885, collection Nancy Miller Arnn.

    Ezra Gilliland's September 17, 1885, letter to TAE with Edison's marginal notations is in ENHS, as is Edison's September 30, 1885, handwritten proposal letter to Lewis Miller. I am grateful to Leah Burt, former curator of Glenmont, the Edison estate in West Orange, New Jersey, for discovering and bringing to my attention the inscribed presentation copy of *Character* by Samuel Smiles.

CHAPTER 14

161 Correspondence in the collection of the Charles Edison Fund, East Orange, New Jersey (hereafter referred to as CEF). The author extends grateful acknowledgement to the following persons at the Fund for their kind assistance in allowing access to material in the Charles Edison Fund archives: Paul J. Christiansen, president; David Schantz, treasurer; Alberta Ench, secretary; and trustees Nancy Miller Arnn and J. Tom Smoot, Jr.

The incoming correspondence was collected into cumulative letter books by Mina Miller, later to become the second Mrs. Edison. The earliest letters examined by the author extend back to 1877, when Mina was a mere twelve years old. The following are cited or drawn upon for anecdotal information in the beginning of this chapter. All letters are to Mina Miller unless otherwise indicated. TAE:, November 5, 1885; Jane (Jennie) Miller:, July 27, 1877, and n.d.; Lewis Miller (brother):, December 12, 1884; Jennie Miller:, November 21, 1885; Jennie Miller, December 17, 1885.

162 During my field visit to the city of Akron, Ohio, in addition to assistance and guidance from Mr. George Ball, formerly of the University of Akron, I also interviewed George W. Knepper, Distinguished Professor of History at the University of Akron and past president of the Ohio Historical Society, and have made use of his definitive work, *Ohio and Its People* (Kent, Ohio, 1989). Another important reference work on the early history of the city is Abe Zaidan, *Akron, Rising Toward the Twenty-First Century: A Contemporary Portrait*, produced in association with the Akron Regional Development Board, 1990. Also: Karl H. Grismer, *Akron and Summit County* (Akron, Ohio, 1989); and George W. Knepper, *Akron, City at the Summit* (Tulsa, Okla., 1981), for the Summit County Historical Society.

163 For a guided tour of the Miller home, Oak Place, overlooking the city of Akron, I am grateful to Dr. Bill E. Thomas, who is in the process of restoring the house to its former grandeur. For a history of the house, the following articles from the files of the Akron *Beacon Journal*: "Lewis Miller Mansion Now Apartment House," September 8,

1940; "Honor Miller House," April 13, 1976; "Much New Light Is Being Shed on Akron's Old Miller Mansion," March 13, 1979; and "Edison's Akron Nuptial Site," November 4, 1979. For contemporary anecdotes on life in the Miller household, Hendrick, *op. cit.*, and George E. Vincent, *Theodore W. Miller, Rough Rider* (privately printed, Akron, Ohio, 1899); correspondence, TAE: John and Theodore Miller, December 24, 1885, ENHS. Excerpts from the Miller boys' weekly newspaper, *The Jumbo*, courtesy Nancy Miller Arnn.

166   The literature on Llewellyn Park is uncollected, and in diverse periodicals. For the purposes of this introductory study, the following were useful: Thomas J. Schlereth, "Chautauqua, A Middle Landscape of the Middle Class," in "Chautauqua" number of *Henry Ford Museum and Greenfield Village Herald,* op. cit.; Samuel Swift, "Llewellyn Park, The First American Suburban Community," *House and Garden*, vol. 3, June 1903; Jane B. Davies, "Llewellyn Park in West Orange, New Jersey," *The Magazine Antiques*, January 1975; and Richard Guy Wilson, "Idealism and the Origin of the First American Suburb, Llewellyn Park, New Jersey," *The American Art Journal*, October 1979.

167   The informing spirit of Glenmont, its roots in the Queen Anne aesthetic, and the intellectual origins of Henry Hudson Holly are defined brilliantly in Vincent J. Scully, Jr., *The Shingle Style and the Stick Style: Architectural Theory and Design from Downing to the Origins of Wright* (revised edition of 1955 publication, New Haven, Conn., 1971). Henry Hudson Holly's books, *Country Seats* (1863) and *Modern Dwellings* (1878), reissued together in 1977 by the Library of Victorian Culture, Watkins Glen, New York, were brought to my attention by Leah Burt, former curator of Glenmont. Ms. Burt's exhaustive *Glenmont Historic Furnishings Report* (1991–92) discusses room-by-room in narrative form the original and then changing decor and environment of the home, as well as the landscaping alterations from the moment of the Edisons' occupancy through to the 1940s.

169   Chronicle of Edison's activities in the month preceding his wedding: laboratory notebooks, January and February 1886; Charles Batchelor's Diary, February 18–26, 1886; correspondence, TAE:Mina

Miller, February 4, 12, and 20, 1886, ENHS. Description of the Edison-Miller wedding, Akron *Beacon Journal*, February 25, 1886, microfilm collection, University of Akron, courtesy John V. Miller, Jr., director, archival services. The diagnosis of the congenital origins and degenerative nature of Edison's deafness are by Dr. Linn Emerson, who was his doctor for a time in the 1910s, and are taken from the ENHS card file index to the Edison papers.

172 Edison's metaphysical speculations, "Edison's Religious Belief," Salt Lake City *Democrat,* July 16, 1885; Fort Myers notebooks, March 20, 26, 30, April 13, 16, 18, 21, and 28, 1886, ENHS. Edison's conceptualization of Fort Myers estate and sketch of grounds, Fort Myers memo book, April 1886, are in the collection of *Edison Winter Estate* and are courtesy of Robert Halgrim, Jr., director, Edison Winter Estate.

174 Mina's conflicts regarding the Edison children, correspondence, Jennie Miller:, February 26 and March 5, 1886; Thomas A. Edison, Jr.:, March 1, 1886; Lewis A. Miller (brother):, March 2, 1886; Lewis Miller (father): (n.d., spring 1886). CEF.

CHAPTER 15

176 The new domestic scene at Glenmont, and Mina's emotional adjustments, are reflected in a series of letters to Mina Miller Edison, all CEF: Mary Miller:, May 16, July 22, and July (n.d.), 1886; Jennie Miller:, September 22, 1886; Ed Miller:, October 1, 1886; Mina Miller: Samuel Insull, August 30, 1886.

177 The strike in New York City shops and the move to Schenectady: Batchelor's *Diaries*, May and June 1886, and clipping therein from *The World of Today*, "Edison's Men on Strike, The Machine Shop Closed Down for an Indefinite Period. . . . What Is Said on Both Sides on the Subject," ENHS. "Walks and Talks in the Grove," an interview with TAE in the Chautauqua *Assembly Herald*, August 16, 1886, E-MC. *The Edison Era, The General Electric Story, A Photo History,* vol. 1 Schenectady, 1976).

PAGE

179   Thomas A. Edison, "The Air-Telegraph, System of Telegraphing to Trains and Ships," *The North American Review*, March 1886, 285–291.

181   On Edison's spring 1887 trip to Fort Myers, Grismer, *op. cit.*; Fort Myers *Press*, February 19, 24, March 3, 10, April 14, and May 5, 1887, in Marietta, op. cit. Edison's medical problems are detailed in Batchelor's Diaries. Batchelor paid Edison a brief visit in Fort Myers and witnessed his physical condition firsthand. Correspondence, TAE:Batchelor, April 6, 1887, ENHS. Mina's "expectation" is discussed in sister Jennie's letter to her of March 8, 1887: "I hope though my dear you will have an easier time from now on. If you are happy in the thought of having a dear little one I am [happy], but if you regret it I am sorry. I think when it comes you will be happy with it."

183   Andre Millard, *Edison and the Business of Innovation*, Johns Hopkins Studies in the History of Technology (Baltimore, 1990). This is the definitive story of Edison as entrepreneur-businessman, with particular emphasis on the West Orange laboratory-factory complex from its genesis in the mid-1880s through to Edison's death in 1931. Edison's concepts for his new laboratory are also detailed in correspondence TAE: "Friend [W. L.] Garrison," August 13, 1887, ENHS; and TAE: J. Hood Wright, a partner of J. P. Morgan, November 15, 1887, cited in Vanderbilt, op. cit. The remarkable similarity in the language of these two letters leads one to believe that Edison may have been sending out a series of "pitches" to excite and attract investors at this time. The precipitous dismissal of Henry Hudson Holly is described in Batchelor's Diaries.

184   Raymond Wile, "Edison and Growing Hostilities," *Archives of Recorded Sound Journal*, XXII.1, Spring 1991, 8–34. This is an exhaustively researched account of the controversy surrounding the reintroduction of Edison's phonograph in 1887 and following, based upon Wile's labors in AT&T and Bell Archives, the Tainter papers at the National Museum of History and Technology, the Villard papers at Harvard, ENHS, court transcripts and depositions, especially *American Graphophone Company* vs. *Edison Phonograph Works*, and other primary sources. See also: Gilliland's handwritten notes, "Commenced work on the standard Phonograph . . . ," October 5, 1886.

Correspondence, George Gouraud:TAE, July 2, 1887; TAE:Gouraud, July 21, 1887; Gouraud:TAE, November 30, 1887; Uriah Painter: TAE, December 4, 1887; Gilliland:TAE, December 16, 1887. ENHS.

CHAPTER 16

188 The account of the continuing, tangled imbroglio over the phonograph continues with the chronicle traced in Raymond Wile, op. cit. Edison's handwritten assessment of his financial record with the major patents; his 14-page deposition before the Court of the Southern District of New York in the matter of "Thomas Edison against Ezra T. Gilliland and John C. Tomlinson"; and his handwritten diatribe against Edward H. Johnson are in ENHS.

189 The rift with Gilliland was permanent and irrevocable—despite impassioned interventions on the part of Lewis Miller—and extended to the disposition of Gilliland's adjoining property in Fort Myers. See, for example, correspondence, A. O. Tate: TAE, January 15, 1889, where TAE instructs Tate to write to Gilliland on the matter; and S. B. Eaton as attorney for Edison asking in a letter of December 14, 1889, whether he should "push the taking of testimony vigorously or not" in the pending litigation, to which Edison replies, "Yes—*push*"; also, correspondence William E. Hibble (caretaker of Edison's Fort Myers property): TAE, during 1889–1890, in which Hibble seeks final confirmation from TAE to "cut the pipes" leading from the windmill on Edison's property which had been supplying Gilliland's house with water. ENHS.

192 The saga of the "perfected phonograph" is reconstructed from the following sources: "A Wonderful Workshop, The Resources of Edison's Laboratory, What the New Phonograph Is Expected to Do," New York *Post*, October 25, 1887; "The New Phonograph," *Scientific American*, December 31, 1887; correspondence, Gilliland: TAE, March 19, 1888 ("The personal end of the Phonograph business needs my personal attention badly and I think from this time forward

I shall give it all of my time."); Thomas A. Edison, "The Perfected
Phonograph," *North American Review*, June 1888, 641–650; "Edison's
Perfected Phonograph," *Illustrated London News*, July 14, 1888; "Reproduction of Articulate Speech and Other Sounds," *Scientific American*, July 14, 1888; "Method of Recording and Reproducing Sounds,"
U.S. Patent Office registrations, July 31 and December 4, 1888 ("As
is well understood, the phonograph invented by me, and the various
modifications of my early instruments that have been made by others
as well as by myself, operate to record speech, music and other
sounds."); TAE:Clement Studebaker, July 23, 1888 ("Phonos soon to
be ready in new factory"); "Edison's Aides, The Great Inventor's
Right Hand Men, The Lieutenants Who Have Aided Him in Producing His Wonderful Inventions," Detroit *Free Press*, August 5,
1888; "Mr. Edison and His New Phonograph," New York *Daily
Graphic*, August 18, 1888.

192   On suspicion of Painter's "spies," see A. C. Couper:TAE, September
10, 1888: "From questions asked me a few days ago, I think that
some one connected with the graphophone people has ascertained
who was working in your laboratory at the time when the principal
improvements were made in the phonograph. . . . Last Friday I was
followed into a restaurant by a gentleman who took a seat at the same
table and after some general conversation began talking about the
phonograph with the details of which he was thoroughly familiar. . . .
I observed that he appeared to be writing down what I said although
his hands being below the top of the table I could not see either pencil or paper. . . . I changed the conversation somewhat but he went on
fishing for information. . . . In response to a question he said he was
slightly acquainted with Painter." This tip is followed in the ENHS
file by a quick note from Tate to Edison: "I have positive information
that Painter has a man at work on the phonograph. We ought to ex-
clude visitors from the Laboratory, *absolutely*." Edison replies, "Get
gates in order and then put up sign, [Edison encloses sketch of sign in
a frame] VISITORS NO LONGER ADMITTED."

192   Grateful acknowledgement to Neil Maken for archival copies of *Scientific American* from this period exhaustively recording new developments in the phonograph-graphophone competition: "The Grapho-

phone and the Phonograph," October 27, 1888; "The Scientific Use of the Phonograph," by George M. Hopkins, serialized in March 8, April 19, and August 16, 1890.

194 The summary of activities in the West Orange laboratory continues to rely upon the excellent work of Andre Millard, op. cit.; also, Horace Townsend, "Edison, His Work and His Work-Shop," *Cosmopolitan,* April 1889, 598–607; Edison's handwritten list of "the things I propose working [*sic*]," ENHS; "Edison's Phonographic Doll," *Scientific American* front-page article, April 26, 1890, accompanied by a full page of illustrations; Edison's proposal to expand the "Toy Doll biz" is a 2-page memo headed "Tate write Dyer as follows," ENHS; the ethnographic applications of the phonograph are described in "The Edison Phonograph in the Preservation of the Languages of the American Indians," *Scientific American*, May 24, 1890. The analysis of the transformed social function of dolls in post-bellum American society is drawn from the essay "Sugar and Spite: The Politics of Doll Play in Nineteenth-Century America," by Miriam Formanek-Brunell, in Petrik and West, op. cit.

197 Lewis Miller's cautionary letter to Mina, March 22, 1888; and Mary Miller's letter, August 26, 1888, CEF. For the ethos of work in late-1880's America, Wiebe, op. cit., especially his argument on "biological organicism and philosophical idealism" with respect to the writings of sociologist Franklin H. Giddings. An extended discussion of Adriano Tilgher's theories related to the decline of the work ethic in late-nineteenth-century America is in James B. Gilbert, *Work Without Salvation: American Intellectuals and Industrial Alienation, 1880–1910* (Baltimore and London, 1977).

198 Edison's dicta on "work" and "working": *Scientific American*, September 14, 1889, "We hear wonderful stories of your working. You have the reputation of being able to work twenty-three hours a day for an indefinite period." "Oh! [Edison replied] I have often done more than that, haven't I, Gouraud? As a rule, though, I get through twenty hours a day. I find four hours sleep sufficient for all purposes."

198 Edward Bellamy, *Looking Backward* (New York, 1986), edited and with an introduction by Cecelia Tichi. Edison's handwritten notes on his futuristic fantasy, ENHS. As late as June 24, 1891, George Par-

sons Lathrop was imploring Mina Edison to return the drafts for *Progress* "which I sent to Mr. Edison about six weeks ago—which, as you remember, were lost sight of for two weeks, and have not yet been read by Mr. Edison."

CHAPTER 17

200   Margaret Cheney, *Tesla: Man Out of Time* (New York, 1981) is the most current biography of the underappreciated Nikola Tesla. Cheney's account of Tesla's short-lived relationship with Edison draws in turn upon two earlier biographies, John O'Neill, *Prodigal Genius* (New York, 1944); and Inez Hunt and W. W. Draper, *Lightning in His Hand* (Hawthorne, Calif., 1964).

201   The chronicle of the AC-DC wars is reconstructed from the following: "Edison Electric Light Company," [summary of current and pending patent litigation, full-page notice] *The Electrical World*, November 24, 1888; "New Death for Murderers, The Alternating Current Tested in Edison's Laboratory, Two Calves and a Horse Killed," New York *Telegram*, September 27, 1888; "Electricity," Asheville (N.C.) *Citizen*, February 10, 1889; "Is Electricity Sure to Kill?" Pittsburgh *Dispatch*, July 13, 1889; "Killed by 1,000 Volts," New York *Herald* (n.d.), clipped and sent to TAE in Paris by S. B. Eaton, summer 1889, with note, "By the way, one of their [Westinghouse's] local officials was killed by the alternating current yesterday." Correspondence, Eugene H. Lewis: S. B. Eaton, re terminology of *westinghouse*, June 1, 1889; S. B. Eaton: TAE in Paris, September 3, 1889, reports that "the enemy . . . making effort, notably Westinghouse, to get patents and control all convertor systems. When you come back, I shall ask you to determine what our attitude is to be. . . . Shall we fight on our own line alone?" *Clippings:* Batchelor Scrapbooks; *correspondence:* ENHS.

202   Thomas Edison, "The Dangers of Electric Lighting," *North American Review*, November 1889. Collection New York Historical Society.

PAGE

203 The ongoing account of Henry Villard's efforts to reconfigure the Edison interests is in Chandler, Clark, Hughes, and Insull, op. cit.

203 Edison in Paris: Two days before his departure, Edison drew up what appears to be an exhaustive "to-do" list for Batchelor, Kruesi, and Ott at the West Orange shop, including specific instructions to continue work in his absence on "the clock phonograph, cheap doll phono., electric brake, electric rock drill, mailing phonograms, telephone induction balance, dipping machine for veneering musical cylinders, model glass blowing machine for lamp bulbs, sapphire-making machinery, plans etc for [phonograph] cabinet shop, rhigolene process for preserving meats, fruits, etc, non-flammable insulation," and other items, to which Batchelor has added, "Design new street car motor."

203 "A Talk With Edison," ("When I was on shipboard coming over [to Le Havre], I used to sit on the deck . . . "), Akron *Beacon*, October 2, 1889; correspondence, Charles Batchelor:TAE August 29, 1889 ("I knew what the feeling was there"); "With Mr. Edison on the Eiffel Tower," *Scientific American*, September 14, 1889, excerpted from R. H. Sherard in the *Pall Mall Budget,* London; "Thomas Alva Edison," *Scientific American*, October 12, 1889, translated from *La Nature*, Paris. *Scientific American*: collection of Neil Maken; *all others:* Batchelor Scrapbooks, ENHS.

204 John A. Kouwenhoven, "The Eiffel Tower and the Ferris Wheel," (essay) originally published in *Arts* magazine, February 1980, reprinted in *Half a Truth Is Better Than None: Some Unsystematic Conjectures About Art, Disorder, and American Experience* (Chicago, 1982); Brian N. Morton, *Americans In Paris, An Anecdotal Street Guide* (Ann Arbor, 1984).

206 The refreshing new look at Buffalo Bill is one section of "The White City and the Wild West: Buffalo Bill and the Mythic Space of American History, 1880–1917," chapter 2 of Richard Slotkin, *Gunfighter Nation, The Myth of the Frontier in Twentieth-Century America* (New York, 1992).

207 The account of Edison's ongoing family problems is drawn from correspondence, all CEF: Mina Edison:her mother, August 28, 1889 [Mina misdated the letter for "September" and then corrected her

error, an indication of the depth of her desire to return home]; Mary Miller: Mina Edison (n.d.); the twelve-page report on Dot's European behavior was sent from Spa, Switzerland, and is dated "[18]'89," and from internal evidence was written in the late spring before Edison and Mina came to Paris.

208   Edison and Marey: Christopher Rawlence, *The Missing Reel: The Untold Story of the Lost Inventor of Moving Pictures* (New York, 1990), blends social history and popular science to tell the story of the French inventor Augustin Le Prince's suppressed role in formulating the early motion picture projector, and discusses Edison's early work, although Rawlence heavily debunks Edison's pretenses at originality; Stephen Kern, *The Culture of Time and Space, 1880–1918* (Cambridge, Mass., 1983) contains a cogent analysis of "atomized movement" by Marey; Siegfried Giedion, *Mechanization Takes Command: A Contribution to Anonymous History* (New York, 1948; 1975), documents the history of the evolution of machines and their function in society from the early nineteenth century; Francois Dagognet, *Étienne-Jules Marey, A Passion for the Trace*, tr. Robert Galeta, with Jeanine Herman (New York, 1992), cites Governor Leland Stanford's interest in Marey's early writings as the stimulus for his contract with Eadweard Muybridge. See also, "The Scientist Who Took Pictures," Andy Grundberg's review of *A Passion for the Trace*, as well as Marta Braun, *Picturing Time,* a new biography of Marey, in *The New York Times Book Review*, March 18, 1992 ("Marey, like Thomas Edison and the Lumières, was only one of several 'fathers' of the cinema.").

209   Edison and Muybridge: James L. Sheldon, biographical essay, "A Man Beyond His Time," in Jock Reynolds and James L. Sheldon, *Motion and Document, Sequence and Time: Eadweard Muybridge and Contemporary American Photography*, exhibition catalog, the Addison Gallery of American Art, Phillips Academy (Andover, Mass., 1991). A fascinating introduction to the life of this iconoclastic figure. George C. Pratt, *Spellbound in Darkness, A History of the Silent Film* (Greenwich, Conn., 1966; 1973). Contains excerpts from contemporary accounts of Muybridge's work and lectures, from *Scientific American Supplement*, January 28 and July 29, 1882; and New York *Times*, November 18, 1882. Muybridge's description of his meeting with Edison in late

February 1888 is from *Animals in Motion*, cited in Pratt and corroborated by Musser, below, with reference to different sources. The Muybridge–Edison correspondence during 1888 pertaining to Edison's receipt of *Animal Locomotion* is in ENHS.

211   Edison and Dickson, the early days: Charles Musser, *Before the Nickelodeon, Edwin S. Porter and the Edison Manufacturing Company* (Berkeley, 1991). Paul C. Spehr, *The Movies Begin, Making Movies in New Jersey, 1887–1920* (Newark, N.J., 1977). Spehr, former assistant chief of the Division of Motion Pictures, Broadcasting and Recorded Sound, Library of Congress, is skeptical of Dickson's claims to have demonstrated a perfected kinetoscopic show for Edison in the fall of 1889. He aptly points out in personal communications with the author and in the notes to his essay, "Edison Films in the Library of Congress," that "The Edison companies were engaged in a number of projects at this time [late 1880s]. Dickson spent at least part of this period working on experiments in ore separation which were entirely unrelated to the work on the motion picture. . . . All of this distracted attention from the development of the motion picture." See also Edison's introduction to W. K. L. Dickson and Antonia Dickson, "*Edison's Invention of the Kineto-Phonograph*," *Century Magazine*, vol. 48, no. 2, June 1894, in which he updates his "idea" for the machine.

212   The seminal 1936 essay "The Work of Art in the Age of Mechanical Reproduction" is in Walter Benjamin, *Illuminations*, selected essays edited and with an introduction by Hannah Arendt (New York, 1969).

CHAPTER 18

213   Edison and ore-milling: There have been three important articles published on this gargantuan enterprise. A richly documented, photographically illustrated contemporary source, including a geographic landscaping plan of the factory complex, is "The Edison Concentrating Works," *The Iron Age* magazine, Thursday, October 28, 1897 (n.a.). A more recent historical approach, with an update on the existing situation based upon a visit to Ogden in the early 1980s, is

PAGE

Karl Anderson, "Edison's Venture in Iron," *The North Jersey High-lander, A Quarterly Publication of the North Jersey Highlands Historical Society*, Newfoundland, New Jersey, vol. XVI.2, summer 1980. The most comprehensive scholarly approach to date is W. Bernard Carlson, "Edison in the Mountains: The Magnetic Ore Separation Venture, 1879–1900," in Norman Smith, ed., *History of Technology*, no.8, New York and London, 1983.

See also, Edison's 21-page handwritten private memorandum on his "industrial undertaking" (n.d., but from internal evidence early summer 1889). Correspondence, John Birkinbine: TAE, July 13, 1889, with Edison's handwritten reply on Birkinbine's letter, "I have firmly resolved to waste no time . . ." John Birkinbine and Thomas A. Edison, author's edition, "The Concentration of Iron-Ore," a paper read before the American Institute of Mining Engineers, New York meeting, February 1889. TAE, U.S. patents: Magnetic Separator, January 15, 1889; Ore Separator, March 26, 1889; Crushing-Roll (Improvements), December 19, 1893: TAE and W. K. L. Dickson, U.S. patent: Magnetic Separator, August 19, 1890. All ENHS.

219   Continuing narrative of development of cinematic techniques: Benjamin, Dagognet, Dickson, Musser, Rawlence, Pratt, and Spehr, all op. cit., chapter XVII.

220   Handwritten memorandum, TAE:Sherbourne Eaton, covering the period January–March 1891, with reference to Lippincott and the phonograph; "Edison's Kinetograph and Cosmical Telephone," *Scientific American*, June 20, 1891; "Edison's Kinetograph," *Harper's Weekly*, June, 1891; "Moving and Talking Pictures," *Scientific American*, January 18, 1913 [retrospective references to sequence of patents filed before and after Edison's]. United States Patent, Kinetographic Camera, August 24, 1891; assignment of patent from W. K. L. Dickson to TAE, October 31, 1891. ENHS.

221   The troubled children, Marion, Tom, Jr., and Will: Correspondence, Mary Miller (sister):Mina, Rome, November 17, 1889, CEF; telegram, Marion in Germany to TAE in New York requesting wiring of funds, August 1, 1891, ENHS; correspondence, Tom, Jr.:

"Jhonney," September 24, November 15, and December 30, 1891, ENHS; Tom, Jr.:"My dear Papa," January 16, 1892; Tom, Jr.: "My dear Mamma," April 25 and November 18, 1892, CEF. Correspondence, William Leslie:"Dear Father" (n.d., 1890–91); William Leslie:"Dear Mamma" [n.d., 1890–91]; William Leslie:"All the Family," October 31, 1890; William Leslie:"Dear Mamma," October 31, 1890, CEF.

223 William Pitt Edison: Correspondence, WPE: TAE, September 28, 1888, asking for money to keep up the mortgage payments on his farm. TAE replies with emphatic note in corner of letter, "If I pay it this time I want Pitt to understand that I shall never do so again. His farm must go before I will do it." Telegram, January 13, 1891, re Pitt's death and Edison's travel plans for the funeral. Also, correspondence, Nellie Edison (sister-in-law): TAE, January 26, February 6, 1891. ENHS.

223 The travels and adventures of Samuel Ogden Edison are chronicled in the faithful letters (and requests for additional funds) from his friend and companion James Symington. Symington: TAE, January 20, February 23, March 19, April 15, 18, 28, May 1, July 26, 29, September 10, 1891; and February 1, 1892, with enclosed newspaper clipping, n.d., "Edison's Father, The Venerable Parent of the Great Inventor on a Visit to New Orleans," stating that "In personal appearance Mr. Edison [senior] strikingly resembles James G. Blaine." At the end of the piece, Samuel Ogden declares his intention to "go to Amsterdam, the city of his father's birth, and spend the next year or so in a tour through Holland." ENHS.

224 Mina Miller Edison and family: Correspondence, Lewis Miller:Grace (daughter), September 29, 1889; May 23, 1890; Lewis Miller:Kate Kimball, October 12, 1892; Lewis Miller:Bishop John Heyl Vincent, October 12, 1892; Lewis Miller:George E. Vincent, December 1, 17, 22, and 27, 1892; Lewis Miller:Mina, May 14 and June 28, 1892. E-MC. Mary Valinda Miller:Mina, July 15, 1888; March 2, 1890; January 11, February 18, November 15, 1891; February 24, 1892. CEF.

225 Redecorating of Glenmont and environs: Leah Burt, "Home Furnishings Report and Cultural Landscape Report," op. cit.

CHAPTER 19

227   The description of Menlo Park's deterioration is from correspon-
       dence, S. B. Eaton:TAE, February 17, 1892, with TAE reply of Feb-
       ruary 18 written upon it; and from Marion Edison: John Randolph,
       January 18, 1893, ENHS. The story of the formation of General Elec-
       tric is in Chandler, Clark, General Electric Research and Develop-
       ment Center, Hughes, and Insull, op. cit. An intriguing footnote to
       Edison's reaction to the corporate identity of GE is retold in "How
       Edison Lost Power," letter to the editor, *Wall Street Journal*, February
       9, 1989, by William Copulsky, associate professor of marketing,
       Baruch College, Collection Edison Electric Institute. A new biogra-
       phy, *Steinmetz: Engineer and Socialist*, by Ronald R. Kline (Baltimore,
       1992), is reviewed with salient bibliographical material in *Technology
       and Culture,* The International Quarterly of the Society for the His-
       tory of Technology, XXXIV.34, no.3, July 1993.

230   Edison's orientation toward the Panic years and their overall impact
       in American society are chronicled in Cashman, Gilbert, Millard,
       Palmer, Vatter, and Welch, op. cit. Andrew Carnegie's uniquely op-
       timistic philosophy is well expressed in Joseph Frazier Wall, editor,
       The Andrew Carnegie Reader (Pittsburgh, 1992), especially the essays,
       "The Gospel of Wealth," "The Bugaboo of Trusts," and "The A B C
       of Money." Carnegie also wrote to TAE on February 3, 1891 (ad-
       dressing him as "My Dear Wizard") in response to Edison's query
       about the usefulness of nickel alloys in steel production, ENHS. The
       chronology of Edison's activities during the onset and duration of the
       Panic is derived from correspondence, TAE:R. D. Benson, treasurer,
       Magnetic Iron Ore Company of New York, September 7, 1892; TAE:
       Rush Young, Boulder Smelting Company, February 8, 1893; TAE:
       J. T. Boyd, April 19, 1893 ("The mill is not now running"); and
       TAE: Haldeman, Grubb and Company, June 20, 1893, ENHS. The
       Chautauqua *Assembly Herald* notes the "arrival of Mr. Thomas A. Edi-
       son and family" July 22, 1893, E-MC. Lewis Miller's correspondence,
       eighteen letters to various family members, spans January 3, 1893
       through October 31, 1894. Correspondence, Ira Miller:TAE, July 31,
       1894, discussing the terms of their loan, E-MC; W. K. L. Dickson:

A. O. Tate (for TAE), February 3, 1893; and TAE:A. B. Dick, February 17, 1893, ENHS, reflects TAE's anxieties about Dickson's nervous condition and leave of absence while the kinetoscope was supposed to be in production.

234 Study of the great 1893 Chicago World's Columbian Exposition provides the welcome opportunity for access to a colorful mixture of literature from the period as well as more recent scholarship. The interview with Edison in which he estimates his net expanse of inventions at "twenty-five acres" was published in the New York *World*, January 17, 1892. General background on the Fair is contained in Kouwenhoven, Smith, and Trachtenberg, op. cit. The pervasive role of electricity at the Fair is broadly if romantically described in "Electric Wonders, Strange Things to be Done at the World's Fair," Chicago *Times*, November 5, 1891. Three excellent recent scholarly works are the following: Lawrence Levine, *Highbrow/Lowbrow: The Emergence of Cultural Hierarchy in America* (Cambridge, 1988), with its discussion of "Order, Hierarchy and Culture" as manifested in the schism between the White City and the Midway Plaisance. Carolyn Marvin, *When Old Technologies Were New: Thinking About Communication in the Late Nineteenth Century* (New York, 1988), contains a thorough description of the electrical highlights of the Fair and the contemporary popular response to them; and Robert W. Rydell, *World of Fairs, The Century-of-Progress Expositions* (Chicago, 1993) places the Chicago Fair in the context of its fellows with particular emphasis on "visions of empire" the Fairs extolled. For contemporary intellectual and popular interpretations of the Fair, we have delved into Henry Adams, *The Education of Henry Adams*, Library of America Edition (New York, 1990), especially chapters 22 and 23, "Chicago (1893)" and "Silence (1894–1898)"; William Dean Howells, *Letters of an Altrurian Traveller (1893–94)*, facsimile edition, ed. Clara and Rudolf Kirk (Gainesville, Fl., 1961), essays that originally appeared in *The Cosmopolitan Magazine* at the time of the Fair; and Dave Walter, ed., *Today Then: America's Best Minds Look 100 Years Into the Future on the Occasion of the 1893 World's Columbian Exposition* (Helena, Mont., 1992), which contains the excerpts from the cultural survey in which Thomas Edison was so favorably cited. Steinmetz's anecdotal tribute

to Edison was published by The Edison Lamp Works of The New York Edison Company on October 21, 1914, as part of a tribute to TAE "by the present lighting industry."

238   Continuing the tale of the kinetoscope, Dickson, Musser, Pratt, and Spehr, op. cit. Charles Musser has written the definitive work on early cinema, *The Emergence of Cinema: The American Screen to 1907* (New York, 1990). This month-by-month analysis of the technological and theatrical developments by Edison and others provides the underlying chronology for this chapter. W. K. L. Dickson's description of the perfections still needed in the kinetoscope, including a schematic drawing of his new shutter device, are contained in his letter to patent attorneys Dyer and Seeley, January 3, 1894; Edison's reply to Muybridge in draft (dated at top, "02-08-94") and final typewritten form dated "Feb 1894"; also of interest on this theme is a handwritten copy in ENHS letter books of a letter to H. Ericksen, a reporter for the Detroit *Journal* from TAE, June 20, 1894, "I beg to state that the difference [between the kinetoscopic image and Muybridge's] is that the record of the motion is pure, no intermittency as in the old Zeotrope [*sic*] as well as Muybridge and Anscheutz. The whole difficulty was in getting a pure motion without which the illusion is destroyed." ENHS. The amusing anecdote about the opening of the first kinetoscopic parlor in New York City is taken from A. O. Tate, op. cit. The response of the New York press to the kinetoscope ("Edison's latest toy"), was summarized by an anonymous writer for *The Critic*, XXIV.638 (May 12, 1894), reviewing the films shown in the debut month of the parlor.

243   The ongoing story of Ogden/Edison is drawn from Anderson, Carlson, Hughes (1976), *The Iron Age* (1897), Nye, and Vanderbilt, op. cit. The description of Marion Edison's final two days in America before her departure for Germany is drawn from a letter from Lewis Miller to his wife written while he was staying as a guest at Glenmont in March 1894. E-MC. Charles Batchelor made an entry in his journal for November 14, 1894, "Ogden Mine changed to Edison, N. J.," ENHS. The visual descriptions of the Ogden site are based upon contemporary panoramic photographs of the period, ENHS.

"Faith in production as an end in itself" at the closing of the nineteenth century is noted in Giedion, op. cit. See also, Frederick W. Taylor's essay, "A Piece-Rate System," *Transactions of the American Society of Mechanical Engineers*, XVI (1895); and "He Employs Women—Good Work Done at Edison's by Nimble Fingers," by Cynthia M. Westover in the New York *Recorder*, November 15, 1894.

CHAPTER 20

247 Kenneth Goldstein, transcripts of interviews with Mrs. Madeleine Edison Sloane, Llewellyn Park, New Jersey, December 1, 1972, and March 13, 1973, Columbia University Oral History Project Archives and ENHS. Grateful acknowledgement is made to Ronald L. Grele, director, COHPA, for his kind assistance in providing access to this material. Also, Helen Henry, "Edison: A Loving Father, A Tease, Too: One of the Inventor's Daughters Reminisces," Baltimore *Sun* magazine, October 19, 1969 (interview with Madeleine Edison Sloane), collection Newark Public Library.

248 The epistolary chronicle of TAE's letters from the Ogden Mine to his wife at Glenmont are grouped in one folder of photocopies at ENHS. Thanks to George Tselos, archives curator, for locating these materials. Most are undated except for the days of the week ("Sunday," "Friday," "Tuesday") but are positioned in the narrative on the basis of internal evidence. There are four dated letters: August 9, August 16, 1895; March 8, December 1, 1898. The citation from William James, *The Will To Believe* (1897) is in Perry Miller, op. cit. Edison's journal entry on cobalt ("I have been reading up on cobalt.") is in ENHS, as is a response to a letter from George Foster Peabody of Spencer Trask & Company, March 28, 1898, "I am only at the Laboratory Mondays and sometimes Tuesday, balance of the time at the Mine. I could not find time to come to New York." For an updated view of the situation at Ogden/Edison, see also the magnificently illustrated "The Edison Magnetic Concentrating Works," *Scientific*

American, January 22, 1898, collection of Neil Maken. The anecdote "A Game of Billiards" is from an apparently unpublished manuscript by W. S. Mallory, typescript in ENHS.

253   An excellent overview of Thomas Edison and his work with the X ray is Allen Koenigsberg, "Edison's Brain: An Inside Look at the Discovery of X-Rays and Recorded Sound," *Antique Phonograph Monthly*, IX.2 (n.d., 1992). A focused succession of events surrounding the X-ray work is documented in ENHS, as follows: TAE, telegram to A. E. Kennelly, "We could do alot before others get their second wind," January 27, 1896. Correspondence, Dr. James Burry (Rookery Building, Chicago): TAE, February 4, 1896. "It is necessary for me to procure Rontgen [*sic*] photographs for hospital purposes at once," and TAE immediate response, "The thing is too new"; Sigmund Bergmann: TAE, February 7, 1896, enclosing "Rontgen [*sic*] photograph of a hand . . . sent to me from Germany"; TAE handwritten reply to A. L. Benedict of Buffalo, New York, regarding X-ray tube inserted in stomach, February 10, 1896; TAE handwritten note to Walter E. Woodbury, editor, *The Photographic Times*, February 13, 1896, "If you could come over and help us further would be obliged;" handwritten note from TAE February 18, 1896, regarding "perfecting the apparatus, the effect of the ray on the Bacterial germ of Cancer will soon be know[n]"; handwritten note from TAE, February 18, 1896, "There is some profound mystery going on in these high vacuum tubes. . . . I am trying to run these to earth."

254   Death of Samuel Ogden Edison. Site visit by the author to Lakeside Cemetery in Port Huron, Michigan, spring 1992. Correspondence, Marion Edison Page:Mina Edison, March 1896; telegrams, Mrs. W. P. Edison:TAE, February 26, 1896; TAE:Mrs. W. P. Edison, February 26, 1896; TAE:B. H. Welton, Port Huron, Michigan; and Welton:TAE, February 26, 1896; Marion Edison Page:TAE, and TAE:Marion Page, February 27, 1896, ENHS. Correspondence, James Symington:TAE, March 20, 1893 and May 27, 1894, re travels and health of S. O. Edison, ENHS. Minneapolis *Tribune* article on SOE, January 12, 1893, reprinted in Fort Myers *Press*, collection of Leslie Marietta; George Parsons Lathrop, "Edison's Father," *Once a Week Magazine*, January 20, 1894, collection of Neil Maken.

PAGE

256 Marion "Dot" Edison, 1893–98: Correspondence is in CEF unless otherwise noted; Marion to Mina letters are n.d., but are positioned with internal evidence. Mina Edison:"My darling Mother and sisters," January 7, 1898 (not sent?); Karl Oscar Oeser:TAE, July 23, 1894; Marion:TAE, July 1894, August 24, 1894; April 1895 and November 16, 1895; J. P. Morgan:TAE regarding Marion's monthly allowance, April 3, 1896; Marion:TAE, April 10, 1896; Mrs. Eugenie Oeser:TAE, May 28, 1896, ENHS; Marion:Tom, Jr., December 12, 1896, ENHS; Marion:TAE, August 27, 1897.

257 Thomas A. Edison, Jr., 1894–98: Jane Miller (sister):Mina, January 29, 1893, "How is it with Thomas? It is too bad he would not go to school. Some day he will regret it. . . . Mr. Edison feels it is all right for him to go to Ogden. . . . Every boy I suppose of that age wants to stop school and gets discouraged about his studies," CEF; Tom, Jr.: Mina, August 12, 1894; April 29, May 27, June 7, September 5, 1895; March 8, August 18, October 11, November 24, December 1, 1896; September 24, and October 3, 1897 ("When I came down from Edison, and you had left for the west, I never dreamed that one could feel so blue and despondent. I never expected to feel so low, never in the world. I immediately went up to the house and then to the Laboratory. I lost my appetite entirely and I felt horrible. I felt that as no single soul cared for me and I had no idea what to do, being useless in this world as I am and always was, Oh! Mother, do not think I am crazy, but the awful thought came to me that it is best for me to leave the world, for I never was very happy, and as the prospects for ever being so were very slim, I thought that this was best. But no—darling Mother I couldn't die. I was not prepared to die."), CEF; Tom Jr.: Mina, November 1894; May 15, August 6, 1897, ENHS. The author gratefully acknowledges photocopies of 1896–98 correspondence from Tom, Jr., to his friend Ed Redington, sent by Mr. C. W. Weintraub in response to a query in the *New York Times Book Review*. Copies of these letters are also in ENHS. Tom, Jr.: "My dear Ed," April 1, 1896; Tom, Jr.:W. S. Mallory, October 11 and October 26, 1896, draft letters not sent, ENHS; Tom, Jr.:Ed Redington, February 21, March 14, April 11, May 7, 23, 1897, ENHS; August 19, September 13, 1897; January 8, January 18, 29, February 18, 22,

March 10, April 18, May 21, 29, July 24, 1898; Tom, Jr.:TAE, January 14, 1898, "I feel that I never have pleased you in anything I have ever done"; Tom, Jr.:William Leslie Edison [January 1898]. CEF.

261 William Leslie Edison, 1894–98: Correspondence, Will:Mina, markings by TAE, back to Will [1894], CEF; J. M. Hawkins:Mina, March 2, 1894; February 25, 1896, CEF; WLE:TAE, April 25, 1898; I. S. Atkinson, Counselor-at-Law:TAE, May 26, 1898; J. M. Ceballos & Co.:W. S. Mallory, October 21, 1898, ENHS.

262 Lewis Miller and family, 1894–98: LM:Mina, August 28, 1893; January 1894, CEF; Lewis Miller (brother):Mina, January 3 and 6, 1894; July 12, 1895, CEF; Mary Valinda Miller:Mina, February 27, October 21, 25, 1894, CEF; Aultman, Miller & Co. to the secretary of the treasury, regarding invention of reaper and mower, January 1, 1897, E-MC.

263 Lewis Miller and Chautauqua, from 1896: All manuscript and archival materials in Chautauqua Library, Chautauqua, New York. Alfreda L. Irwin, op. cit.; William Rainey Harper, D. D., "John H. Vincent, the Founder of the Chautauqua Movement," *The Outlook Magazine,* September 26, 1896; John H. Vincent to George Vincent, "Confidential," handwritten note, April 28, 1896; Correspondence, JHV:George Vincent, April 30, October 29, November 21, 1896; Mina's handwritten draft on Chautauqua, "Religion is the foundation of a great many good things and Chautauqua is one of its monuments"; Lewis Miller, multiple handwritten pencil drafts of letter to William Rainey Harper [November], 1896.

265 Theodore Westwood Miller, 1894–98: Ellwood Hendrick, op. cit., is source for citation of Theodore as "his father's Benjamin." See also George Vincent, op. cit., privately printed memorial biography, *Theodore Westwood Miller, Rough Rider.* An excellent single-volume history of the Spanish-American War, from which the quotations from Theodore Roosevelt and geographical and anecdotal details on the Battle of San Juan Hill have been drawn, is G. J. A. O'Toole, *The Spanish War, An American Epic, 1898* (New York, 1984). Correspondence, Lewis Miller:Theodore, April 26, 1896, May 29, 1897; April 19, 24, 30, 1898; LM:Grace Miller, May 7, 1898, E-MC; Theodore Miller:Mina, "On Board the 'Yucatan,' " June 11, 1898, CEF; Robert

Miller:Mina, June 15, 1898, "I enclose Theodore's card. . . . He seems to be having a fine time of it so far but I cannot say I am delighted he went. I hope he will come out with high honors and have no doubt he will merit them but some how I don't understand how he happened to go." CEF. Excerpts from Theodore's letters to his mother of April 20, 26, May 28, and July 7, 1898, and the details of his wounds on the battlefield and final days are in George Vincent, op. cit. Details of the news of Theodore's death arriving within one day subsequent to the birth of the Edison baby boy are from the Newark *Daily Advertiser*, July 12, 1898, "Joy and Grief in Edison's Home, A Son Is Born and Wife's Brother Killed at Santiago," CEF, and a related story in the New York *Morning Journal*, July 12, 1898, "Grief and Joy in Edison Home, Son Born to the Inventor, and Brother Lost to His Wife, Wizard's Day of Anxiety," ENHS.

CHAPTER 21

268 John Miller's letters home from the Spanish-American War were collected and transcribed by George Vincent in preparation for his book on Theodore Miller, op. cit., and photocopies are in the E-MC at Chautauqua Institution. The Glendale Cemetery records and inscriptions from the flyleaf of the Miller family Bible were provided by Mr. George Ball of Akron. Correspondence, George Vincent:William Rainey Harper, August 24, 1898; Lewis Miller:John Vincent, October 24, 1898, E-MC.

269 Correspondence, Thomas A. Edison, Jr.:Edward Redington, regarding his relationship with Miss Twohey [*sic*], August 24, 1898, ENHS. "Wed Casino Girl, Son of Inventor Edison Marries Secretly, Kept It Quiet for Months, His Bride Has Left 'La Belle Helene' to Keep House," New York *News,* February 17, 1899; "May Take His Wife to Orange," Newark *Daily Advertiser,* February 17, 1899; Certificate and Record of Marriage, State of New York, regarding Thomas A. Edison, Jr., and Marie Louise Toohey, signed and witnessed, February 19, 1899; correspondence, William Leslie

Edison:TAE, November 24, 1900, regarding the irregular behavior of Marie Edison, and TAE handwritten note to T. C. Martin, "He never could go to school," ENHS.

270 Descriptions of the illness and death of Lewis Miller are taken from scrapbook clippings in E-MC, dated February 18, 1899, including "He Raced with Death and Lost," New York *Herald;* "Lewis Miller Died Under the Knife," New York *World;* "Blizzard Kills the Founder of Chautauqua," New York *Journal;* "Death of Lewis Miller," New York *Sun;* "Chautauqua Founder Dead," New York *Evening Journal;* and "Hon. Lewis Miller," *The American Thresherman* (Madison, Wisconsin), vol. 1, no. 10. Medical diagnosis of Miller's illness, David S. Baldwin, M.D., personal communication. Also, see Ellwood Hendrick, op. cit., for tributes to Lewis Miller from Akron friends and neighbors. Mina Edison was perennially haunted by the memories of the deaths of her brother and father, and every year on these anniversaries shared her grief: "My darling Mother," she wrote on July 8, 1901, "I would love to be just near you where in silent sympathy we might sit together on the old Oak Place piazza. Ah, how often father darling comes before me seated in one of those large porch chairs on the corner of the porch trying to rest and get cool after a busy day. There where we sat the summer of the sad return of 'our soldier boy.' "

272 The ongoing story of Thomas Edison and the cinema continues in Musser (1990 and 1991), Pratt, and Spehr, op. cit. Edison's tirade against the Circuit Court was published as an excerpt from the New York *Sun* in *Scientific American,* January 8, 1898. The Biograph case, including Edison's deposition, is exhaustively documented in ENHS, Legal Department and Related Records, Boxes 73–77: Complainant's Record, Brief for Complainant, and Closing Argument.

277 TAE and cement: The citation from Brooks Adams, *The New Empire,* 1902, is in Perry Miller, op. cit. For background information on the history of cement, I am grateful to Richard H. Lucas, communications supervisor, and editor, *PCA Progress,* at the Portland Cement Association, Skokie, Illinois. Information on David O. Saylor's pioneering work on Portland cement was obtained in a visit to the Saylor Park Cement Industry Museum in Coplay, Pennsylvania. TAE's

notes on cement go back to the early 1880s. For the purposes of this chapter, an 1895 handwritten note to "Johnny" [Randolph] sent from the Ogden Mines indicates that "I am now designing a small crushing machine similar to our 3 High Rolls at the Mine, for crushing cement . . . *I feel sanguine that this machine will be an epoch in the Portland Cement Industry.*" TAE's 51-page handwritten prospectus to Herman E. Dick, May 11, 1900, including schematic diagrams of the cement plant and the geological strata of the area near Stewartsville; and list of "Jobs that can be done now," October 4, 1900, ENHS. The best description of Edison's long cement kiln will be found in Appendix XVII to Frank Dyer and Thomas C. Martin, with William H. Meadowcroft, *Edison, His Life and Inventions* (New York, 1929), which also contains Walter Mallory's assertion that the mill was planned by TAE in one day of intensive thought and scribbling.

280 The progress of the phonograph at West Orange from 1896–1900 is drawn from "The Phonograph," *Scientific American,* July 25, 1896; and "The Manufacture of Edison Phonograph Records," ibid., December 22, 1900, which includes six detailed photographs on the front page portraying the technical steps from casting the blank records to testing the machines. Collection of Neil Maken.

281 Edison and the storage battery: Hughes (1976), and Vanderbilt, op. cit., provide excellent overviews. The contemporary documentation begins with Arthur E. Kennelly, "The New Edison Storage Battery," publication of his paper presented before the American Institute of Electrical Engineers, May 21, 1901; Thomas A. Edison, "The Storage Battery and the Motor Car," *The North American Review,* vol. 175, no. 1 (July 1902). Information on Edith Wharton, in Anita Brookner's "About the Author," at the conclusion of *The House of Mirth* (New York, 1987); Ray Stannard Baker, "Edison's Latest Marvel: The New Storage Battery," *Windsor Magazine,* November 1902; E. S. Foljambe, "The Edison Storage Battery—Its Inception and Method of Manufacture, A Trip Through the Plant at Orange, N. J.," *The Commercial Car Journal,* January 13, 1914; Dyer, Martin and Meadowcroft, 1929, ibid., Appendix XVIII, "Edison's New Storage Battery;" and W. Bernard Carlson, "Thomas Edison as a Manager of R & D: The Case of the Alkaline Storage Battery, 1898–1915," Feature Article, *IEEE*

Technology and Society Magazine, vol. 7, no. 4 (December 1988). These articles are kept in Curator's Subject Files at ENHS.

284   Life at Glenmont: Thorstein Veblen, *The Theory of the Leisure Class,* intro. Robert Lekachman (New York, 1979). Madeleine's 13th birthday party, correspondence, Mina Miller Edison: Mary Valinda Miller, June 4, 1901, CEF; Madeleine's anecdotal reminscences, Columbia Oral History Interviews and Baltimore *Sun* magazine, op. cit; "Acoustiguide" tour of Glenmont, late 1960s. In her interviews, Madeleine refers to sister Marion's memories of childhood also. Charles Edison's life is documented through oral history transcripts in John D. Venable, *Out of the Shadow, The Story of Charles Edison,* published privately by the Charles Edison Fund, East Orange, New Jersey, 1978. The changing decor and elaborate menus at Glenmont as well as other details about the domestic situation, including the number and variety of servants, is in Leah Burt, op. cit. The increasing prominence of Edison as a man of means had its complications for his family. In May 1901 he received letters threatening to kidnap Madeleine for ransom. Many of his wealthy neighbors in Llewellyn Park were also endangered. He was forced to hire Pinkerton guards to patrol his estate: "Edison Threatened by Kidnapping and Arson, Police Are Guarding His Children," Chicago *American,* May 28, 1901. ENHS.

288   Edison returns to Fort Myers: "Two Years' Rest for Edison," the Philadelphia *North American* (n.d., 1901 from internal evidence). Articles in the Fort Myers *Press,* February 28, March 7, 14, April 4, 11, June 27, August 22, November 14, 1901. Collection of Leslie Marietta.

CHAPTER 22

290   Glenmont and Fourth of July vignettes: Leah Burt, Madeleine Edison Sloane, John D. Venable, op. cit. Regarding Theodore's early years: correspondence, Mary Valinda Miller:Mina Miller Edison, November 15, 1903 and May 19, 1906, CEF. Mina:Mother [c. 1904–06], CEF.

Feature article on Theodore Edison with two photographs, Newark *News,* July 10, 1904. ENHS. Theodore Edison, interview with the author, June 19, 1992. Charles and TAE competitive fishing, "Wizard Lands Two Tarpon," Fort Myers *Press,* March 31, 1904, including chart, "Tarpon Record for 1904," listing TAE's catch at 40 pounds and Charles's at 100 pounds.

292 The account of TAE and family motor car excursion is drawn from a journal kept by Margaret Miller (Newman), August 15–29, 1905, provided to the author by Nancy Miller Arnn.

293 TAE and children from first marriage: Correspondence, Tom, Jr.:TAE, December 29, 1902; January 1, July 21, 1903, with TAE handwritten reply on verso; Anna M. Miller:John Randolph, September 8, 1903; Charles Supe:TAE, regarding Tom, Jr. and the sham "Vitalizer," December 16, 1903, with TAE handwritten disclaimer re "crooked" concern; Tom, Jr.:John Randolph, December 17, 1903; Charles Stilwell:John Randolph, December 18, 1903; newspaper articles, "Edison, Jr., Held for Mail Fraud," New York *Herald,* October 6, 1904, and "Humbugs and Swindles, Thomas A. Edison, Jr.'s Electric Vitalizer," *Bee Culture,* October 15, 1904. John Randolph:Tom, Jr., November 23, 1905; Tom, Jr.:TAE, November 22, 1905; February 9, 1906; May 5, June 26, September 27, November 27, 1907; transcript of article in New York *Times,* "Edison, Jr., Weds Quietly in Jersey," July 10, 1906; Beatrice Heyzer handwritten autobiographical sketch, September 30, 1929; Tom, Jr.:John Randolph, August 22, 1906; ENHS. Tom, Jr.:Mina Edison, February 6, 1909, CEF. Correspondence, William Leslie Edison:Mina Edison, July 8, 1900; January 28, 1902, CEF. WLE:Walter Mallory, July 14, 1902; WLE:TAE, August 12, October 12, 1903, with TAE handwritten replies, October 13 and 14; December 16, 1903, "Your letter addressed to my wife came duly to hand"; December 16, 1905; Blanche Edison:TAE, December 12, 1903, and TAE handwritten reply, same day; October 23, 1907; Samuel H. Scoggins:TAE, December 27, 1903; WLE:John Randolph, February 16, 1905; Blanche Edison:John Randolph, September 30, 1907; March 30, November 16, 1907; October 3, 1907; James P. Whelan:TAE, November 13, 1907; WLE:Frank Dyer, October 2, 1908; Frank Dyer:WLE, Octo-

ber 3, 1908; WLE:Frank Dyer, November 24, 1908. ENHS. Correspondence, Marion Edison Oeser:Mina Edison, August 14, 1904 and (n.d.) 1906. CEF.

296 M. A. Rosanoff, "Edison in His Laboratory," *Harper's Magazine,* May 1932. The article is a lengthy and affectionate look back over twenty-nine years. Edison fired Rosanoff after one year in the laboratory, in a note to William Gilmore at the time calling him "dangerous and of the same type of [*sic*] K. L. Dickson only even more impracticable. He has worked over a year and practically got nothing done . . . He is all talk and not a practical chemist." T. C. Martin's analogy is drawn from his article in *Electrical World,* "The Edison of To-Day" (n.d., but from internal evidence written at the time of the concrete house innovation).

297 *The Great Train Robbery,* videotape in collection ENHS. Also, see Musser, Pratt, and Spehr, op. cit.

298 The cement house: "Means for Operating Motors in Dust-Laden Atmospheres," U.S. patent application filed January 9, 1903, and accepted March 31, 1903, number 724,089. "Process of Constructing Concrete Buildings," U.S. patent application filed August 13, 1908, and accepted March 13, 1917, number 1,219,272. *The Romance of Cement,* published by Thomas A. Edison, Inc. and the Edison Portland Cement Company (1925). "Edison's Latest Discovery," Chicago *Times,* May 28, 1901. William A. Radford, "Edison Poured Cement House," *American Carpenter and Builder,* July 1909. ENHS. Courtesy Portland Cement Association: Peter Collins, *Concrete: The Vision of a New Architecture* (New York, 1959); "Thomas Edison's Concrete House," *Concrete Construction* magazine, June 1965; Raymond C. Heun, "Thomas A. Edison's Adventures in Concrete," *American Concrete Institute Journal,* August 1975, reprinted with emendations in *Concrete International Magazine,* March 1979. Collection Edison Electric Institute: "Cement Firm Started by Edison in 1898 to Close," New York *Herald Tribune,* April 18, 1942.

300 Storage Battery: Correspondence, Mina Edison:"My darling Mother," dated "Laboratory, Sunday, 8 p.m."; memorandum, "500 Cells Daily," and handwritten letter to "[Herman] Dick," ENHS. Mina Edison:"My darling Mother," October 23, 1904, CEF. "New-World

Alchemist," New York *Daily Graphic,* February 19, 1910, collection of Neil Maken. Of particular interest in tracing developments in the production of the storage battery over a decade of experimentation is a detailed letter from Jonas Aylsworth, Edison's chief chemist, to Dr. T. J. Parker at the General Chemical Company in New York City, September 20, 1915, in which he reviews "the details of the inventions in which Mr. Edison has made use of chemistry." The letter concludes with three pages on the battery alone, including the pivotal discovery of nickel flake. Aylsworth concludes with characteristic modesty echoing other Edison disciples, disavowing any "credit for this or that invention or accomplishment. After close association with Mr. Edison for the past twenty-seven years I can say that this is not the case with his important achievements during my time with him . . . His personal genius and generalship is responsible for all he has accomplished." ENHS. On the battery, see also Harold H. Smith, "The Edison Storage Battery in Service," a paper read before the Electric Vehicle Association of America, Third Annual Convention, Boston, Massachusetts, October 8–9, 1912, and Baker, Carlson, Foljambe, Meadowcroft, and Vanderbilt, op. cit.

301 The cultural setting for Henry Ford is admirably laid down in Loren Baritz, *The Good Life: The Meaning of Success for the American Middle Class* (New York, 1982). Baritz cites the important announcement by Henry Ford of his "car for the great multitude," taken from Warren Susman's essay, "Culture Heroes: Ford, Barton, Ruth," originally published in 1980 and anthologized in *Culture As History: The Transformation of American Society in the Twentieth Century* (New York, 1984). For the story of Henry Ford's childhood, youth and success up to the Model T, I have turned to the words of the industrialist himself in *My Life and Work,* written in collaboration with Samuel Crowther (New York, 1922); as well as the following selected classic biographies: the first volume of Allan Nevins and F. E. Hill's epic, *Ford: The Times, The Man, The Company* (New York, 1954); Keith Sward's more critical *The Legend of Henry Ford* (New York, 1948); Roger Burlingame, *Henry Ford,* (New York, 1957); and more recently, Peter Collier and David Horowitz, *The Fords, An American Epic* (New York, 1987). William and Marlys Ray, *The Art of Inven-*

tion, Patent Models and Their Makers (Princeton, N.J., 1974), was useful in establishing the chronology for the internal combustion engine. Jeanine M. Head and William S. Pretzer, *Henry Ford, A Pictorial Biography,* Henry Ford Museum and Greenfield Village (Dearborn, Mich., 1990), beautifully conveys the pictorial flavor of Ford's early days. The story of the first meeting of Edison and Ford is told differently in Conot and Josephson, but the undeniably influential gist is consistent, certainly according to Ford and his undying adulation for Edison.

304   The philosophical dialectic at the conclusion of this chapter is staked out in Henry Adams, *The Education of Henry Adams,* especially chapters 26–35; and William Graham Sumner, "The Nineteenth Century's Legacy to the Twentieth," in Albert G. Keller and Maurice R. Davie, eds., *Essays of W. G. Sumner,* vol. I (New Haven, Conn., 1934). See also Nathan Reingold and Ida H. Reingold, eds., *Science in America, A Documentary History, 1900–1939* (Chicago, 1981); and for a clear description of the establishment of Thomas A. Edison, Inc., Andrew Millard, op. cit.

CHAPTER 23

306   Charles Edison's play, "Prayer Meeting Night," is an 8-page handwritten manuscript in CEF. Charles's pastimes in Hotchkiss recounted in letters to Madeleine ("Dear Lin"), n.d, his academic performance documented in periodic, dated grade reports; and Madeleine's letters home to her mother from Bryn Mawr, and Mina's descriptions of the campus and her daughter's residence are also CEF, especially Madeleine:Mina, December 4, 1906; and Mina:"My darling mother and two sisters," n.d., [Fall 1906]. The chronology of Henry and William James's activities during this period is found in the Library of America editions of their works; Peggy James's adolescence and beyond, as well as correspondence about the Edisons are detailed in R. W. B. Lewis, *The Jameses, A Family Narrative* (New York, 1991), especially 542 ff. and 618–621. The account of Madeleine's

"debut" party at Glenmont in June 1908 is drawn from Irene Seiber-
ling Harrison (b. 1890), interview with the author held at the Stan
Hywet Hall Gate House, Akron, Ohio, June 27, 1992. Mrs. Harrison, third child of rubber industry magnate Francis Augustus Seiber-
ling and his wife, Gertrude, was present at the luncheon described.
The party was also described in the Fort Myers *Press*. Theodore's letters home from Camp Pasquaney, June 20, 1907; July 11, and August 17, 1909; and a report from the headmaster of Montclair Academy, December 23, 1912, are in CEF. Aunt Grace Miller's postcard
directions to Monhegan from the mainland, July 14, 1908, are in
CEF. For descriptions of Monhegan Island, Maine, "Monhegan Associates: An Experiment in Private Conservation," Sierra Club *Bulletin*,
August 1968; and the *New Monhegan Press*, a "Nearly Monthly Letter" published at Monhegan Island, IV.8 (November 1992) and
IV.10 (January 1993). I am grateful to Curtis Harnack for putting
me in touch with Jan and William McCartin, longtime residents of
Monhegan, and for their personal reminiscences at their home in
New York City.

311 Madeleine's social activities after leaving college, her summer at the
James cottage, and her springtime 1912 trip to Fort Myers with
Peggy James: Correspondence, Madeleine:Mina, August 5, 8, September 6, and November 20, 1909, CEF; R. W. B. Lewis, op. cit.;
and Fort Myers *Press*, March 14, 1907–December 12, 1912, inclu-
sive, collection of Leslie Marietta.

314 Charles Edison at MIT and the Edisons' European trip: Correspondence, Charles:"Dearest parents," "Dearest Both," "Blessed Mother,"
"Dearest Mother" {n.d., 1909–1911}; Mina: "My darling, darling
mother," June 10, 1911; Grace Miller: "My darling Mina"; Marion
Oeser: "My darling father," June 6, 1911, and Marion Oeser,
Voicewritten Reminiscences, March 1956, CEF. The shipboard
meeting of Edison and Henry James is recounted by Charles Edison
in Venable, op. cit., and also in R. W. B. Lewis, op. cit. The Edisons'
itinerary is detailed in Venable and Madeleine Edison Sloane oral history interviews, op. cit. The reunion with Francis Jehl is recollected
in Francis Jehl:Marion Oeser, October 15, 1913, and Marion:Mina
attached, October 21, 1913. ENHS.

317   Edison and the Diamond Disc phonograph: Millard, Schlereth, Sloane, and Edison patents, op. cit. The comprehensive and definitive work on Edison's involvement with the "modern" phonograph is John Harvith and Susan Edwards Harvith, eds., *Edison, Musicians and the Phonograph: A Century in Retrospect* (Westport, 1987). This is a fascinating compendium of interviews with recording artists who worked with Edison during his "intense involvement with every aspect of his record company, including the selection of artists and repertoire." The definitive works on the Disc phonograph and records are George L. Frow, *The Edison Disc Phonographs and the Diamond Discs: A History With Illustrations*, self-published by Frow (Sevenoaks, Kent, England, 1982) and Ronald Dethlefson, *Edison Blue Amberol Recordings, 1912–1929*. See also, *The Phonograph and How to Use It: Being a Short History of Its Invention and Development*, originally published by The National Phonograph Company (West Orange, N.J., 1900) and reprinted in a facsimile edition by Allen Koenigsberg (New York, 1971).

321   Edison and Ford in 1912 and the evolving myth of TAE:Chandler, Collier, Ford, Hughes (1989), Ray, and Rosanoff, op. cit.; TAE:Correspondence, J. Soissons, August 5, 1911; TAE:David T. Dickinson, November 4, 1911; and Richard Cole Newton, M.D., *Harper's Magazine*, February 1912, ENHS.

CHAPTER 24

327   The literature on John Burroughs has most recently been enriched by a loving and thorough biography, Edward J. Renehan, Jr., *John Burroughs: An American Naturalist* (Post Mills, Vt., 1992), the first full account of the naturalist's life to be published since 1925. It draws upon a wealth of manuscripts, journals and letters. Renehan is also the editor of *A River View*, an anthology of Burroughs's nature writings. For Burroughs's aesthetic and personal debts to Emerson and Whitman, I have turned to Gay Wilson Allen, *Waldo Emerson* (New York, 1981) and *The Solitary Singer, A Critical Biography of Walt Whitman* (New York, 1967); as well as the Chronology section in the Library of America Edition of *Emerson's Essays and Lectures* (New

York, 1983). Burroughs's own critical perceptions of his friend the Good Gray Poet are in the uniform set of the naturalist's oeuvre first published in 1896 and 1904 by Houghton Mifflin Company, especially vol. 10, *Whitman: A Study.* The depth of Burroughs's involvement with birds and birding is most evident in his biography of John James Audubon (Boston, 1902; reissued in a facsimile Woodstock, N.Y., 1987). The chronicle of Burroughs's life and observations in upstate New York where he lived for most of his life are found with great verve and color in one of his last books, *Under the Maples* (Boston, 1921). This volume contains the essay referenced directly in this chapter, "A Strenuous Holiday." See also a facsimile edition of essays collected as *In the Catskills* (Boston, 1910; Cornwallville, N.Y., 1986). Two local histories were also helpful: *The Town of Esopus Story* (1979); and Elizabeth Burroughs Kelley, *The History of West Park and Esopus* (Hannacroix, N.Y., 1978). I am grateful to Jonathan Plutzik and Lesley Goldwasser for their hospitality during a weekend of research in their Tiffany Lodge, guided hiking through the surrounding unspoiled forests of West Park including a visit to Slabsides, and access to their rare research materials on John Burroughs.

330 John Burroughs's notebooks containing chronicles of his camping trips with Edison, Ford, and Firestone are housed in the Berg Collection of English and American Literature at the New York Public Library; and on microfilm at the Vassar College library, Poughkeepsie, New York. The Berg Collection holds the original, handwritten notebook for August 4–13, 1919. I am grateful to Nancy McKechnie, curator, special collections, and her staff at the Vassar College library for access to microfilm and photocopied excerpts of the following Burroughs notebook entries: September 3, 1913; February 25, March 10, 1914; May 6, September 29, 1916; August 3, 1919; November 16, 1920; and January 17, 1921.

331 Two published articles on the Burroughs, Edison, Firestone, and Ford camping trips added authentic color to my account: R. J. H. De-Loach, "In Camp with Four Great Americans," *The Georgia Review,* Spring 1959. Professor DeLoach also sent a 10-page typewritten transcript of the detailed journal he kept during the August 1918 camping trip to Thomas Edison on October 30, 1918; the transcript

PAGE

is in ENHS. Dorothy Boyle Huyck, "Over hill and dale with Henry Ford and famous friends," *Smithsonian Magazine*, June 1978. Harvey Firestone's version of the camping trips is in his memoir, written in collaboration with Samuel Crowther, *Men and Rubber, The Story of a Business* (New York, 1926), which I was fortunate to discover in a secondhand bookshop in Akron, Ohio. See especially chapters XIII–XV. On Henry Ford and the Jews, see Albert Lee's book of that name (New York, 1980); Lee also discusses Edison's clarification of the Detroit *Free Press* interview in November 1914 that disturbed Jacob Schiff. TAE's correspondence with Schiff and Isaac Markins is in ENHS. Two films produced by the Ford Motor Company ["Edison, Ford, Firestone, Burroughs"], 1915, and "A Week in the Open," 1918, are in the Collections of the Library of Congress, Motion Picture, Broadcasting and Recorded Sound Division. Grateful acknowledgement to Paul Spehr, Barbara Humphrys, and Patrick Sheehan of the Division for their kind assistance in allowing access to viewing this rare footage. Independent scholar Norm Brauer of Dalton, Pennsylvania, has assembled a compendium of accounts tracing the entire decade of camping trips and loaned me a typescript copy of his research.

335 The negotiations leading to Henry Ford's purchase of his home at Fort Myers, Florida, are described in related correspondence accompanying transcripts of newspaper articles in Leslie Marietta's collection of the Fort Myers *Press*, March 7, 1916–December 30, 1916. Grateful acknowledgement to Mary Anna Carroll, housekeeper-caretaker, and Diane Burris, guide supervisor, at the Edison–Ford Winter Homes in Fort Myers, for their informative tour of the house and environs in November 1991. The Ford home was restored by Buffy Donlon and Foxhill Design through a $350,000 project conducted under the auspices of the city of Fort Myers, and formally opened to the public on January 28, 1990. The project received the John MacArthur Award for Excellence in Museum Quality Conservation.

336 The most detailed anecdotal description of the fire at the Edison plant on December 9, 1914, is in the "Special Fire Edition" of the *Edison Works Monthly* III.1 ("Notes of Interest on the Great Fire Which Destroyed Most of the Edison Phonograph Works"), a 20-page corporate report illustrated with photographs and charts, pub-

PAGE

lished at West Orange, collection of Nancy Miller Arnn. Follow-up stories appeared in *Our Plant* (formerly *The Edison Works Monthly),* III.2 (March 1915). Also, "Edison Fire at Its Height," Newark *Evening News,* December 10, 1914, ENHS.

337 Continued detail on the evolution of the phonograph and the Edison Tone Tests: Frow and Harvith, op. cit. On Anna Case: Thomas A. Edison, Inc., corporate brochure, 1912, "A list of some of the Operatic Artists who have made or will make Records for the Edison Phonograph" (Facsimile edition, Syracuse University Library Audio Archives, edited and with an introduction by Walter L. Welch, 1968); contains a photograph and biographical sketch of Miss Case; collection of Charles Hummel. Also, K. J. Kutsch and Leo Riemens, eds., *A Concise Biographical Dictionary of Singers* (Philadelphia, 1969). Grateful acknowledgment to Jean Bowen, chief; and Charles Eubanks, special collections librarian, Music Division, the New York Public Library at Lincoln Center, for access to the "Anna Case clipping files," and particularly the following articles: "Edison Concert at Carnegie Hall Demonstrated" (description of Tone Test), New York *Telegraph,* March 14, 1920; "Thomas A. Edison Hears Anna Case Sing Over 3,000 Miles of Telephone Wire," New York *Musical Courier,* October 28, 1915; "America's Most Charming Singer Arrives in City, Miracle Girl of Opera Makes Debut in Gate City of Northwest," Fargo (N.D.) *Extra,* October 19, 1916; "Miss Anna Case" (Liberty Bond pose), *Theatre Magazine,* 1916; "Anna Case Sings Star-Spangled Banner for Soldiers," ibid., October 1917; "Anna Case," *Musical America* (n.d.), 1917; "Anna Case Says Good-Bye to Caruso," New York *Telegram,* May 20, 1917; "Tell me," a friend once asked Anna Case, "about all the hard knocks of your life." [No newspaper cited. Extended interview], 1918; "Clarence H. Mackay Wed to Anna Case," New York *Times,* July 19, 1931.

338 On Edison's debt to the work of Hermann von Helmholtz: David Cahan, ed., *Hermann von Helmholtz and the Foundations of Nineteenth-Century Science* (Berkeley, Calif., 1993), a landmark "collective intellectual biography," the first book-length study of Helmholtz in ninety years and the first ever to analyze Helmholtz's accomplishments by critically assessing his published and unpublished writings.

Of special pertinence in this definitive volume including fifteen essays, comprehensive bibliography, and historical chronology are two essays: Gary Hatfield, "Helmholtz and Classicism: The Science of Aesthetics and the Aesthetics of Science," primarily concerned with an analysis of the tenets in Helmholtz's *Tonempfindungen* (*Sensations of Tone*), the book which had such an influence on Edison's thinking in the 1870s; and David Cahan, "Helmholtz and the Civilizing Power of Science," defining the subject's preeminent stance in the context of nineteenth century philosophy and the extent to which he embraced and incorporated larger moral and artistic issues into the scientists' realm.

339   For an outspoken discussion (in his own words) of his likes and dislikes in contemporary music and art, and his standards for music appreciation, Thomas A. Edison, "New Aspects of the Art of Music by the Great American Inventor and Scientist," *The Etude,* April 1917; for a candid interview with Edison on the decline of his cinema activities, *Motion Picture World*, XXI (July 11, 1914). Both publications are in ENHS.

CHAPTER 25

342   On Henry Ford, World War I, and the Peace Ship: See William E. Leuchtenburg, *The Perils of Prosperity, 1914–1932* (Chicago, 1958), for a good general background on the causes of the war. Roger Burlingame, *Henry Ford: A Great Life in Brief* (New York, 1955), provides a workmanlike account. The argument that Henry Ford's earliest attitudes toward Jews may have been shaped by what he read in the McGuffey Readers of his childhood is found in Allan Nevins, *Ford: The Times, The Man, The Company* (New York, 1954). Charles E. Sorensen's eyewitness account of the meeting between Ford and Mme. Schwimmer is in *My Forty Years with Ford*, written with Samuel T. Williamson (New York, 1956). For an extended discussion of the Jewish problem see Lee, op. cit. Also, Collier and Horowitz, Sward, and Ford/Crowther, op. cit.

343   Thomas Edison and "preparedness": The essay, "On Atomic Energy," 1922, in Dagobert D. Runes, *Diary of Edison,* op. cit. Edison's pro-

nouncements on America's readiness for war have been traced back to early interviews, "Edison Sees the Value of a Great War, Says Nothing Short of the Almighty Can Prevent a Great Conflict," Fort Myers *Press*, August 10, 1914; "Edison Sees Good Year for Industries," ibid., January 7, 1915; "Two Years' War, Declares Edison, Submarines Are Effective," St. Augustine *Record*, February 22, 1915; "If We Are Prepared, War Will Be Averted, That Is Edison's Creed," Scranton *Tribune-Republican*, March 5, 1916; "Must Be Prepared or War May Come," Brooklyn *Citizen*, March 14, 1915; "Edison Would Not Fight for the Lusitania," Detroit *Free Press*, May 22, 1915; "Edison Holds U.S. War Talk Absurd," Indianapolis *Tribune*, May 23, 1915; "Secretary of the Navy Daniels Hopes for Disarmament," Detroit *Free Press*, May 23, 1915; "An Afternoon with Thomas B.[*sic*] Edison in His Laboratory," Houston *Chronicle*, June 6, 1915; "Utilizing Waste" (Edison on recycling of benzol), Bisbee (Ariz.) *Review*, June 15, 1915; "War Tools Should Be Stored, Says Noted Inventor," Beaver (Penn.) *Times*, June 15, 1915; "If Edison Should Go to War," Lancaster (Penn.) *Journal*, June 23, 1915; "Train Soldiers, Then Send them Back," Portland *Times*, June 23, 1915; "If War Hits U. S., Edison Will Make Science Master," Atlantic City *Press*, June 24, 1915. Selections from Fort Myers *Press*: collection of Leslie Marietta; all others: ENHS.

345   Edison and the Naval Consulting Board: Reingold, 1981, op. cit., discusses the precedent for the Board during the Civil War, and includes Robert S. Woodward's statement on the history of the Board as included in the minutes of the board of trustees meeting, Carnegie Institute of Washington, December 13, 1918. In its "Tattler" column, "Notes from the Capital," of October 28, 1915, *The Nation* approved of "the choice of Thomas Alvah [*sic*] Edison to head the new Naval Advisory Board" as "the natural thing under existing conditions, the navy being in particular need of new inventions, and Edison being to-day the foremost American inventor." Microfilm Collections, the New York Public Library. From the Collections of Charles Hummel, I am grateful for access to the following documents: "Mr. Edison's War Work," typescript narrative account of TAE's plan for the Naval Consulting Board; "List of War Subjects upon which Mr.

Edison worked in 1917 and 1918," checklist of 48 projects, by name; "Mr. Thomas A. Edison, September 30, 1919," 20-page typescript transcription of oral history interview with Edison in which he runs down the list of projects named above and describes them in detail. Includes his statements of disillusionment about the lack of interest expressed by the Navy as time went on: "I was right next door to the Navy Board [in Washington] but not one had the curiosity to ask what we were doing"; two handwritten letters to "Darling Billy" (n.d., but from internal evidence, summer 1917 and after).

346   "Great Promise in New Invention," Fort Myers *Press*, May 9, 1917; and "Edison Devices Aided U. S. Navy to Defeat U-Boats" (interview with Josephus Daniels excerpted from the New York *American*), November 12, 1920, collection of Leslie Marietta, provide corroboration and additional detail on Edison's navy projects, as does "Edison War Plan Revealed by Son, Inventor 25 Years Ago Urged Preparation After Conflict Against Any Invasion" (interview with Charles Edison), New York *Times*, February 12, 1942, Collection Edison Electric Institute. Edison's "projectile" plans, one of the first combat items he developed, were registered for patent in January and February 1916, as number 1,296,294.

347   Marion, Tom, Jr., and Will: Marion's oral history memoir, March, 1956, op. cit.: "War is a terrible tragedy and my husband used to say that if the old men around the council table who decide on war had to fight in the front lines and be killed or maimed there would be no more wars"; Correspondence, Marion:TAE and Mina, January 19, 1912; May 9, 1922 ("Now I know that Oscar wanted to deceive me in 1912 when he tried to seduce Fritz's wife and about a year later had an affair with a woman in Mullhausen"); and December 18, 1914; July 7, 1915; July 2, 1916; June 29, October 2, 1917; December 2, 1918; April 23, August 26, 1920. Madeleine's letter to her mother warning of the dangers of sending shoes to Marion is undated, but from internal evidence is 1917–18. All letters cited above are CEF. Correspondence, Tom, Jr.:Father, April 22, October 29, 1911; Dr. E. R. Mulford:Frank Dyer December 6, 1911, on behalf of Beatrice Edison, conveying the parlous mental and physical state of Tom, Jr.: "Yesterday he took some 'Booze' and has been very dis-

agreeable especially to Mrs. Edison"; accompanied by note from Beatrice Edison to Dyer, December 5, 1911; Tom, Jr.:TAE, December
21, 1911; November 17, 30, 1913; Tom, Jr.:Mina, September 13,
1913; April 13, 1915, "my sudden attack of neuralgia yesterday suffered me great pain"; January 16, 1917; Beatrice Edison:Mina, October 19, 1917, and Thanksgiving, 1917. Tom, Jr., correspondence
cited above is ENHS, except letters to Mina, which are CEF.

348   Correspondence, William Leslie Edison: TAE, January 28, 30, 1911;
[March], April 27, 1913; Blanche Edison: TAE, [n.d.] 1918; WLE:
TAE, regarding help with discharge from the army since armistice
has been signed, November 20, 1918; TAE: WLE, December 18,
1918. ENHS.

348   Madeleine Edison Sloane: Oral history interview with Kenneth K.
Goldstein, op. cit., regarding details of her wedding, early married
life, and first two babies; correspondence, MES:Mina, January 14, 15,
February 15, March 15 [August], 1918; August 13, 1919. CEF.

349   Charles Edison: John D. Venable, op. cit., provides the outline for
Charles's Western trip and his early experiences working for T.A.E.,
Inc. On his final months at college, correspondence, Charles:Mina,
January 4, 1913, CEF. Very little documentation remains regarding
Charles's "bohemian days" in Greenwich Village: "The Mexican Border," his booklet-length poem published by Guido Bruno in 1915;
an autobiographical portfolio made up of cut-and-paste newspaper
headlines, "The Adventures in The Life of Charles Edison, Hero of
To-Day;" an assortment of handbills from the "[Little] Thimble Theatre" for productions of *Miss Julia* [*sic*] and *The Stronger*, by Strindberg, and *The State Forbids*, a play in two scenes by Sada Cowan; and a
snapshot of Charles posing stiffly in front of the "Thimble Theatre"
curtain adorned with paintings of a peacock with its tail unfurled.
His firsthand account, "My Experiences Working for Father" appeared in an article under Charles's byline in *The American Magazine*,
1917, accompanied by a photograph of Charles, intently serious of
mien, pipe in one hand and telephone receiver in the other, over the
caption: "Today, because Thomas A. Edison is devoting all his time
to government work, Charles Edison is Operating Manager and Chief
Executive of all the Edison organizations. . . . It was not until he had

proved that he could work for other men that he went into his father's business. He is not an inventor. His abilities lie rather along business lines." This narrative was amplified in the *Literary Digest* article cited below (chapter 26), "Edison's Laboratory Tests for Human Nature," March 9, 1918. ENHS. Charles's wedding to Carolyn Hawkins as they "pledged their vows in simple manner under the beautiful palms," was described in the Fort Myers *Press,* March 27, and their return north was reported, April 2, 1918. Madeleine's reaction to the suddenness of her brother's ceremony appears in her correspondence with Mina, n.d. [April 1918], CEF.

CHAPTER 26

355 Thomas Edison's hiring practices at West Orange: The Motion Picture, Broadcasting and Recorded Sound Division, Library of Congress, holds rare footage of Edison at work in his later years. The film *A Day with Thomas A. Edison* is part of this archive. Paul Kasakove's extensive memoir, *Reminiscences of My Association with Thomas Alva Edison*, covers the years 1920–31, and was provided to the author by Robert B. Halgrim, Jr., director of the Edison-Ford Winter Estates in Fort Myers, Florida. "Edison's Laboratory Tests for Human Nature," an unsigned article, appeared in *The Literary Digest*, March 8, 1918. Collection New York Public Library.

357 Edison and resistance to radio: The important citation from the works of Buckminster Fuller is found in Susan Sontag's essay, "One Culture and the New Sensibility," published originally in *Mademoiselle*; then collected in *Against Interpretation* (New York, 1966). Edison's notes on the *Ave Maria* sound tests: "Edison's Experiments—Musical Instruments—No.3," Notebook, collection of Charles Hummel. As far back as December 1911, Edison had written, "I have now personally listened to all our records made in the last seven years (3,600 all together)." The rationale for Edison's shift away from intuitive hiring and toward the standardized questionnaire is chronicled in A. L. Shands, "The Real Thomas A. Edison," *The Haldeman-Julius*

Monthly, VIII.3 (August 1928); acknowledgment to George Tselos, archivist, Edison National Historic Site, for drawing this piece to my attention; Roderic Peters, "Tom Edison's Sales Techniques," *Nation's Business*, March 1971, Collection Edison Electric Institute (Peters worked in the Edison Phonograph sales department, 1922–1926); "If You Can't Answer These Questions You Are Surely Ignorant—At Least That Is What Mr. Thomas A. Edison Says," Fort Myers *Press*, May 16, 1921; "Mr. Edison's Questions and Answers," itemized listing of the entire exam, ENHS; Edward Marshall, authorized interview with TAE, "Youth of To-Day and To-Morrow," *The Forum*, January 1926 ("As he approaches his eightieth birthday, the Nestor of American invention, unlike his Homeric prototype, takes a cheerful view of his juniors."); Marshall's piece contains a section on "The Famous Questionnaire"; Dagobert D. Runes, Diary of TAE, op. cit., presents "sundry observations" pertaining to education, youth, memory, and the values of true (experiential) knowledge; Reginald D. Archambault, ed., *John Dewey on Education: Selected Writings* (Chicago, 1974), provides many points of commonality between Edison's organic view of education and Dewey's faith in empirical "knowledge of the actual."

360 Charles Edison and TAE in conflict: Undated questionnaire from mid 1920s in ENHS presents Edison's views on the relative merits of radio and the phonograph. To test Edison's strong reaction against radio in the context of the early 1920s, see Frederick Lewis Allen, *Only Yesterday, An Informal History of the 1920s* (New York, 1931); and the best recent single-volume history of radio, Tom Lewis, *Empire of the Air—The Men Who Made Radio* (New York, 1991). Charles Edison's three letters to his mother bracketing the beginning and end of this section, [n.d.] 1921; March 9, 1924; and February 22, 1926, CEF. The "paper trail" of handwritten notes and memoranda cited extends from April 25, 1921 to November 27, 1923. Courtesy Charles Hummel.

362 Mina Edison: "New Civic Organization Is Formed at the Home of Mr. and Mrs. Edison," Fort Myers *Press*, March 24, 1920; four articles from the *Chautauquan Daily*, "Bird and Tree Club—Mrs. Thos. A. Edison Addressed First Meeting," July 1922; "Mrs. Edison Elected

Trustee," August 2, 1922; "Cottage Owners' Meeting," August 9, 1922; and "Chautauqua Garden—Mrs. Edison Beautifies Miller Property," August 21, 1922; pamphlet, "History of the Bird and Tree Club of Chautauqua, N.Y., 1913–1941," E-MC. Acknowledgment to Nancy Miller Arnn for a personal guided tour of the Miller Cottage at Chautauqua, including the garden and patio. "Mrs. Mina M. Edison's District Plan," one-page flyer, ENHS. Two-page curriculum vitae prepared by Theodore Edison, "Mrs. Thomas A. Edison," dated 1947, E-MC.

365   The two most significant sources for TAE's "likes and dislikes" are James R. Crowell, "What It Means to Be Married to A Genius—Mrs. Thomas A. Edison Tells You in This Intimate and Sparkling Story," *The American Magazine*, February 1930; and Milton Marmor, "Interview with Mrs. Thomas A. Edison," January 10, 1947 (15-page unedited typed transcript, beginning, "What were the trials and tribulations and compensations of being the wife of a genius?"). On TAE and MME, see also Madeleine Edison Sloane, op. cit., especially the following exchange: *"Sloane:* Of course, you were always a woman and [there were] certain things women did not do. *Q:* What did [Edison] consider that a woman shouldn't do? Ride a horse? *Sloane:* Ride a horse, yes. But she did what her husband told her to, period. *Q:* That came through in [his] letters, 'I'm glad you did what your mama asked you to.' *Sloane:* Yes. We were supposed to be good kind creatures and take care of our husbands and fathers." The citation from Mary Valinda Miller was written in a pocket notebook kept by Mary Miller Nichols, Mina's sister, collection of Nancy Miller Arnn. On Edison's particular preference in cigars, see telegram, Mina in Fort Myers: Theodore in West Orange, "Have Meadowcroft cancel Key West cigars. No good. Hoffman cigars in mine or linen closet large box under Father's table living room. Please send one immediately," March 10, 1925. On Mina's moments of self-doubt, interview, NB with Irene Sieberling Harrison, June 1992, op. cit. Mina made the "worm in the dust" comment directly to Mrs. Harrison; correspondence, Mina:Madeleine, March 1, 1928 and March 16, 1946, ENHS.

367   Mina and Theodore: "The little golden-headed boy" letter was written by Mina to her mother in the summer of 1901, when Theodore

was three years old. Correspondence, Mina:sister Mary April 28, 1908, refers to having "times alone" with Theodore. Both letters collection of Nancy Miller Arnn. Theodore Edison:Mina, January 11, July 4, 1920; January 14, June 2, July 1, November 22, December 1, 1921, CEF. Mina:Theodore, May 2, 10, 19, June 1, 15, October 22, November 19, 1922, collection of Nancy Miller Arnn. Madeleine Edison Sloane, op. cit., cites her June 14, 1923 letter to the editor of the New York *Times* defending TAE's decision to send his youngest son to college even in view of his skeptical opinion of the viability of a college education. "Edison, Jr., Has Views of His Own" (re Theodore's graduation from MIT and his decision to continue on with his studies), Fort Myers *Press*, June 25, 1923. Correspondence, Theodore:Mina, May 28, September 13, December 8, 1923 (in which he tells his mother he wants a slide rule for Christmas); January 28, 1924 (describes first meeting Miss Osterhout); February 17, March 8, April 11, 24, 1924; March 23, 1925 ("My goodness, what a lot of questions there were in your last letter! . . . I have burned *a lot* of your letters"). CEF.

CHAPTER 27

369   "Edison Bats .500 in Tryout with Mack's Men—Inventor Solves Gleason's Shoots," Fort Myers *Tropical News*, February 25, 1926; "Charles Edison Made President of Edison Industries," ibid., September 16, 1926. On "TAE voted greatest living American," and other phenomena of the age, Wyn Wachhorst, *Thomas Alva Edison, An American Myth* (Cambridge, 1981), also contains one of the most comprehensive recent bibliographies on literature by and about TAE. The Constitution and By-Laws of the Edison Pioneers and information on the eventual physical ruin of Menlo Park are CEF. The Francis Upton Papers at the New Jersey Historical Society, Newark, contain evocative late correspondence between TAE and his old friend.

371   The Edison "Dicta": Excerpts from TAE's pronouncements on a variety of subjects are drawn from Runes, ed., op. cit.; also, Edward Mar-

shall, "Youth of To-Day and To-Morrow, An Authorized Interview With Thomas A. Edison," *The Forum*, January 1927; "Edison on Trusts," New York *Commercial*, December 14, 1911; Edwin R. A. Seligman, *Currency Inflation and Public Debts, An Historical Sketch* (New York, 1921), Edison's personal copy, with extensive marginalia critiquing the Federal Reserve system; "Thomas Edison on the Money Subject," pamphlet published and circulated by Alfredo and Clara Studer, January 1947, Poetry/Rare Books Collection, SUNY at Buffalo, courtesy Robert J. Bertholf, curator; TAE handwritten draft for letter in response to editorial in the Dearborn (Mich.) *Independent*, December 17, 1921 on theories to revise the Federal Reserve; and Thomas A. Edison, "A Proposed Amendment to the Federal Reserve Banking System, Plan and Notes," (West Orange, February 10, 1923). Edward Marshall, "The Scientific City of the Future, An Authorized Interview with Thomas A. Edison," The *Forum*, December 1926. Correspondence, Edwin E. Slosson:TAE, September 8, 1925, with accompanying questionnaire; "Edison and Slosson Talked of Many Things," *Science News Letter*, October 24, 1931. All ENHS.

375   Edison and the monads: John Meurig Thomas on Faraday, op. cit.; Ralph Waldo Emerson, *Essays and Lectures* (New York, 1983) re the definitive text for "Representative Men"; Lewis Spence, *Encyclopedia of the Occult,* op. cit.; Sylvia Cranston on Madame Blavatsky, op. cit. Thomas A. Edison, "Our Brains Are Like Records on the Talking Machine," Gary (Ind.) *Gazette*, January 2, 1911; Edward Marshall, "Thomas A. Edison on Immortality," *The Columbian Magazine*, III.4, (January 1911); "Time Is Defied by Wizard," Chicago *Tribune*, April 23, 1911. Broca's brain manuscripts, two handwritten drafts, labeled "A" and "B," ENHS. "Thomas Alva Edison, The Record of a Life," form filled out by Theodore Edison after his father's death. Under the category, "Religious Affiliations," is written, "He believed in a Supreme Intelligence but had no creed. His religion was the Golden Rule." ENHS. On May 1, 1976, Theodore presented a lecture for the MIT Club on "The Monoid [or Monad?] Concept": "Father never ceased to marvel at Nature's grand design, and now and then he speculated on the basic features of life itself." "Views of Edison on Religion," Fort Myers *Press*, September 1, 1923: "I am convinced that the

body is made up of entities, which are intelligent." Edward Marshall, "Has Man An Immortal Soul? An Authorized Interview with Thomas A. Edison," *The Forum*, November 1926. ENHS.

378 Edison and his children in the mid and late 1920s: Correspondence, Marion:TAE, April 2, 1923; Marion:Mina, January 24, 1921; August 9, September 24, 1924; from Norwalk [n.d.]; December 31, 1929; February 24, 1930. CEF. Tom, Jr.:Mina, March 4, 1929. ENHS. Tom, Jr.,'s obituary, *Newsweek*, August 31, 1935. The New York Public Library. William Leslie: "Dear Mother" [n.d.]; to William Leslie: Meadowcroft, April 8, 1925; William Leslie: TAE, March 23 and June 8, 1925; Meadowcroft (on behalf of TAE): Norman MacLeod, Wilmington *Evening Journal*, November 30, 1926: "William L. Edison has had an uneventful life, spending nearly all of it as a farmer and a breeder of birds, chickens, pheasants, etc." Meadowcroft (on behalf of TAE): Charles W. Duke, Curtis-Martin Newspapers, Philadelphia, February 26, 1926: "I am sorry that we have no photograph of William L. Edison." ENHS. Madeleine:Mina, April 4, 1923, and then TAE, April 11, 1923, asking for loan; Madeleine: "Darling Mother" regarding her daughters-in-law, n.d.; Madeleine: TAE and Mina, February 20, 1925; Madeleine:Mina, February 26, May 28, 1928. CEF. Charles:Mina, "You are systematically fooling yourself into thinking that you are 'alone.' " [n.d., 1925]. CEF. Wedding announcement for Theodore and Ann, Fort Myers *Press*, April 25, 1925. Re Theodore and Ann on their honeymoon in Monhegan Island, *The New Monhegan Press*, IV.10, January 1993. Theodore: Mina, March 21, 1927; Ann:"Mother Edison," March 26 and April 26, 1927. Theodore:Mina, February 29, 1928, mentions the work of Tom, Jr., at the Lab, amd April 7, 1928 with postscript by Ann. CEF.

380 Edison and the search for domestic rubber: The citation from Lewis Mumford is in Hughes, 1989, op. cit.; see also Clark, Firestone, and Vanderbilt, op. cit. The citation from Andre Siegfried's *America Comes of Age* (1927 and 1929) is in Richard Guy Wilson, Dianne H. Pilgrim, and Dickran Tashjian, *The Machine Age in America, 1918–1941* (New York, 1986). For an important contemporary perspective on the period, Harold E. Stearns, ed., *Civilization in the United States, An*

Inquiry by Thirty Americans (New York, 1922); and Charles A. Beard and Mary R. Beard, *The American Spirit, A Study of the Idea of Civilization in the United States,* vol. 4, *The Rise of American Civilization* (New York, 1942). I am grateful to James Newton, author of *Uncommon Friends* (New York, 1987), for his firsthand anecdote about Sunday afternoon rubber searches with TAE in Fort Myers in 1927 and years following. The best single account of Edison's rubber endeavors is a monograph by Loren G. Polhamus, Crops Research Division, U.S. Department of Agriculture, "Plants Collected and Tested By Thomas A. Edison As Possible Sources for Domestic Rubber" (Agricultural Research Service, Publication 34-74, July 1967). Early newspaper accounts of Edison's interest in rubber can be traced back to "Edison Looks into the Future," New York *Times,* October 9, 1913. The chronology then proceeds: "Rubber Experts Are Interested in Lee County," Fort Myers *Press,* September 4, 1923; "Mr. Firestone and His Experts Confer with Thos. A. Edison," ibid., March 14, 1924; "Another Great Industry," ibid., June 24, 1924; "Climate Right for Rubber, Says Expert," ibid., March 19, 1925; "Ford, Firestone and Edison Unite," ibid., April 7, 1925; "Firestone May Use Florida Rubber to Break Monopoly," ibid., January 18, 1926; "Edison Celebrates Birthday Today," ibid., February 11, 1927; "Edison Sees Success for Rubber Experiment," ibid., February 27, 1927; "Guayule Rubber Growers Visit Edison Experiment," ibid., March 29, 1927. Grateful acknowledgement to Leah Burt, curator of Glenmont, for allowing me to examine Edison's annotated copy of L. H. Bailey, *The Standard Cyclopedia of Horticulture* (New York, 1927). Edison's correspondence with the authorities at the New York Botanical Garden appears to have begun in May 1927. See correspondence, Marshall A. Howe:TAE, May 12, 1927, and TAE:Howe, the same day; Howe: TAE, May 14, 1927; TAE:Howe, May 17, 1927; TAE:Roland M. Harper (potential "field man"), May 17, 1927; TAE:Howe, June 3, 1927; Howe:TAE, June 4, 1927, enclosing lists of *Euphorbia* and *Ficus*; Meadowcroft:Howe, June 8, 1927. This correspondence is in the Edison Botanic Research Corporation (EBRC) file at ENHS. I am grateful to Prentiss M. Howe, son of Marshall Howe, who responded to my author's query in *The New York Times Book Review* with helpful information about his father's sessions with Edison at the New York Botanical Garden in June 1927. For guidance in my on-site research at the New York Botanical Garden library, grateful acknowledgment to Susan Fraser, head of information services, and her staff. I reviewed the New York Botanical Garden library's Accession No. 233 (1964), typed transcriptions of TAE's "Rubber Studies" Notebooks at West Orange. A covering letter to the transcriptions, dated August 10, 1964, by Harold Moldenke, at that time director of the Trailside Nature and Science Center, Mountainside, New Jersey; and a conversation at the New York Botanical Garden with Steven E. Clemants, taxonomist, Brooklyn Botanic Garden, led me to a successful search for the Edison Herbarium at the ENHS. I am grateful to Douglas Tarr, archivist at ENHS, for allowing me access to the Herbarium. Mr. Tarr also provided me with an excerpt from the *Edison Phonograph Monthly* for January 1916, illustrating the Edison Disc Vault (Building 32), which houses the Herbarium. Correspondence between Edison and the New York Botanical Garden resumed on June 17, 1927, with the first letter of many from Dr. John K. Small, curator of the Herbarium; Meadowcroft: Howe, June 20, 1927. ENHS. "Edison Still Busy on Rubber Plans, Visits New York Botanical Garden Looking for New Specimens," Fort Myers *Tropical News*, June 21, 1927. Memorandum, Charles Edison:TAE, "Dept. of Rubberology," June 21, 1927; Meadowcroft:Small, June 23, 1927; Small: Meadowcroft, June 24, 1927; Howe:Meadowcroft, July 6, 1927, with photograph from the New York *Times* rotogravure section showing Edison with Drs. Britton and Howe; Meadowcroft:Howe, July 7, 1927; Small:Meadowcroft, July 25, 1927. ENHS. "Edison Loads Up on Rubber Seed and Data," Fort Myers *Tropical News*, July 27, 1927; "Edison Forms Company to Back Rubber Product," ibid., July 31, 1927; John V. Miller:J. K. Small, September 23, 1927; Small: Miller, September 26, 1927, ENHS. Edison interview with *Popular Science Monthly* cited in Fort Myers *Tropical News*, November 3, 1927; Madeleine's quotation from her father re his race with time, inside flyleaf of copy of Dyer, Martin, and Meadowcroft, op. cit. at ENHS; Small: Edison and Edison draft reply to Small on same sheet, November 10, 1927, ENHS; "Edison Coming In January to Push Rubber

Work," ibid., November 30, 1927; "Edison Aides Rushing Work on Laboratory," ibid., December 31, 1927; "Edison's Right Hand Man Here to Make Ready for Wizard," ibid.., January 5, 1928; "Edison Here for Working Vacation," ibid., January 14, 1928.

CHAPTER 28

386 From Fort Myers to Chautauqua and back: Emil Ludwig, "Ludwig Calls Edison Secret King of U.S.," Premier Syndicate, March, 1928. ENHS. Miscellaneous articles on Edison's sojourn at Fort Myers, February 1928, Fort Myers *Tropical News*. Mrs. Edison's views on prohibition were explicitly spelled out in, among other places, her lecture before the Women's Chautauqua Temperance Union on August 18, 1934, where she referred to her mother's experiences in Akron. E-MC. "Edison to Receive Medal Awarded by Congress Tomorrow," Fort Myers *Tropical News*, October 19, 1928; account of argument with Charles over "getting into radio" is from Venable, op. cit.; the story of the Edison relationship with the Splitdorf Company is drawn from Frow, op. cit.; "Thomas Edison to Arrive Tonight on His 46th Annual Visit," Fort Myers *Tropical News*, January 16, 1929; "Hoover Visits Edison on His Birthday," ibid., February 12, 1929; "Special Stamp Coming," and "Cohan Writes Song," ibid., June 1, 1929; TAE's rubber processing notebook, N 29-06-02, June 2 and 3, 1929, ENHS; George M. Cohan, "Thomas A. Edison, Miracle Man," copyright© VosBurgh's Orchestration Service, New York, publishing agent for Light's Golden Jubilee committee; correspondence, Madeleine:Mina, n.d., "I wish you would give up going to Chautauqua—this summer—and go where you could enjoy yourself as a human being and not as a public figure." CEF. John H. Vincent: Mary Valinda Miller, "Christmas Day," 1910; John H. Vincent obituary, May 10, 1920; Arthur E. Bestor:Mina, with Mina's handwritten commentary before she passed the letter on to Louise Igoe Miller, April 9, 1929; "Chautauqua Movement Now Fifty Years Old," New

York *Times*, July 1924; draft letter, Mina: "Dear, Dear George [Vincent]," [1930s]; Mina speech to board of trustees, Chautauqua, October 1939, "Gentlemen—I warn you that I thoroughly resent the treatment of my father in regard to Chautauqua"; program pamphlet, "Centenary of Lewis Miller, Wednesday, July 24, 1929," published at the Chautauqua Institution. All E-MC. Nancy Miller:"Auntie Ga" n.d. (c. 1928). CEF, and Nancy Miller Arnn, personal communications. Alfreda Irwin, op. cit., for background history on the celebration. "Edison Better After Serious Attack of Pneumonia," Fort Myers *Tropical News*, September 2, 1929.

392 Light's Golden Jubilee: General background, see Baritz, Head and Pretzer, Leuchtenburg, Pretzer, ed., and Nye, op. cit. Reynold M. Wik, *Henry Ford and Grass-roots America* (Ann Arbor, 1972) provides a sympathetic account of Ford's progress toward the concept of Greenfield Village. The "history is bunk" story is told in Sward, op. cit., and discussed by Michael Kammen in his essay contribution, "Why Study History? Three Historians Respond," in Paul Gagnon, op. cit. For bringing Eric Hobsbawm's resonant term "invented tradition" to my attention, I am grateful to Lawrence W. Levine, op. cit. Henry Ford's collecting mania leading to his special focus on Edison's laboratory and old equipment, gestating over three years, is described in Fort Myers *Press*, February 19, 1925; February 18, July 25, 1926; June 7 and 19, 1928. Madeleine Sloane's opinion of Ford and Firestone was expressed in her oral history interview of December 1, 1972, with Kenneth Goldstein, op. cit. Mina Edison:Theodore, August 5 and 21, 1923, mentioning Henry Ford, courtesy Nancy Miller Arnn. Mina's private comments the night before the Jubilee ceremony are in ENHS, dated "Sunday—Oct. 20—1929," and were discovered and photocopied by Norman R. Speiden, previous curator of the Edison Archive. For a firsthand reminiscence of events leading up to the Jubilee, I am grateful to Edward Bernays, who spoke with me by telephone on January 18, 1992. Mr. Bernays designed the public relations campaign for the Jubilee and met with Edison several times. The Jubilee was chronicled assiduously through wire service dispatches published in the Fort Myers *Tropical News* on September 24,

October 16, 19, 20, and 22, 1929. For an insightful tour of the reconstructed Menlo Park laboratory and surrounding property on April 27, 1993, I am grateful to John Bowditch, curator of industry at the Henry Ford Museum in Dearborn, Michigan. Terry Collins, research associate on the Edison Papers Project, provided me with a research seminar paper (April 8, 1993) by John M. Staudenmaier, S.J., of the University of Detroit, Mercy Campus, "Clean Exhibits, Messy Exhibits: Henry Ford's Technological Aesthetic," presented at the Hagley Museum and Library in Wilmington, Delaware. Staudenmaier brilliantly analyzes Henry Ford's perfectionist, controlling vision of the Museum and Village, and his relationship to the place as a "refuge." The definitive, popular account of the vast Dearborn site is Harold K. Skramstad, Jr., and Jeanine Head, *An Illustrated History of Henry Ford Museum and Greenfield Village* (Dearborn, Mich., 1990), which is a revised and abridged version of Geoffrey C. Upward, *A Home for Our Heritage* (Dearborn, 1979).

397 Into and Out of Radio and Phonograph: Correspondence, Theodore: Mina, January 30 and April 22, 1939, describe the transition to radio through Splitdorf. CEF. Also see Frow, op. cit., final chapter, "Bitter End," and New York *Times*, "Edison to Suspend Phonograph Making," November 8, 1929, in addition to New York *Telegram* piece cited above. Meadowcroft's announcement of the resolution with goldenrod appeared in Fort Myers *Tropical News*, December 6, 1929, over the subhead, "Ends Long Search." Edison's self-referential discussion of "the patience of Job" appeared in Crowther, *Saturday Evening Post,* op. cit. *The New Yorker* profile, December 28, 1929; January 4, 11, 1930, is in the New York Public Library microfilm collection. "Edison on Birthday Feels More Than 83," Fort Myers *Tropical News*, February 12, 1930, mentions Mina's wheelchair gift. The *Plant Patent Bill* is described in the issue of May 30, 1930. Edison's final known notebooks, N 30-01-06 through 30-08-02 (January–August 1930) are in ENHS. A typescript draft for Remsen Crawford's *Saturday Evening Post* article, "Patents, Profits and Pirates," dated August 12, 1930, is ENHS, along with a tearsheet of the published version.

CHAPTER 29

400 TAE, December 1930—April 1931: Last physical descriptions of Edison, William H. Crawford, "Things Never Told About Thomas Alvin [*sic*] Edison," St. Louis *Globe-Democrat*, May 16, 1926 ("He is no Beau Brummel"). "Thomas Edison Talks on Invention in the Life of Today," *The Review of Reviews*, January 1931. ENHS. The long tales by TAE at Grace Miller Hitchcock's birthday dinner party were taken down in shorthand by the family social secretary, Miss Barrett, and then retold in an 18-page letter, Louise Igoe:"Aunt Sade," written from the Hitchcock home, December 3, 1930. CEF. Details surrounding the birth and naming of Michael Sloane are in Madeleine Edison Sloane, oral history interview, op. cit., and Madeleine:"Darling Mother and Father," February 13, 1931. CEF. Articles in the Fort Myers *Tropical News* from January 1–April 26, 1931, describe Edison's return for his final stay. Also, Mary Miller Nichols:Mina, February 25, 1931. CEF. Harvey Firestone's birthday interview with TAE as well as his June 11 radio remarks to the National Electric Light Association Convention are included in the recording, "Edison Speaks," Michael Biel, Ph.D., associate producer, Mark56 Records, Anaheim, California, 1985, courtesy Neil Maken.

403 Theodore Edison leaves the company: This convoluted tale can be traced back to correspondence, February 14, 1929, Madeleine:TAE and Mina, "I think Father should command [Theodore's] presence or something. He might just as well leave the business now instead of being thoroughly sick and having to leave it for a longer time." Madeleine's characterization of Theodore as a "maverick" is in her oral history interview, op. cit., of March 13, 1973. Ann:Mina, February 28, 1930; March 14, 1931; Theodore:Mina, March 25, 1930; January 30, March 8, 1931; Madeleine:Mina, April 1930; formal memorandum Theodore:Charles, April 1, 1931; Theodore:TAE, April 5, 1931, enclosing his memo to Charles; Madeleine:Mina, April 19, 1931, "I hope you will not feel too badly about Theodore's resigning at the Lab. He looks like a different person already"; Theodore:Mina, April 21, 26, 1931. CEF.

406   Illness, Death, Funeral: Daily bulletins appeared in the Fort Myers
Press and *Tropical News* from June 6–October 20, 1931. These blow-
by-blow news releases on Edison's condition were carried by the wire
services in most American daily newspapers. On October 12, 1930,
Edison filled out a Longevity Inquiry sent to him by Professor Irving
Fisher of Yale University, in the course of which he provided a complete medical history of himself, his parents and siblings, as well as
information about sleep habits, diet, weight, and so on. Late in 1927,
Dr. Hubert Howe wrote a note to TAE describing conjectures on his
recalcitrant patient's medical problems and urging TAE to submit to
an X ray of his digestive tract, which Edison refused. ENHS. The il-
lustrative anecdote about Edison's automobile rides through the Orange countryside during his illness is taken from Robert Washburn's
feature article in the New York *Post* of September 10, 1931, "Edison
Supervises Doctor, Studies Medicine He Takes," collection Consolidated Edison Company of New York Library. Grateful acknowledgment to Steven Jaffe, librarian at Con Ed, for his kind assistance. On
a regular basis beginning August 2, Dr. Howe and Charles Edison
provided updates usually twice daily to the press. Copies of these
typewritten bulletins are in ENHS, as is Charles's rigorous listing of
procedures to be followed by the servants at the house respecting incoming phone calls, newspaper reporters, and visitors. Edison's last
words were originally cited in Francis T. Miller, *Thomas A. Edison,
Benefactor of Mankind* (New York, 1931). James Newton wrote in *Uncommon Friends,* op. cit., that Mina personally told him "near the end
her husband had come out of his coma" and uttered the same sentence. Further citation occurs in Berthold Eric Schwarz, M.D., "The
Telepathic Hypothesis and Genius: A Note on Thomas Alva Edison,"
Corrective Psychiatry and Journal of Social Therapy, XIII.1 (1967). Photographs of TAE's death mask and copy of interview with James
Earle Fraser published in the Newark *Star-Eagle* were provided to the
author by Neil Maken. Description of the collections housed in Edison's library at the ENHS are drawn from a guided tour of the library
provided by archivist George Tselos on August 9, 1991. The procedures to be followed at the factory after Edison's death were outlined
in a memo by R. H. Allen and C. S. Williams, Jr., of T.A.E., Inc.

ENHS; and correspondence, William Wallace Nichols (brother-in-law): Mina, September 18, 1931, CEF. Henry Ford's comment at the lying-in-state is in Newton, op. cit. The newsreel "Edison Dead!" is in the Library of Congress Recorded Sound Archives. The compendium of editorial commentary and cartoons in the days immediately following Edison's death are in ENHS. For background on Walter Lippmann's columns in the New York *Herald Tribune,* Ronald Steel, *Walter Lippmann and the American Century* (New York, 1981); Lippmann's essay on Edison is anthologized in Gilbert A. Harrison, ed., *Public Persons by Walter Lippmann* (New York, 1976). There is voluminous documentation pertaining to Edison's funeral and interment in ENHS: "Seating plan desired by M.M.E.," with Charles's original ideas; lyrics and sheet music for "I'll Take You Home Again, Kathleen," by Thomas P. Westendorf (New York, 1904) and "Little Grey Home in the West," by D. Eardley-Wilmot and Hermann Lohr (New York, 1911); Charles's handwritten draft notes for the funeral service including proposed musical compositions to be played, and final typescript; summary of funeral service and interment as documented with notes by Reverend Herben and sent to Mina on May 28, 1934; and a complete overview of arrangements for TAE's passing set forth by Madeleine, Charles, Theodore, and John V. Miller, then sent to Mina for her review and annotation on September 17 and 18, 1931. All ENHS.

CHAPTER 30

411   For resolution of the rubber experiments, and chronology of Charles Edison's career, Vanderbilt and Venable, op. cit. For analysis of Thomas Edison's will and subsequent contesting, John Miller:Mina, November 21, 1928, CEF; "Edison Left Two Sons Bulk of $12,000,000," New York *Times,* October 30, 1931 (including complete text of will and codicils); and "Edison Sons Avoid Fight Over Estate," ibid., February 26, 1932, ENHS. Correspondence, Marion: Mina [October 1931], "I think that since Father's will was read you

have found out more about my real character. . . . I from the very first never considered for a moment contesting his will." CEF.

412   The death of Tom, Jr.: "T. A. Edison, Jr., Dies; Son of Inventor, 59. Refused to Contest Will," New York *Times*, August 25, 1935; "Hotel 'Alias' Hides Death of Edison Jr.," New York *Post*, August 26, 1935, ENHS; "Thomas Alva Edison, Jr.," obituary, *Newsweek*, August 31, 1935, New York Public Library. Correspondence, Marion:Mina [September 1935], "I am half sick from insomnia for Tom's death. In the circumstances it has been a terrible blow." CEF. Death of WLE: Marion:Mina, August 17, 1937, "Billy's passing was not the shock Tom's was although it was a blessing in disguise." CEF, and "I would like my ashes . . ." [n.d.].

413   Mina Edison's last years: Interview with Jeannette (Mrs. Chesley) Perry, Fort Myers, November 25, 1991. Mrs. Perry was Mina Edison's personal secretary in Fort Myers, 1935–40, and remained friendly with her until the end. "Mrs. Edison Dies at 82 Years of Age, Chautauquans Sing in Her Honor Not Knowing of End," Newark *Evening News*, August 25, 1947, E-MC.

415   Theodore Edison: Interview with Whitney Landon, Esq., one of the executors of Theodore's estate, at his home in Llewellyn Park, December 8; 1992. Interview in New York City with Sylvia Alberts, painter, and friend of Theodore and Ann's for thirty years on Monhegan Island, December 29, 1992. Certificate of Organization and By-Laws of Monhegan Associates, organized September 8, 1954, ENHS. "Vietnam—What Should We Do?" Pamphlet reprint of Theodore's full-page advertisement in the New York *Times*, October 16, 1966. Courtesy Nancy Miller Arnn. "Diversity Unlimited, The Creative Work of Thomas A. Edison," talk given by Theodore to the MIT club of northern New Jersey, January 24, 1969; "Progress Toward Points of No Return," letter by Theodore submitted for the record of a hearing sponsored by the New Jersey Commission on Open Space Policy, reprinted from *The Living Wilderness*, Winter 1970–71; Talk on Glenmont presented by Theodore at a meeting of the MIT club of northern New Jersey, November 10, 1973. Courtesy Nancy Miller Arnn, who also arranged my visits to Theodore Edison on June 19, 1992 and Ann Edison on November 26 and 27, 1992; as

well as my attendance at the Memorial Service for Ann Edison held in the conservatory at Glenmont on April 17, 1993. "Theodore M. Edison, An Illustrious Father Guided Inventor, 94," obituary by Eric Pace, New York *Times*, November 26, 1992; "Last Edison, Inventor's Son, 94, Dies in West Orange," Newark *Star-Ledger*, November 26, 1992. "Ann Osterhout Edison, Pharmacist, Dies at 91," obituary, New York *Times*, January 30, 1993.

PHOTO CREDITS:

Pages 8, 27, 33, 52, 70, 79, 81, 82, 96, 107, 113, 119, 136, 151, 170, 204–205, 233, 240, 243, 249, 276, 281, 283, 286, 287, 288, 291, 294, 298, 306, 317, 324, 325, 330, 352, 355, 359, 366, 367—U.S. Department of the Interior, National Park Service, Edison National Historic Site (ENHS); Pages 9, 164, 207, 248, 406—Edison Birthplace Museum, Milan, Ohio. Photo by Tom Koba; Page 14—Milan Free Library. Photo by Tom Koba from Henry Howe's "History of Ohio"; Pages 18, 53—From the Collection of the Henry Ford Museum & Greenfield Village, Dearborn, Michigan; Page 36—Edison Birthplace Museum; Page 88—*Harper's Weekly*, January 3, 1980. Courtesy of ENHS; Page 112—*Frank Leslie's Illustrated Newspaper*, January 10, 1880; Page 117—*The Daily Graphic*, January 3, 1880. Courtesy of ENHS; Pages 147, 156–157, 232, 387, 391, 393—From the Collection of Photographs in the Chatauqua Institution Archives; Page 338—"A List of Some of the Operatic Artists Who Have Made or Will Make Records for the Edison Phonograph" (Thomas A. Edison, Inc., West Orange, New Jersey, 1912), p. 6.

ACKNOWLEDGMENTS

The actual writing of a book such as this must be a solitary act. But along the way, I was fortunate to receive countless kindnesses—ranging from research assistance and leads on interviews to hospitality while on the road—from many people: Amanda Aaron; Sylvia Alberts; Kim Arnn; Victoria Azara, Ph.D.; David S. Baldwin, M.D.; George W. Ball in Akron, Ohio; Robert Halgrim, Jr., director, Mary C. Fitzpatrick, assistant director, Gary Thomas, horticulturist, Diane Burris, guide supervisor, Robert Beeson, guide, and Mary Anna Carroll, housekeeper, The Edison-Ford Winter Homes in Fort Myers, Florida; Anne Bernays; Edward Bernays; Peter Boehmer, editor, *The New Monhegan Press*; Howard Blum; Norm Brauer; Tom Bridwell and Marilyn Kitchell; Patty Brune; Charles and Barbara Cantalupo; Tom Carling Design; Richard Chalfin; Ron Chernow; Carolyn Cohen; Robert Creeley; Mary Cross; Patricia Crown; John Dale; Bernese Davis; Martin Duus; Jane Faus; Colleen Fehrenbach; Lisa Frigand; George and Beth Frow of Sevenoaks, Kent; Ellen Gilchrist; Thomas F. Gillen; Marjorie and Howard Ginaven; Robert Halgrim, Sr.; Curtis Harnack; Frances P. Harpst; Irene Sieberling Harrison; Karen Heffner; Steven Hicks; Alice O. Howell; Edith F. Hunter; William M. Jackson; Haig Kafafian; Susan and Michael King; James W. Kitchell, vice president, Turner Broadcasting System, Inc.; Carole Klein; George W. Knepper; Tom Koba; Fred Kobrak; Allen Koenigsberg, *Antique Phonographs Monthly Magazine*; Elizabeth Kopley; S. Whitney Landon; R. W. B. Lewis; Jan and William McCartin; Susan Magrino; Neil Maken; Sarita Wilson Martinez; Andy and Allison Merriman; Andrew Millard; Howard and Ruth Miller; Margaret Nagy; Gloria Naylor; James Newton; Nancy Novogrod; Chesley and Jeannette Perry; Jonathan and Lesley Goldwasser Plutzik; Ned Polsky; Susan Regan; Charles and Peggy Robbins; David G. Robinson; Roger Rosenblatt; Mrs. John E. Sloane; Ann and J. Tom Smoot, Jr.; Dorothy Stevens; Catharine R. Stimpson; Bill E. Thomas; Frank T. Waters; T. H. Watkins; Charles Weintraub; Dan White; and Sean Wilentz.

For scholarly help with access to special collections and archives, I am grateful to Tasha Mabry, index researcher, ASCAP; John Baker, chief, Preservation Division, and Edward Koeppel, New York Public Library; Cheryl Hurley and Gila Bercovitch, The Library of America; Frank Mattson, Curator, the Berg Collection, New York Public Library; Mary Bowling, curator of manuscripts, New York Public Library; Norman Harris and Yvonne Smith, British Telephone Archives and Historical Information Center, London; Kathleen Burk, department of history, University College, London; Leah Burt, curator (retired) at Glenmont, Edison National Historic Site; Microfilm Division, Clifton Public Library, New Jersey; Reese V. Jenkins, director/editor, and Paul Israel, Tom Jeffrey, and Terry Collins, The Edison Papers Project, Rutgers University; Ronald L. Grele, Director, Columbia University Oral History Project, Butler Library; Steven Jaffe, librarian, Consolidated Edison Archives; Beth Diefendorf, chief, General Research Division, New York Public Library; Larry Russell, curator, The Edison Birthplace, Milan, Ohio; Paul J. Christiansen, president, David Schantz, secretary-treasurer, and Alberta Ench, secretary, The Charles Edison Fund, East Orange, New Jersey; Jack Young, senior vice president, and Ethel Tiberg, manager, Library Services, Edison Electric Institute, Washington, D.C.; Jill B. Street, curator, Edison Plaza Museum, Beaumont, Texas; Seth Feldman, Associate Dean, Faculty of the Arts, York University, Toronto; George J. Fluhr, township historian, Shohola, Pike County, Pennsylvania; Robert Fox, professor of the history of science, Oxford University; Laura Linder, archivist, Hall of History Foundation, Schenectady, New York; Roger Harlan, Tucson, Arizona, Community Foundation; Charles Hummel, museum coordinator, Charles Edison Fund, East Orange; Alfreda Irwin, historian, and Margaret Wade, Barbara Haug, and Nathalie Leonard, curators, Chautauqua Library; Paul C. Spehr, assistant chief (retired), and Barbara Humphrys and Patrick J. Sheehan, Motion Picture, Broadcasting, and

Recorded Sound Division, Library of Congress, Washington, D.C.; Leslie H. Marietta, historian, Edison-Ford Winter Estates, Fort Myers, Florida; John V. Miller, Jr., director, Archival Services, University of Akron; David Morgan, director, Genealogy and History Collection, Milan (Ohio) Public Library; Robert Rubik, Precision Chromes, New York City; David Wright, archivist and registrar, Pierpont Morgan Library, New York City; Nancy Blankenhorn and Jessica Peters, New Jersey Historical Society; Charles Cummings and George Hawley, Newark Public Library; Paul Bunten and Bill Landers, New York Academy of Medicine; Jean Ashton, librarian, New York Historical Society; Jean Bowen, chief, and Charles Eubanks, the Music Division, New York Public Library; Nancy Cotter, Olcott Library and Research Center, Theosophical Society of America, Wheaton, Illinois; Stephen H. Paschen, executive director, Summit County Historical Society, Akron, Ohio; Edward Pershey, director, Tsongas Center, Lowell, Massachusetts; Christian C. Braig, associate librarian, Philadelphia Electric Company; Richard H. Lucas, editor, *PCA Progress*, Portland Cement Association, Skokie, Illinois; Andrew Perry, archivist, Royal Post Office Archives, London; (Mrs.) I. M. McCabe, librarian/information officer, and Susan Devlin and Judith Moak, Royal Institution of Great Britain, London; Howard Dodson, chief, Schomburg Center for Research in Black Culture, New York Public Library; Richard B. Stamps, director, and Nancy E. Wright, historian, Edison Archaeological Project, Oakland University, Rochester, Michigan; William S. Pretzer, director of exhibitions, and Robert H. Casey, Luke Swetland, and Jeanine Head, Henry Ford Museum and Greenfield Village, Dearborn, Michigan; Jane Hartye, curator, Lieb Collection of DaVinciana, Stevens Institute of Technology Library, Hoboken, New Jersey; Lenore Symons, archivist, Institution of Electrical Engineeers, London; Catherine Tierney, Chief Librarian, Akron *Beacon-Journal*; Nancy McKechnie and Melissa O'Donnell, Special Collections, Vassar College Library, Poughkeepsie, New York; Steven Williams, director, Museum of Arts and History, Port Huron, Michigan; Susan Fraser, director, New York Botanical Garden Library, Bronx, New York; and Judith Ann Schiff, chief research archivist, Manuscripts and Archives Division, Yale University, New Haven, Connecticut.

My agent, David Chalfant, has been by my side with steadfast moral support, encouragement, stamina, and sheer belief in the project from start to finish. Christopher Little and Patrick Walsh avidly represented my literary interests overseas. Jenny Cox came up with the original idea for this book when she was an editor at the then-fledgling Hyperion, and acquired it with enthusiasm and alacrity.

Without George Tselos and Douglas Tarr, archivists at the Edison National Historic Site, my research never would have been completed; devoted guardians of the Edison legacy, they helped me through many seemingly impenetrable reference thickets. Kristin Herron, the new curator at Glenmont, unearthed some rare photographs at the very end of my research. Ken Chandler, collections manager at the Edison National Historic Site, was also helpful, especially when guiding me through the mysterious Third Floor of the Edison site.

Rick Kot's perfect pitch and linguistic precision, coupled with his unfailing respect for the written word and courteous yet always firm approach, made him my ideal editor. David Cashion, Victor Weaver, Lisa Kitei, Bob Miller, and Michael Lynton of the Hyperion family provided support in many ways.

Without the allegiance of the families and descendants of Thomas and Mina Edison, this book would not possess the substance of the man beneath the myth. My deepest thanks to David Edison Sloane; Madeleine Edison Sloane; Nancy Miller Arnn; Theodore and Ann Edison; and Robert and Linda Wheeler.

Finally, as always, to my wife, Roberta, and my children, Nicholas and Allegra—an apology, for the endless hours of upstairs silences, doors closed, and weekends missed; and a hope, that they will accept the result of all this time spent necessarily in the "craft and sullen art."

INDEX

Neil Baldwin grew up in New York City and received his Ph.D. in Modern Poetry from the State University of New York at Buffalo. From 1974 to 1982 he edited and published *The Niagara Magazine*, a journal of contemporary poetry. Dr. Baldwin has taught literature and creative writing at City College of New York, Hunter College, Baruch College, The New School, Fordham University, and New York University. Since 1989 he has been Executive Director of the National Book Foundation.

Neil Baldwin is the author of several volumes of poetry, including *On the Trail of Messages* (Salt Works Press, 1982). His other works include *To All Gentleness: William Carlos Williams, the Doctor-Poet* (Atheneum, 1984); *Man Ray: American Artist* (Clarkson N. Potter, 1988; reissued 2001 by Da Capo Press); *Edison: Inventing the Century* (Hyperion, 1995; reissued 2001 by the University of Chicago Press), and *Legends of the Plumed Serpent: Biography of a Mexican God* (Public Affairs Press, 1998). *Henry Ford and the Jews* will be published in 2001 by Public Affairs Press.